AF753312

PLANT SCIENCE RESEARCH AND PRACTICES

THE PROMISING FUTURE OF *JATROPHA CURCAS*

PROPERTIES AND POTENTIAL APPLICATIONS

PLANT SCIENCE RESEARCH AND PRACTICES

Additional books in this series can be found on Nova's website under the Series tab.

Additional e-books in this series can be found on Nova's website under the eBooks tab.

PLANT SCIENCE RESEARCH AND PRACTICES

THE PROMISING FUTURE OF *JATROPHA CURCAS*

PROPERTIES AND POTENTIAL APPLICATIONS

MAIRA RUBI SEGURA-CAMPOS
AND
DAVID BETANCUR-ANCONA
EDITORS

nova publishers
New York

Copyright © 2016 by Nova Science Publishers, Inc.
All rights reserved. No part of this book may be reproduced, stored in a retrieval system or transmitted in any form or by any means: electronic, electrostatic, magnetic, tape, mechanical photocopying, recording or otherwise without the written permission of the Publisher.

We have partnered with Copyright Clearance Center to make it easy for you to obtain permissions to reuse content from this publication. Simply navigate to this publication's page on Nova's website and locate the "Get Permission" button below the title description. This button is linked directly to the title's permission page on copyright.com. Alternatively, you can visit copyright.com and search by title, ISBN, or ISSN.

For further questions about using the service on copyright.com, please contact:
Copyright Clearance Center
Phone: +1-(978) 750-8400 Fax: +1-(978) 750-4470 E-mail: info@copyright.com.

NOTICE TO THE READER
The Publisher has taken reasonable care in the preparation of this book, but makes no expressed or implied warranty of any kind and assumes no responsibility for any errors or omissions. No liability is assumed for incidental or consequential damages in connection with or arising out of information contained in this book. The Publisher shall not be liable for any special, consequential, or exemplary damages resulting, in whole or in part, from the readers' use of, or reliance upon, this material. Any parts of this book based on government reports are so indicated and copyright is claimed for those parts to the extent applicable to compilations of such works.

Independent verification should be sought for any data, advice or recommendations contained in this book. In addition, no responsibility is assumed by the publisher for any injury and/or damage to persons or property arising from any methods, products, instructions, ideas or otherwise contained in this publication.

This publication is designed to provide accurate and authoritative information with regard to the subject matter covered herein. It is sold with the clear understanding that the Publisher is not engaged in rendering legal or any other professional services. If legal or any other expert assistance is required, the services of a competent person should be sought. FROM A DECLARATION OF PARTICIPANTS JOINTLY ADOPTED BY A COMMITTEE OF THE AMERICAN BAR ASSOCIATION AND A COMMITTEE OF PUBLISHERS.

Additional color graphics may be available in the e-book version of this book.

Library of Congress Cataloging-in-Publication Data
Names: Segura-Campos, Maira Rubi, editor. | Betancur-Ancona, David, editor.
Title: The promising future of Jatropha curcas : properties and potential applications / editors: Maira Rubi Segura-Campos and David Betancur-Ancona.
Other titles: Plant science research and practices.
Description: Hauppauge, New York : Nova Science Publishers, [2016] | Series: Plant science research and practices | Includes index.
Identifiers: LCCN 2016010058 (print) | LCCN 2016011133 (ebook) | ISBN 9781634849890 (hardcover) | ISBN 9781634850155 ()
Subjects: LCSH: Jatropha. | Jatropha--Utilization.
Classification: LCC QK495.E9 P73 2016 (print) | LCC QK495.E9 (ebook) | DDC 583/.69--dc23
LC record available at http://lccn.loc.gov/2016010058

Published by Nova Science Publishers, Inc. † New York

Contents

PREFACE

The purpose of this book is to provide an insight into the possibilities of applying *J. curcas* for commercial purposes. Emphasis is placed on the biology, chemistry and potential uses of *J. curcas* and its derived ingredients. The book has thirteen chapters: (1) Seeds of *Jatropha curcas* L.: Diversity, Yield, Quality, Storage and Germination. This chapter discuss some important aspects of the *J. curcas* seeds in Tunisia as their diversity and yield, the effect of the season on seed quality and the percentage of aborted seeds as well as the effect of storage in plastic bag under ambient laboratory conditions on the germination rate. (2) Chemical Composition and Biological Activities of *Jatropha curcas*. This chapter summarizes the most important aspects of *J. curcas* species, in order to enable handle information to the reader. A brief description of the most distinctive taxonomy features, worldwide distribution and biosynthetic pathways of the secondary metabolites commonly isolated from it has been described, as well as a review of different biological activities reported for this species. (3) *Jatropha curcas* L. Biomass Raw Materials: The Current State and Future Potentialities. This chapter deals to discuss different part of *J. curcas L.* biomass with its present applications and future potentialities. (4) Phytochemical Aspects of *Jatropha curcas* and its Application in Health. In this chapter the phytochemical studies in order to know the main metabolites are presented; as well as some functional and biological aspects of *J. curcas*. (5) Physic Nut (*Jatropha curcas* L.) Extracts and Active Compounds with Biological Activity Against Pest Insects and Pathogenic Fungi. This chapter document some work related to the use of *J. curcas* and its active compounds, contained in both the oil and the plant, to control some pests and microorganisms of agriculture interest. (6) Functional Potential of *Jatropha curcas* L. in Prevention and Treatment of Non Communicable Diseases. This chapter document some studies *in vitro* and *in vivo* that describe the antioxidant and anti-inflammatory potential of *J. curcas* as alternative in prevention and treatment of non communicable diseases (7) Biotechnological Applications of Protein Hydrolysates and Curcin from *Jatropha curcas* L. Seeds. This chapter open avenues for new applications of the proteins in the *J. curcas* plant, thereby enhancing the financial viability and sustainability of a *Jatropha*-based biodiesel industry. (8) Morphoanatomy and Ontogeny of the Fruit and Seeds of *Jatropha curcas* L.: A Promising Biofuel Plant. The purpose of this chapter is to describe the evolution of the development and main structures formed by *J. curcas* seeds and fruits four days after anthesis (DAA) to the completely ripe fruit at 58 DAA and to provide evidence that the seeds of this species are not waterproof and therefore dormant, as has widely described in the literature. (9) *Jatropha curcas*: A sustainable feedstock for biodiesel synthesis. In this chapter, the authors present the significance of *J. curcas* oil and the dependence of the feedstock to production route selection. The use of various catalytic approaches was explored and the trends in the production capacities in different locations of the world. (10) *Jatropha curcas* seed oil and biofuels. This chapter document the *J. curcas* potential to be used as feedstock to produce green diesel for the transportation fuel. (11) Perspective of Bioenergy Crop *Jatropha curcas* Improving Agriculture and Mitigation of Global Climate Change. This chapter addresses perspective of *J. curcas* towards improving soil biological health, carbon sequestration and greenhouse gas emission reduction. Manuscript highlights different bioactive molecules like organic acids, amyrin, flavonoids, phytates, etc present in different parts of the *J. curcas* and perspective improving agriculture. This chapter also highlights C sequestration potential of *J. curcas* in different ecological scenarios and usefulness rehabilitating degraded lands recommendation of *J. curcas* cultivation and approaches managing agriculture, ecosystem and global change issues. (12) Agribusiness Potential Use of *Jatropha curcas* in The World: A Review. This chapter explore the possibilities of potential exploitation of *J. curcas* for various applications as well as its importance for the production of biodiesel. These issues are dealt herewith to observe its future scope to mitigate energy crisis, environmental management and sustainable productions. The information about the detoxification methods, the cake, phytoremediation and some other topics are collected and discussed. (13) Can Jatropha Have a Second Chance? Insights Based on Previous Experiences in México. This Chapter presents a synthesis related to the current situation of *J. curcas* projects in some Mexican states where the commercial plantations for biofuel production have been established.

We are grateful to all internationally expertize and renowned authors for their state-of-the-art compilation of recent rapid developments in this field, which made possible the publication of this book: *E. Saadaoui, J.J. Martín, N. Ghazel, N. Massoudi, E. Cervantes, Lucero Méndez, Janne Rojas, N. A. Sri Aprilia, H.P.S. Abdul Khalil, M. Dani Supardan, Adisalamun, Amri Amin, M.K. Mohamad Haafiz, M.I. Syakir, M.R Nurul Fazita, Ludy C. Pabón B., Patricia Hernández-Rodríguez, Liliana C. Córdova-Albores, Silvia Bautista-Baños, Laura L. Barrera-Necha, Miguel Salvador-Figueroa, Ma. de Lourdes Adriano-Anaya, Isidro Ovando-Medina, Edwin Enrique Martínez Leo, Maira Rubi Segura Campos, L.P. Jaramillo-Quintero, A.L. Martínez-Ayala, N. Corte-Real, L. Endres, K. P. O. Santos, R. C. B. Figueirêdo, E. C. P. Arruda, C. Ulisses, M. F. Pompelli, Nurudeen Ishola Mohammed, Nassereldeen Ahmed Kabbashi, J. García-Davila, Santosh Ranjan Mohanty, Bharati Kollah, Betty Jarma Arroyo, Alfredo Jarma Orozco, Arturo Pérez-Vázquez, Jorge Martínez-Herrera, Ofelia Andrea Valdés-Rodríguez, Leobigildo Córdoba Tellez, and Odilon Sánchez Sánchez.*

We thank members of the production team of NOVA for their time, effort, advice and expertise.

Maira Rubi Segura-Campos
David Betancur Ancona

In: The Promising Future of *Jatropha curcas*
Editors: Maira R. Segura-Campos et al.

ISBN: 978-1-63484-989-0
© 2016 Nova Science Publishers, Inc.

Chapter 1

SEEDS OF *JATROPHA CURCAS* L.: DIVERSITY, YIELD, QUALITY, STORAGE AND GERMINATION

E. Saadaoui[1], J. J. Martín[2], N. Ghazel[1], N. Massoudi[3] and E. Cervantes[2,*]

[1]University of Carthage, Regional Station of Gabes, LGVR-INRGREF, Tunisia
[2]IRNASA-CSIC, Salamanca, Spain
[3]Regional Commission for Agricultural Development, Gabes, Tunisia

ABSTRACT

Jatropha curcas L. or physic nut (Euphorbiaceae) is a perennial, deciduous, stem-succulent shrub native of Central America and cultivated in America, Africa and Asia. Drought tolerant, it grows in tropical and subtropical regions and in degraded soils having a significant agronomic, environmental, medicinal and economic potential. *J. curcas* is an oleaginous plant and the seeds are characterized by high oil content (30-50%). Eight accessions of *J. curcas* from four countries (Brazil, Suriname, Mozambique and Tanzania) were introduced in Tunisia and tested in a plantation of 200 plants cultivated in Gabes (south of Tunisia), in arid bioclimate and irrigated with wastewater. High phenotypical variability among the eight provenances is observed, essentially in leaf traits and size and shape of seeds, but low molecular diversity is showed. Three fructification periods were observed in 2011 and 2012, four periods in 2013 and six periods in 2014. Seed yield is 35, 171, 398 and 804 g/individual in 2011, 2012, 2013 and 2014 respectively. Maximum seed yield registered for an individual is 1800 g, a quantity that would represent 3 ton/hectare. Seed yield is high in summer and low in autumn and winter. Yield of fixed oils varies from 24 to 32%. The study of seed diversity, yield, quality, germination and storage is a crucial step before plant cultivation. We studied diversity and seed yield of *J. curcas,* the effect of the season (summer, autumn or winter) on seed quality and the percentage of aborted seeds. The effect of storage in plastic bag under ambient laboratory conditions on the germination rate was also investigated. Results obtained in 2013 and 2014 showed high diversity, moderate seed production of *J. curcas* in Tunisia, low percentage of aborted seeds in summer and high percentage in

* Corresponding author: E. Cervantes. IRNASA-CSIC. Cordel de Merinas 40. E-37008 Salamanca, Spain.

winter. The percentage of aborted seeds is 22.4, 10.31 and 40.75% in summer, autumn or winter of 2013 respectively. In 2014, percentages are of 6.45, 9.41 and 22.79%, respectively. Germination rate is 76.25% for fresh seeds, and 37.25, 21.25, 1.6, 0 and 0% for seeds stored for 1, 2, 3, 4 and 5 years, respectively. These results showed that seed yield is low compared with other regions; the choice of better adapted genotypes with the Tunisian climate is recommended. Seed quality varies depending on the year, seasons and storage periods.

Keywords: *Jatropha curcas* L., yield, aborted seeds, germination, storage, diversity

INTRODUCTION

The name *Jatropha* derives from the Greek word jatròs (doctor) and trophé (food) perhaps due to its well-known medicinal properties. Worldwide *Jatropha curcas* seeds are known by many different names, the most popular one is "physic nut" (Camaggio and Amicarelli, 2012). *Jatropha curcas* is a perennial, deciduous, stem-succulent shrub/small tree up to 5 to 7 m (Figure 1). The plant is able to remain productive for up to 50 years. It originated from Central American regions having a sub-tropical humid climate with no marked winter; even the coldest month averages are above 18 °C, with summers of 27 to 28°. Optimum rainfall for seed production is thought to be in the range of 1,000 to 1,500 mm, but high yields on trial sites have been reported in very high rainfall areas (2,300 mm) in Asia.

Similarly the plant is reported to perform well in drier regions of about 400-800 mm where irrigation is provided (Hawkins and Chen, 2015). After the plantation of *J. curcas* in different sites in Burkina Faso, Bayen et al. (2015) noted that this crop may not be as drought-resistant as suggested by the literature: most plants died between April and May, in the peak of the dry season. Propagation of *J. curcas* is easy by seeds and vegetative cutting. The vegetative propagation (cuttings) is the best alternative for the rapid production of fruit while the generative propagation (direct seeding) is interesting for a high biomass production (Nguema et al., 2013). *In vitro* propagation of *J. curcas* has been achieved on medium supplemented with benzyladenine and indole-butyric acid but the concentration of growth regulators varies with the genotype and explants (Divakara et al., 2010).

Jatropha curcas has the potential of retaining marginal and degraded soil by re-anchoring the soil with substantial roof and reducing possibility of soil erosion and it is targeted to marginal land where soil nutrient is low. The roots of *J. curcas* harbor arbuscular mycorrhizae like *Acaulospora* sp., *Gigaspora* sp., *Glomus* sp., *Sclerocystis* sp. and *Scutellospora* sp. and support the growth of bacteria in the rhizosphere including the following genera and species: *Azotobacter*, *Rhizobium*, *Pleomorphomonas diazotrophica*, *Bacillus megaterium*, *Bacillus thuringiensis*. *J. curcas* renders a service to the ecosystems by fixing 5,100-6,100 kg/ha C as the aboveground plus belowground biomass (Agbogidi et al., 2013; Dubey et al., 2015). *J. curcas* has a high allelopathic effect in the germination and root growth of some cultivated species, essentially with the extract of leaf and pericarp (Saadaoui et al., 2015b).

Figure 1. *Jatropha curcas* in Gabes plantation: Top: defoliate plant (left) and green plant (right). Bottom: flowers, immature fruits, mature fruits and seeds.

The major attributes proclaimed for the wide scale adoption of *J. curcas* in developing countries were its biodiesel production potential, higher oil content than other biofuel crops, rapid growth, easy propagation, drought tolerant nature, ability to grow and reclaim various kinds of land, plants can grow without much irrigation and agricultural inputs, pest resistance, small gestation periods and suitable traits for easy harvesting (Edrisi et al., 2015). *J. curcas* is an interesting tropical oil plant for biocarburant production, eligible for the production of better quality of biodiesel. Nontoxic *J. curcas* is a good source of antioxidant and metal chelating peptides, which may have a positive impact on the economic value of this crop, as a potential source of food functional components (Tintoré et al., 2015). *J. curcas* has medicinal interests; it has anti-malarial, insecticidal, anti-cancerous and anti-tumor properties. Hexane partition from root extract showed high anti-inflammatory activity; hexadecanoic acid methyl ester, octadecanoic acid methyl ester and octadecanoic acid could be responsible for the anti-inflammatory activity of the *J. curcas* root extract (Agbogidi et al., 2013; Othman et al., 2015). Some genotypes of *J. curcas* are toxic, due to the presence of phorbol esters, alkaloids which cause a laxative effect, stomach pain and other discomfort on humans and curcin, a toxalbumin similar in structure and effects to ricin, the well known poison of *Ricinus communis* (Camaggio and Amicarelli, 2012; Rodríguez et al., 2014).

The assembled genome of *J. curcas* has a total length of 320.5 Mbp and contains 27,172 putative protein-coding genes. The genome sequence information showed the coexistence of 13,887 gene families in *J. curcas* and *Ricinus communis* genomes. Which is 91 and 88.7% of the gene families in the two species, respectively (Wu et al., 2015). For Sato et al. (2010), total length of the non-redundant sequences thus obtained was 285 858 490 bp consisting of 120 586 contigs and 29 831 singlets. They accounted for ~ 95% of the gene-containing regions. The average G + C content was 34.3%. A total of 40,929 complete and partial structures of protein encoding genes have been deduced. Comparison with genes of other plant species indicated that 1,529 (4%) of the putative protein-encoding genes are specific to the Euphorbiaceae family. A high degree of microsynteny was observed with the genome of castor bean and, to a lesser extent, with those of soybean and Arabidopsis thaliana. *J. curcas* genome can advance our knowledge of the evolution of the plant kingdom because of the retention of many ancient gene families (Wu et al., 2015). The small genome size, chromosome number, ease of vegetative manipulation and transformation are favorable features for the use of biotechnological tools for *J. curcas* improvement (Divakara et al., 2010).

JATROPHA CURCAS DIVERSITY

There is wide variation in morphological and agronomic traits of eight provenances of *J. curcas* introduced in Tunisia from Africa and America (Saadaoui et al., 2015a). Osorio et al. (2014) studied the biodiversity of 182 *J. curcas* accessions from 26 countries (Central America, South America, Africa and Asia) and registered high level of phenotypic variation. Biabani et al. (2012) obtained notable differences among the physic nut populations for various traits measured including: flowering, number of inflorescences, number of fruits per plant, number of seeds per plant, seed yield per hectare, seed oil yield per plant, and seed oil yield per hectare. Plant height was positively correlated with seed yield per plant, while seed oil content was negatively correlated with seed yield per plant. The majority of studies showed a low molecular variability (Jingura and Kamusoko, 2015).

Gonu et al. (2015) studied the yield and morphological characteristics of the seeds and kernels of *J. curcas* cultivated in Ghana. Seed length, width, breadth, geometric and arithmetic mean diameter, volume, bulk density, solid density and sphericity were variable between two provinces with the exception of bulk density. Yue et al. (2014) applied 29 microsatellites located on 11 linkage groups to study genetic variation in 276 accessions of *J. curcas* collected from nine locations in five countries in South America, Asia and Africa. They did not detect any genetic diversity at all 29 microsatellite loci. All the 276 accessions are homozygous at all loci and shared the same genotype at each locus, suggesting no microsatellite variation in the genome of *Jatropha curcas*. Montes et al. (2014) studied 70 accessions with Simple Sequence Repeats (SSR) and Single Nucleotide Polymorphism (SNP). High level of homozygosis was registered with high genetic diversity in accessions from Central America and Mexico and low for African, Asian, and South American accessions. Similar results were obtained by Osorio et al. (2014). Trebbi et al. (2015) used 18 genetic markers (10 SNP, 5 SSR and 3 EST-SSR) and obtained extreme high genetic uniformity and homozygosity of *J. curcas* accessions cultivated in several countries of South America, Asia and Africa, including the eight provenances grown in Tunisia, and confirmed the general idea recognizing Central America as the center of origin of the species, identifying regions with increased genetic variability and heterozygosity in Mexico and Guatemala.

Seed Morphology and Diversity

The seeds become mature when the capsule changes from green to yellow. The seeds are black spheroids of about 2 cm long and 1 cm wide (Figure 1). Their unitary weight ranges from 0.4 g to 0.9 g which is equivalent to 1,250-2,000 seeds per kg (Camaggio and Amicarelli, 2012; Osorio et al., 2014). Montes et al. (2015) revealed a large variation among the seeds within a plant and genotype for seed traits (oil content, seed mass, kernel mass, and ratio of kernel to seed mass). Central American accessions showed a much higher genetic and phenotypic variation for the seed traits than accessions from other regions (Osorio et al., 2014). In Eastern India, Kumar and Singh (2014) studied 28 Indian accessions with nine seed traits: Seed length, seed width, 100 seeds weight, 2D surface area, aspect ratio, germination percentage, total carbohydrate, total soluble protein and seed oil content and registered genetic differences in all the seed characteristics among the different *J. curcas* accessions. In Ghana, Gonu et al. (2015) observed high difference for some seed traits between two provenances. Saadaoui et al. (2015a) studied eight provenances from American and African countries introduced in Tunisia and observed high variability in morphology of *J. curcas* seeds; the biodiversity concerns area, perimeter, and shape (circularity index and J index; Figure 2). Seed shape images were adjusted to an ellipse, J index indicating the percent of similarity of the seed image with the ellipse. Reduced seed size in some accessions was associated with lower J index. Seed mass is in positive correlation with seed yield (Saadaoui et al., 2015a) but it is not well associated with oil content (Montes et al., 2015).

Seed Yield

Seed yield of *J. curcas* ranges from 0.4 to 12 ton/ha/year with plants being productive for 30-50 years (Camaggio and Amicarelli, 2012; Jingura and Kamusoko, 2015). In Tunisia, the maximum obtained is 1800 g/ individual, i.e., about 3 ton/ha for six year old plants (Saadaoui et al., 2015c). In Colombia, yields ranged from 4,570 to 8,800 kg/ha of fruits and from 2,430 to 4,746 kg/ha of seeds (Montenegro et al., 2014). In Indonesia, the best recorded yield performance was 6,004 kg/ha at five years of age on a degraded agriculture land (Santoso et al., 2014). In Senegal, dry seed yield obtained is less than 200 kg/ha for plantations aged four or five years (Terren et al., 2013). In South of China, eighty sources of *J. curcas* were collected and planted in a germplasm. Six sources presented the higher oil yield. A maximum value of 783 kg oil/ha, that is 2.7-times the national best yield, was achieved for 3-year-old trees (Yang et al., 2010). On two sites in Tamil Nadu (Southern India), the selective variety (JO S2) produced up to 2.95 ton/ha of dry seeds in the first year and up to 4.25 ton/ha of dry seeds in the second year, much better than the local variety (Yi et al., 2015). With the fertilizer dose of 150 kg/ha N + 120 kg/ha K, fruit production increased by 92% and seed production by 95%, which represents an increase of about 100% in oil production that rose from 947 to 1,900 kg/ha (Montenegro et al., 2014). Expected seed yields of 4-5 ton/ha are considered to be commercially viable for a *J. curcas*-based biofuel program and an average seed yield of 3.75 ton/ha with an oil content of 30-35% and oil yield of 1.2 ton/hectare is better than the yield profile of other important oil seed crops such as soybeans and rape-seed (Singh et al., 2014).

Figure 2. Variability in the morphology of *J. curcas* seeds. Seeds were collected from plants grown in Gabes (South of Tunisia). Left: Accession J4 from Valley of Jequintinhonha (Brasil). Right: Accession J7 from Mozambique. Mean area is 1.43 cm^2 in J4 and 1.30 cm^2 in J7. Mean circularity index is 0.91 (J4) and 0.92 (J7). Mean J index is 0,96 (J4) and 0,94 (J7). Seeds are larger in J4 than in J7. Circularity index is higher in J7 and J Index is higher in J4 (Saadaoui et al., 2015a).

For Biabani et al. (2012), the number of inflorescences, number of fruits per plant, number of seeds per plant, and plant height characters are important traits for selection of productive genotypes. The genetic map and the assembled chromosomes therefore provide a valuable foundation for quantitative trait locus mapping of agronomic traits and the development of better varieties of *J. curcas* for seeds yield and oil content (Wu et al., 2015).

Seed Oil and Fatty Acids

Jatropha curcas oil (JCO) is non-edible because the seeds contain a toxic protein called curcin (lectin) and diterpene esters (phorbol esters), both of which are harmful to humans and most other animals (Hawkins and Chen, 2015). The content of phorbol esters in the JCO is 3.09 mg/g oil, the treatment by ultraviolet irradiation combined with ethanol washing of JCO showed results in a decrease by 100% in phorbol esters (Xiao et al., 2015).

Oil content varies largely among varieties and genotypes. In Tunisia, Saadaoui et al. (2015c) registered 24 to 32% oil content in *J. curcas* seeds for American and African accessions. In India, Waghmare and Naik (2015) obtained oil content of 32.5%. Hawkins and Chen (2015) cited 250 to 350 kg (25 to 35%) of oil after processing one metric ton of *J. curcas* grain. In Mozambique, Rodriguez et al. (2015) registered oil content of 45.7%. The predominant fatty acids of JCO are oleic acid (42.02-43.86%), linoleic acid (34.63-35.38%), palmitic acid (13.51-14.54%), and stearic acid (6.3-6.92%) (Table 1; Xiao et al., 2015; Utomo et al., 2015). The fatty acid composition of *J. curcas* oil differs between regions contrary to oil content (Osorio et al., 2014). Free Fatty Acid (FFA) content ranges from 2.79 to 4.95%; low FFA content is often preferred for direct utilization in the transesterification process during biodiesel production (Derkyi et al., 2014).

Table 1. Fatty acid composition (%) of JCO

Fatty acid	Percentage
Myristic acid	0.04
Palmitic acid	13.51
Palmitoleic acid	0.62
Stearic acis	6.92
Oleic acid	43.86
Linoleic acid	34.63
Linolenic acid	0.13
Eicosanoic acid	0.19
ΣSaturated fatty acid	20.61
ΣUnsaturated fatty acid	79.32

Xiao et al., 2015.

The oil from *J. curcas* seeds is used in helping with rashes and parasitic skin diseases. When the oil is mixed with benzyl benzoate, it becomes effective against scabies and dermatitis. The seed oil is also emetic, laxative, purgative and for skin ailments (Agbogidi et al., 2013). Alonso and Santos (2013) revealed that seed oils exert considerable toxic activity against the leaf-cutting ant *Atta sexdensrubropilosa.*

JATROPHA CURCAS BIOFUELS

In industry, *J. curcas* is applied for biodiesel production or green fuel production. Its oil content ranges from 44.20 to 49.70% and its biodiesel yield ranges from 82 to 90% (Derkyi et al., 2014). Optimum biodiesel yield obtained is at a molar ratio of 1:9-1:13 (oil to ethanol), temperature range of 35-45°C, an immobilized enzyme loading of 3.2-6.2 g, mixing intensity range of 4.5-6 g RCF and reaction time of 20-24 hours (Abdulla and Ravindra, 2015). Flash point of *J. curcas* oil biodiesel is 170°C, kinematic viscosity is 5.64, cloud point is 3°C, specific gravity at 15/15°C is 0.88 and pour point is - 6°C, but physical and chemical properties of *J. curcas* oil biodiesel change during handling and storage (Berchmans and Hirata, 2008; Aransiola et al., 2012). Proper selection of *J. curcas* seed maturity based on color/maturity level (half matured or matured stage) must be ascertained not only to produce higher quantity but also to ensure the highest level of quality of biofuel. The amount of oil produced from the half-matured seeds is about 2.5% less than the matured stage. The matured stage shows 13% better performance in contrast to the half-matured stage. The performance for different matured JOD blends was found to be very close to diesel and can be used as a biodiesel without engine modification (Samsuri and Zoveidavianpoor, 2014).

Biodiesel from *J. curcas* has a much higher performance than current biofuels, relative to oil-derived diesel fuels. *J. curcas* biodiesel allows a 72% saving in greenhouse gas emissions compared to conventional diesel fuel (Ndong et al., 2009). Transesterfied *J. curcas* oil has the highest speed and power compared to other two fuels (petroleum diesel and transesterified rubber seed oil), and have lesser carbon dioxide (CO_2) and carbon monoxide (CO) than the petroleum diesel (Lawrence et al., 2014). The use of biodiesel and its blends has minimal effect on the environment and reduces the consumption of nonrenewable energy sources when compared to the diesel fuel (Kalaivani et al., 2014).

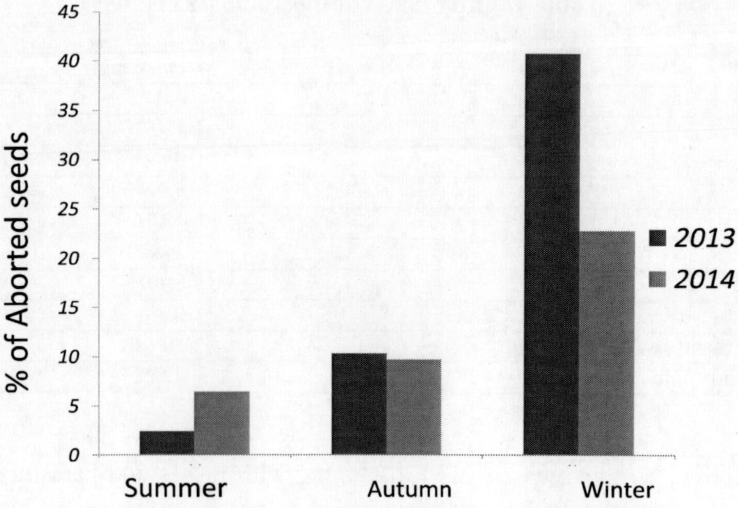

Figure 3. Percentage of aborted seeds in 2013 and 2014 of seeds harvested in summer, autumn and winter.

QUALITY OF SEEDS

In Tunisia, in the experimental station of Gabes, quality of seeds was observed in two years (2013 and 2014) in each period of fructification. The percentage of aborted seeds varied between seasons; in 2013, it was low in summer, high in autumn and very high in winter (2.44, 10.31 and 40.75%, respectively). The percentage of aborted seeds in 2014 was of 6.45, 9.71 and 22.79% in summer, autumn and winter, respectively. The values differ between years, mainly in winter, where the percentage of aborted seeds was higher in 2013. In contrast, in summer, percentage of aborted seeds was higher in 2014 than in 2013 (Figure 3).

These variations are explained by the variability in climatic conditions among the years, mainly the cold that acts directly on fructification and seed filling; temperature being an important factor in growth and productivity of *J. curcas* (Maes et al., 2009). Also, the quality of seeds varies with fruit maturity; seeds separated from fruits harvested at yellow capsule stage show maximum germination efficiency, whereas seeds separated from fruits harvested at brownish black stage yield highest oil content (Ahmad and Sultan, 2015). Seeds frozen (-20°C) showed a significant decrease in free fatty acid (FFA) percentage in comparison to fresh seeds (Sahay and Shyam, 2014).

Seed Storage and Germination Rate

Seeds of *J. curcas* were stored in plastic bag under ambient laboratory conditions at Regional Station of National Institute of Research in Rural Engineering, Waters and Forests (INRGREF) in Gabes (Tunisia), and the germination capacity of different lots was evaluated. The seed lots consisted of seeds harvested in summer 2010, winter 2011, and those collected in summer, autumn and winter for three successive years (2012, 2013 and 2014), as well as fresh seeds harvested in summer 2015. Germination test was conducted in summer 2015 at a temperature of 28°C, evaluating germination capacity in Petri dishes (80 seeds per lot; 10 seeds per dish; 8 Petri dishes per lot).

The germination rate is 0% for the seeds stored for 4 to 5 years and it is very low for the seeds stored during three years (0; 3.3 and 1.25% for the harvests of summer, autumn and winter, respectively). For seeds stored in the laboratory for 2 years, the percentage of germination was 31.25, 26.35 and 6.25% respectively for the harvest of summer, autumn and winter. The seeds collected in 2014 and stored for one year in the laboratory showed a germination rate of 31.25, 25 and 26.25% for summer, autumn and winter harvests, respectively. In the end, we obtained a high germination rate for fresh seeds (76.25%) (Figure 4).

The average of germination percentage for the three seasons (summer, autumn and winter) together in each year, showed that the germination rate decreases with the storage period; it is 37.25, 21.25, 1.6, 0 and 0% for seeds stored for 1, 2, 3, 4 and 5 years, respectively (Figure 5).

Germination percentage for fresh seeds is elevated; In India, Kumar and Singh (2014) obtained a percentage of 52 to 84%. For Moncaleano-Escandon et al. (2013), *J. curcas* seeds have a short viability period, less than 6 months and the increase of storage temperature accelerates the loss of seed germination potential. They obtained germination percentage of 7% and 2% after 12 months in storage at 4°C and 25°C, respectively. *J. curcas* seeds preserve the physiological quality when the fruits were harvested at the appropriate stage of maturation (yellow, ripe and dry fruit on the plant) and by threshing without mechanical damage. *J. curcas* seeds can be stored for 240 days (in multilayer paper bag in control of laboratory condition) without loss in their germination (Veronesi et al., 2014).

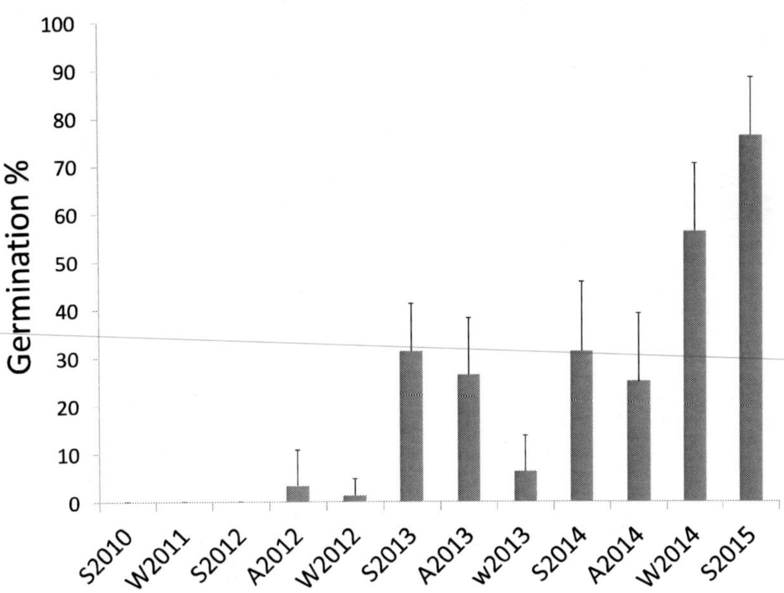

S: summer, A: Autumn and W: winter.

Figure 4. Germination percentage of *J. curcas* seeds harvested from 2010 to 2015 in different seasons and tested in 2015.

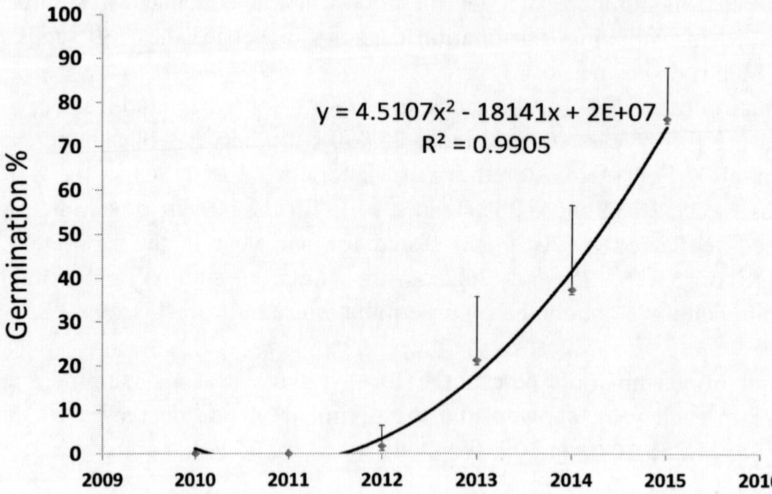

Figure 5. Annual average of germination rate after storage of *J. curcas* seeds (polynomial regression).

Morphology as a Possible Index for Seed Quality

For the application of *Jatropha curcas* in culture, it may be interesting to guarantee that the germplasm used is of the highest quality. In this regard, it is interesting to investigate the correlation between seed shape and germination potential or crop yield. Seed shape of *Jatropha curcas* adjust well to an ellipse. The degree of adjustment is evaluated by means of J index, a magnitude that gives the similarity of the seed image with the geometric figure of the ellipse. In the experiment with seeds obtained from eight accessions, mean values oscillated between 0.94 and 0.96 (Saadaoui et al., 2015a). In some instances, such as seeds obtained from accessions J6 from Suriname and J7 Mozambique, lower J index values were associated with smaller seeds (Figure 2). It remains to be investigated whether this association persists in other instances such as seeds from diverse varieties or stored during variable periods of time under different conditions.

Conclusion

Jatropha curcas is a perennial, deciduous, shrub, which produces seeds rich in oil that is easily convertible into biodiesel. Indeed, *J. curcas* appears to be a perfect crop to produce second generation of biofuels with multitude of environmental benefits. This species is characterized by high genetic variability exclusively in Central America, the origin region of physic nut. *J. curcas* is being cultivated in many parts of Africa including south of the Mediterranean area with variable yields, in general lower than in other regions. In Tunisian bioclimatic conditions, *J. curcas* has a moderate seed yield. Seed oil content varies between 24 and 32% and it's characterized by high percentage of oleic and linoleic acids. Percentage of aborted seeds is elevated essentially in winter harvest (22.79-40.75%). Phenotypic variability concerns essentially seed traits. Percentage germination is up to 70% for fresh seeds harvested in summer but is reduced to 0% after five years of seed storage. These results suggest that the culture of this species is possible in Tunisia and Mediterranean region with elevated risk.

REFERENCES

Abdulla, R. and Ravindra, P. (2015). Process optimization for biodiesel production from crude *Jatropha curcas* oil with immobilized lipase using response surface methodology. *Journal of Applied Science and Agriculture*, 10 (3), 17-20.

Agbogidi, O.M., Akparobi, S.O. and Eruotor, P.G. (2013). Health and environmental benefits of *Jatropha curcas* Linn. *Unique Research Journal of Agricultural Sciences* 1 (5), 76-79.

Ahmad, S. and Sultan, S.M. (2015). Physiological Changes in the Seeds of *Jatropha curcas* L. at Different Stages of Fruit Maturity. *Braz. Arch. Biol. Technol.* 58 (1), 118-123.

Aransiola, E.F., Daramola, M.O., Ojumu, T.V., Aremu, M.O., Layokun, S.K. and Solomon, B.O. (2012). Nigerian *Jatropha curcas* oil seeds: Prospect for biodiesel production in Nigeria. *International Journal of Renewable Energy Research*, 2(2), 317-325.

Bayen, P., Sop, TK., Lykke, A.M. and Thiombiano, A. (2015). Does *Jatropha curcas* L. show resistance to drought in the Sahelian zone of West Africa? A case studyfrom Burkina Faso. *Solid Earth Discuss*, 7, 639-659.

Berchmans, H.J. and Hirata, S. (2008). Biodiesel production from crude *Jatropha curcas* L. seed oil with a high content of free fatty acids. *Bioresource Technology*, 99, 1716-1721.

Biabani, A., Rafii, M.Y., Saleh, G.B., Shabanimofrad, M. and Latif, M.A. (2012). Phenotypic and genetic variation of *Jatropha curcas* L populations from different countries. *Maydica*, 57, 164-174.

Camaggio, G. and Amicarelli, V. (2012). *Jatropha curcas* a no edible versatile cultivar. *Forum WARE International*, 15-24.

Derkyi, N.S.A., Sekyere, D. and Oduro, K.A. (2014). Variations in Oil Content and Biodiesel Yield of *Jatropha curcas* from Different Agro-Ecological Zones of Ghana. *International Journal of Renewable and Sustainable Energy*, 3 (4), 76-81. doi: 10.11648/j.ijrse.2014 0304.11.

Divakara, B.N., Upadhyaya, H.D., Wani, S.P. and Laxmipathi Gowd, C.L. (2010). Biology and genetic improvement of *Jatropha curcas* L.: A review. *Appl Energy*, 87, 732-742. doi:10.1016/j.apenergy.2009.07.013.

Dubey, G., Bharati, K., Ahirwar, U., Mishra, R., Tiwari, S. and Mohanty, S.R. (2015). Prospect of bioactive molecules from *Jatropha curcas* to improve soil and microbial quality for sustainable agricultura. *African Journal of Microbiology Research*, 9 (3), 140-147.

Edrisi, S.A., Dubey, R.K., Tripathi, V., Srivastava, P., Jamil, S., Singh, H.B., Singh, N., Abhilash, P.C. (2015). *Jatropha curcas* L.: A crucified plant waiting for resurgence. *Renewable and Sustainable Energy Reviews*, 41, 855-862.

Fersi, S., Chtourou, N. and Bazin, D. (2013). Energy analysis and potentials of biodiesel production from *Jatropha curcas* in Tunisia. *Int. J. Global Energy Issues*, 35 (6), 441-455.

Gonu, H., Opoku, N., Appiah-Opong, R., Joseph, K., Barnor, M.T. and Quarm, Ph. (2015). Phenotypic traits and chemical properties of *Jatropha curcas* seeds from Northern Ghana. *European Scientific Journal*, 11 (9): 327-340.

Hawkins, D. and Chen, Y. (2015). *Future Harvest.* 21st Century *Jatropha.* Hardman and Co London, United Kingdom, 125p.

Jingura, R.M. and Kamusoko, R. (2015). Utility of Markers for Determination of Genetic Diversity in Jatropha: A Review. *The Open Renewable Energy Journal*, 8, 1-6.

Kalaivani K, Ravikumar G and Balasubramanian N (2014). Environmental impact studies of biodiesel production from *Jatropha curcas*in India by life cycle assessment. *Environmental Progress and Sustainable Energy* 33 (4): 1340-1349, DOI 10.1002/ep.

Kumar, S. and Singh, S. (2014). Variability assessment of seed traits in *Jatropha curcas* L. for improvement of oil yield. *International Journal of Genetics and Molecular Biology*, 6 (1), pp. 8-15.

Lawrence, I., Jim, O.I. and Uwaifo, A.J. (2014). Production of biodiesel from (RSO) rubber seed oil and *Jatropha curcas* seed oil (JCSO). *Standard Journal of Petroleum and Gas Exploration*, 1 (1): 1-4. http://standresjournals.org/journals/SJPGE/index.html.

Maes, W.H., Trabucco, A., Achten, W.M.J. and and Muys, B. (2009). Climatic growing conditions of *Jatropha curcas* L. *Biomass Bioenerg*, 33, 1481-1485.

Moncaleano-Escandon, J., Silva, B.C.F., Silva, S.R.S., Granja, J.A.A., Alves, M.C.J.L. and Pompelli, M.F. (2013). Germination responses of *Jatropha curcas* L. seeds to storage and aging. *Industrial Crops and Products*, 44, 684-690.

Montenegro, R.O., Magnitskiy, S. and Henao, M.C.T. (2014). Effect of nitrogen and potassium fertilization on the production and quality of oil in *Jatropha curcas* L. under the dry and warm climate conditions of Colombia. *Agronomía Colombiana*, 32 (2), 255-265.

Montes, J.M., Technow, F., Martin, M. and Becker, K. (2014). Genetic Diversity in *Jatropha curcas* L. Assessed with SSR and SNP Markers. *Diversity*, 6, 551-566. doi:10.3390/d 6030551.

Montes, J.M., Bulach, A., Martin, M. and Senger, E. (2015). Quantitative Trait Variation in Self- and Cross-Fertilized Seeds of *Jatropha curcas* L.: Parental Effects of Genotypes and Genetic Pools. *Bioenerg Res.* doi 10.1007/s12155-014-9576-8.

Ndong R, Vignoles M, Saint-Girons O, Abrielle B, Pirot R, Domergue M and Sablayrolles C (2009). Life cycle assessment of biofuels from *Jatropha curcas* in West Africa: a field study. *Global Change Biology Bioenergy*, 1 (3): 197-210. http://dx.doi.org/10.1111/j. 1757-1707.2009.01014.x.

Nguema, N.P., Bouanga, E.B., Massounga, Y.C. and Boussiengui, B G. (2013). Étude comparée de trois méthodes de multiplication de *Jatropha curcas* L. dans les conditions climatiques du sud-est du Gabon. *J. Appl. Biosci.*, 65, 4989-4998.

Osorio, L.R.M., Salvador, A.F.T., Jongschaap, R.E.E., Perez, C.A.A., Sandoval, J.E.B., Trindade, L.M., Visser, R.G.F. and van Loo, E.N. (2014). High level of molecular and phenotypic biodiversity in *Jatropha curcas* from Central America compared to Africa, Asia and South America. *BMC Plant Biology*, 14 (77). doi: 10.1186/1471-2229-14-77.

Othman, A.R., Abdullah, N., Ahmad, S., Ismail, I.S. and Zakaria, M.P. (2015). Elucidation of in-vitro anti-inflammatory bioactive compounds isolated from *Jatropha curcas* L. plant root. *BMC Complement Altern Med.* 2015; 15: 11. doi: 10.1186/s12906-015-0528-4.

Rodríguez, O.A.V., Vázquez, A.P. and Gamboa, C.M. (2014). Drivers and Consequences of the First *Jatropha curcas* Plantations in Mexico. *Sustainability*, 6, 3732-3746; doi:10. 3390/su6063732.

Rodrigues, J., Miranda, I., Furquim, L., Gominho, J., Vasconcelos, M., Barradas, G., Pereira, H., Bianchi-de-Aguiar, F. and Ferreira-Dias, S. (2015). Storage stability of *Jatropha curcas* L. Oil naturally rich in gamma-tocopherol. *Industrial Crops and Products*, 64, 188-193.

Saadaoui, E., Martín, J.J., Bouazizi, R., Ben Romdhane, CH., Grira M, Abdelkabir, S., Khouja, M.L. and Cervantes E. (2015a). Phenotypic variability and seed yield of *Jatropha curcas* L. introduced to Tunisia. *Acta Botanica Mexicana*, 110: 121-136.

Saadaoui, E., Martín, J.J., Ghazel, N., Ben Romdhane, Ch., Khouja, M.L. and Cervantes, E. (2015b). Allelopathic Effects of Aqueous Extracts of *Jatropha curcas* L. on the Germination of Six Cultivated Species in *Jatropha curcas*: Biology, cultivation and potential uses. Nova Science Publishers US, 159-169.

Saadaoui, E., Tlili, N., Ghazel, N., Ben Romdhane, Ch., Abdelkebir, S., Grira, M. and Khouja ML (2015c). Growth and seed yield of *Jatropha curcas* L. cultivated in arid region of Tunisia in *Jatropha curcas*: Biology, cultivation and potential uses. Nova Science Publishers US, 117-129.

Sahay, S. and Shyam, M. (2014). Storage conditions to improve the shelf life of *Jatropha curcas* seeds in terms of quality of oil. *J Fundam Renewable Energy Appl*, 4 (2). http:// dx.doi.org/10.4172/2090-4541I000136.

Samsuri, A. and Zoveidavianpoor, M. (2014). Does the maturity of *Jatropha curcas* L. affect the quality and quantity of the yield of oil for biodiesel production? International *Journal of Green Energy*, 11: 193-205, DOI:10.1080/15435075.2013.772055.

Santoso, B.B., Purwoko, B.S. and Jaya, I.K.D. (2014). Yield of *Jatropha curcas* L. Accessions of West Nusa Tenggara During five a year production cycle on a degraded agricultural land. *Journal of Degraded and Mining Lands Management*, 1 (3), 123-130. DOI:10.15243/jdmlm.2014.01.

Sato, S., Hirakawa, H., Isobe, S., Fukai, E., Watanabe, A., Kato, M., Kawashima, K., Minami, C., Muraki, A., Nakazaki, N., Takahashi, C., Nakayama, S., Kishida, Y., Kohara, M., Yamada, M., Tsuruoka, H., Sasamoto, S., Tabata, S., Aizu, T., Toyoda, A., Shin-i, T., Minakuchi, Y., Kohara, Y., Fujiyama, A., Tsuchimoto, S., Kajiyama, S., Makigano, E., Ohmido, N., Shibagaki, N., Cartagena, J.A., Wada, N., Kohinata, T., Atefeh, A., Yuasa, S., Matsunaga, S. and Fukui K. (2010). Sequence Analysis of the Genome of an Oil-Bearing Tree, *Jatropha curcas* L. *DNA Research*, 1-12, (2010) doi:10. 1093/dnares/dsq030.

Singh, K., Singh, B., Verma, S.K. and Patra, DD. (2014). *Jatropha curcas*: a ten year story from hope to despair. *Renew Sustain Energy Rev*, 35, 356-60.

Terren, M., Cissé, C. and Mergeai, G. (2013). Analyse des perspectives de rentabilité de la culture extensive de *Jatropha curcas* L. dans la zone agroécologique de transition orientale du Sénégal: cas de la communauté rurale de Dialacoto (Analysis of the profitability prospects of the extensive cultivation of *Jatropha curcas* L. in the agroecological area of eastern Senegal's transition: the case of rural community Dialacoto). *Cah Agric*, 22: 568-574. doi:10.1684/agr.2013.0674.

Tintoré, S.G., Fuentes, C.T., Feria, J.S., Alaiz, M., Calle, J.G., Ayala, A.L.M., Guerrero, L.Ch. and Vioque, J. (2015). Antioxidant and Chelating Activity of Nontoxic *Jatropha curcas* L. Protein Hydrolysates Produced by *In Vitro* Digestion Using Pepsin and Pancreatin. *Journal of Chemistry,* Volume 2015, Article ID 190129, 9 pages, http://dx.doi.org/10.1155/2015/190129.

Trebbi, D., Papazoglou, E.G., Saadaoui, E., Vischi, M., Baldini, M., Stevanato, P., Cettul, E., Sanzone, A.P., Gualdi, L. and Fabbri, A. (2015). Assessment of genetic diversity in different accessions of *Jatropha curcas. Industrial Crops and Products 2015* (article in press). http://dx.doi.org/10.1016/j.indcrop.2015.06.051.

Utomo, C., Subroto, A.P., Darmawan, Ch., Setyobudi, R.H., Nugroho, Y.A. and Liwang, T. (2015). Construction for Δ-12 Fatty Acid Desaturase (FAD2) Silencing to Improve Oil Quality of *Jatropha curcas* Linn. *Energy Procedia*, 62: 36-41.

Veronesi, C.O., Alves de Souza, C.M., Serra, A.P., Rafull, L.Z.L., da Silva, C.J., Ros, V.V. and Conrad, V.A. (2014). Quality of *Jatropha curcas* L. Seeds Harvested at Different Stages of Fruit Maturation. *American Journal of Plant Sciences*, 5, 2716-2725, http://dx.doi.org/10.4236/ajps.2014.518287.

Waghmare, T. and Naik, G.R. (2015). Chemical analysis of seed and oil of *Jatropha curcus* L., a biofuel plant cultivated in Hyderabad Karnataka region, North Karnataka. *International Journal of Applied Biology and Pharmaceutical Technology*, 6 (1), 186-192.

Wu, P., Zhou, Ch., Cheng, S., Wu, Z., Lu, W., Han, J., Chen, Y., Chen, Y., Ni, P., Wang, Y., Xu, X., Huang, Y., Song, Ch., Wang, Z., Shi, N., Zhang, X., Fang, X., Yang, Q., Jiang, H., Chen, Y., Li, M., Wang, Y., Chen, F., Wang, J. and Wu, G. (2015) Integrated genome sequence and linkage map of physic nut (*Jatrophacurcas* L.), a biodiesel plant. *Plant J.*, 81 (5), 810-21 doi: 10.1111/tpj.12761.

Xiao, J.H., Mao, X., Zhang, H. and Niu, L. (2015). Detoxification of *Jatropha curcas* oil by ultraviolet irradiation combined with ethanol washing. *Grasas y aceites*, 66 (1), e063. doi: http://dx.doi.org/10.3989/gya.0723142.

Yang, Ch., Deng, X., Fang, Z. and Peng, D.P. (2010). Selection of high-oil-yield seed sources of *Jatropha curcas* L. for biodiesel production. *Biofuels*, 1(5), 705-717.

Yi, Ch., Reddy, Ch., Varghese, K., Ngoc, T., Bui, H., Zhang S., Kallath, M., Kunjachen, B., Ramachandran, S. and Hong, Y. et al. (2014). A new *Jatropha curcas* variety (JO S2) with improved seed productivity. *Sustainability*, 6, 4355-4368.

Yue, G.H., Lo, L.Ch., Sun, F., Cao, S.Y., Yi, Ch.X., Hong, X. and Sun, W.B. (2014). No Variation at 29 Microsatellites in the Genome of *Jatropha curcas*. *Journal of Genomics*, 2, 59-63.

In: The Promising Future of *Jatropha curcas*
Editors: Maira R. Segura-Campos et al.

ISBN: 978-1-63484-989-0
© 2016 Nova Science Publishers, Inc.

Chapter 2

CHEMICAL COMPOSITION AND BIOLOGICAL ACTIVITIES OF *JATROPHA CURCAS*

Lucero Méndez[1] and Janne Rojas[2,*]
[1]Polymer and colloids laboratory "POLYCOL," Faculty of Engineering, University of Los Andes, Mérida, Venezuela
[2]Organic Biomolecular Research Group, Research Institute, Faculty of Pharmacy and Bioanalysis, University of Los Andes, Mérida, Venezuela

ABSTRACT

Jatropha curcas L. belongs to Euphorbiaceae family, distributed mainly in México, Central America, India, Ecuador, Perú, Brazil, Bolivia, Paraguay and Venezuela. This species synthesize a variety of secondary metabolites such as diterpenes, triterpenes, coumarins, flavonoids, alkaloids, fatty acids, among others. Several investigations have also revealed a wide range of biological properties like antibacterial, antifungal, antiseptic, diuretic, laxative, antimalarial, anti-cancer, anti-inflammatory, cytotoxic, insecticide and antioxidant. This chapter summarizes the most important aspects of *Jatropha curcas* species, in order to enable handle information to the reader. A brief description of the most distinctive taxonomy features, worldwide distribution and biosynthetic pathways of the secondary metabolites commonly isolated from it has been described, as well as a review of different biological activities reported for this species.

Keywords: *Jatropha curcas*, secondary metabolites, biological activities

JATROPHA GENUS

Family Euforbiaceae is one of the largest and most geographically worldwide distributed. According to its classification, it is grouped into five subfamilies and 52 tribes. These are

* Corresponding author: Janne Rojas Vera. Tel.: + 58 274 2403557; fax: + 58 274 2403455. E-mail address: janner @ula.ve.

divided into Sub-tribes and about 300 genuses containing over 1.100 species are known to date (Webster, 1994). *Jatropha* gender is composed by 170 species worldwide, approximately. Some examples are described below:

J. aconitifolia Miller,	*J. elliptica (Pohl)* Muell. Arg.	*J. integérrima* Jacq.,
J. acotinifolia Miller,	*J. kunthiana* Muell-Arg.	*J. gaumeri* Greenm.
J. hastata Jacq.	*J. macrocarpa* Griseb,	*J glandulifera* Roxb.
J. capensis (L.f.) Sond.	*J. molissima* Muell. Arg.	*J. gossypifolia* Linn.
J. cilliata M. Arg	*J. grossidentata* Pax ex Hoff.	*J. multifida* Linn.
J. chevalieri Beille	*J. angustidens* (Torr.) Müll. Arg.	*J. podagrica* Hook.
J. curcas Linn.	*J. heynii* Balnom Nov.	*J. unicostata* Balf.f
J. curcus Linn.	*J. hieronymi* Kuntze.	*J. urens* Linn. (Webster, 1994)

JATROPHA CURCAS BOTANICAL FEATURES

Jatropha curcas is a shrub (herbaceous plant) perennial, about 2 meters high, grayish-white bark and translucent latex. The plant shows articulated growth, with a morphological discontinuity at each increment. Normally, five roots are formed from seedlings, one central and four peripheral. Leaves are simple, long petiolate, cordate and deep, with 3-7 lobes acuminate with a length and width of 6 to 15 cm, which are arranged alternately. Inflorescences are formed terminally on branches and are complex, possessing main and co-florescences with paracladia. Flowers are male and small (6-8 mm). Each inflorescence, white, produces a beam of about 10 or more ovoid fruits. Fruits are drupaceous and ovoid capsules; initially green, becoming dark brown or black when these enter the stage of maturation. Fruit development needs 90 days from flowering to seed matures. At the beginning are fleshy but when dry, it opens allowing the output of seeds (dehiscent). When fruit changes color from green to yellow, seeds are ripe. Seeds are black, 2 cm long and 1 cm thick (Dehgan and Webster, 1979; Heller, 1996; Berry et al., 1999; Guerrero, S/A). In Table 1 the taxonomic classification of *Jatropha curcas* L. species is presented.

JATROPHA GENUS WORLDWIDE DISTRIBUTION

Jatropha genus is represented by approximately 170 species. It was developed mainly in Mexico, Central America, western India, Colombia, Guyana, Suriname, French Guiana, Ecuador, Peru, Brazil, Bolivia, Paraguay, Venezuela and Argentina. In the United States this genus is spread in sandy, dry forests, hills and sandy fields, mainly on the coastal plain from Virginia to Florida, Hawaii and Texas (Dehgan and Webster, 1979; Jongschaap et al., 2007). Figure 1 shows the global distribution of this genus.

Table 1. Taxonomic classification of *Jatropha curcas* Linn species

Family	Euphorbiaceae
Subfamily	Crotonoideae
Tribe	Jatropheae
Gender	*Jatropha*
Species	*J. curcas* Linn

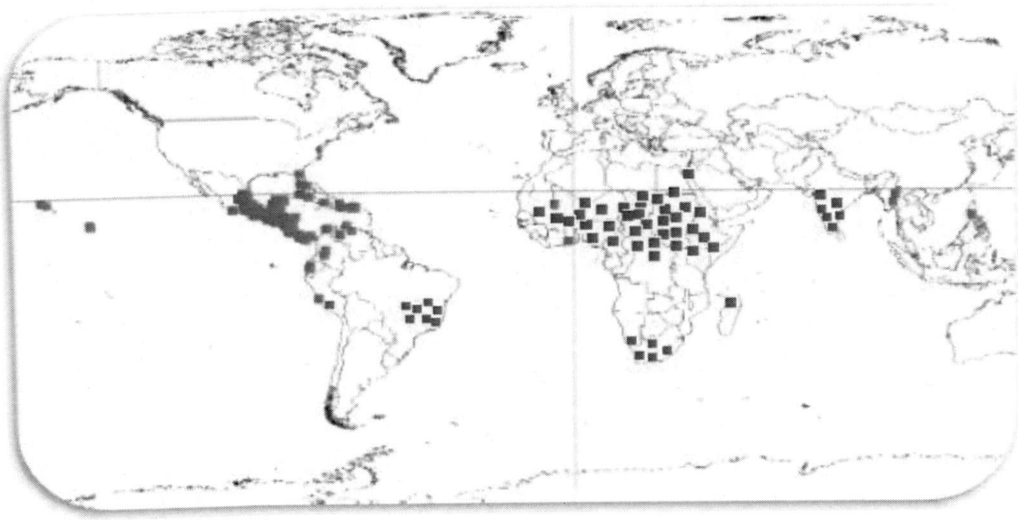

Dehgan and Webster, 1979.

Figure 1. Worldwide distribution of the genus *Jatropha*.

PHYTOCHEMISTRY OF *JATROPHA CURCAS* L.

Plants produce primary and secondary metabolites (Figure 2). Primary metabolites have high molecular weight and immediate utility on the organism that produces it, while secondary metabolites have low molecular weight. Secondary plant metabolites can be divided into four major categories: (a) polyketides, (b) terpenes, (c) phenolic compounds and (d) alkaloids. Compounds classified as terpenes are undoubtedly the largest and diverse class of natural products (Croteau et al., 2000).

Metabolites produced by this genus are vary and complex. There have been reported alkaloids, diterpenes, triterpenes, tannins, cyanogenic glycosides, glucosinolates, flavonoids, steroids, coumarins, steroidal sapogenins, lignans, cyclic peptides and phenolic glycosides. It also contains fixed oils and fatty acids (Odebiyi 1980 and 1985; Aiyelaagbe et al., 2007).

Besides the known biosynthetic pathways (acetate-malonate, mevalonate, shikimate and alkaloids), there are also compounds that are synthesized by the union of two routes (acetate malonate and shikimate), such as flavonoids, several alkaloids and glucosides, which are considered to come from a mixed route. Each biosynthetic pathway and chemical compounds derived from it are described below.

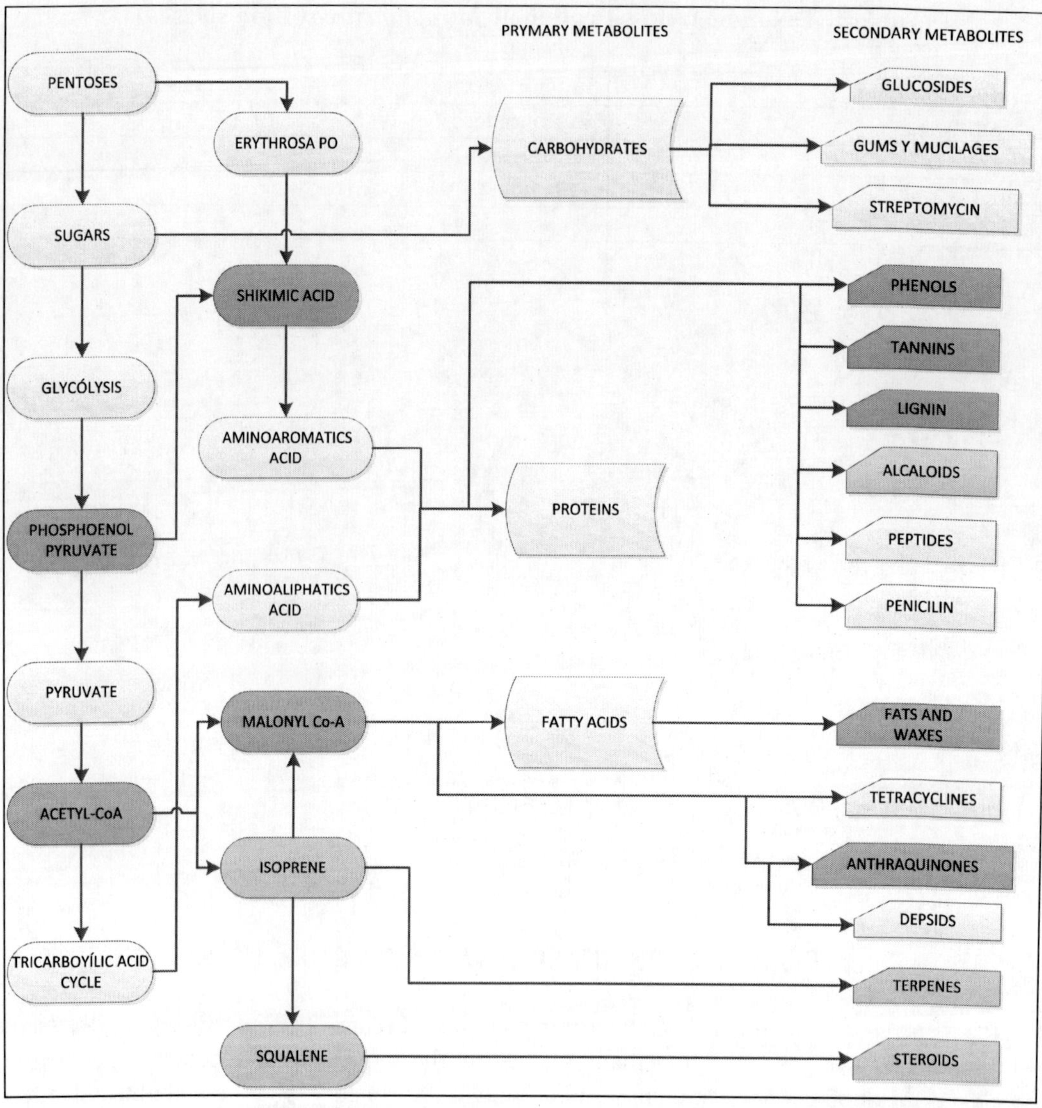

Figure 2. Biosynthetic route followed by plants to produce metabolites (Morales and Negrón, 1988). Biosynthetic pathways are essentially four: acetate-malonate pathway (blue), the mevalonate pathway (green), shikimate pathway (violet) and the alkaloids pathway (yellow).

Acetate-Malonate Pathway

Through acetate-malonate pathway, aliphatic substances such as fatty acids and aliphatic acetogenins polyacetylenes are formed. They are generally lightly branched acyclic structures, which are formed by consecutive condensation of acetyl coenzyme A and malonyl coenzyme A. Figures 3 and 4 show metabolic acetate-malonate pathway for the formation of fatty acids and aromatic polyketides.

Compounds isolated from *Jatropha* species biosynthesized by acetate malonate pathway are shown in Table 2. These are mostly fatty acids, extracted from seeds of *J. curcas* by

Martinez-Herrera et al., 2006, and one anthraquinone, isolated from aerial parts of same species by Ravindranath et al., 2004b.

Mevalonate Pathway

This route is named after the mevalonic acid which is the precursor of isoprene. It generates the full range of terpenes: monoterpenes, sesquiterpenes, diterpenes, triterpenes and derivatives (saponins and steroids) and tetraterpenes. The structural unit of terpenes is the active isoprene, forming isopentenilpirofosfato units (IPPF) and dimetilalilpirofosfato (DMAPF) (Marcano and Hasegawa, 2002). Figure 5 shows schematically the formation of terpene type components.

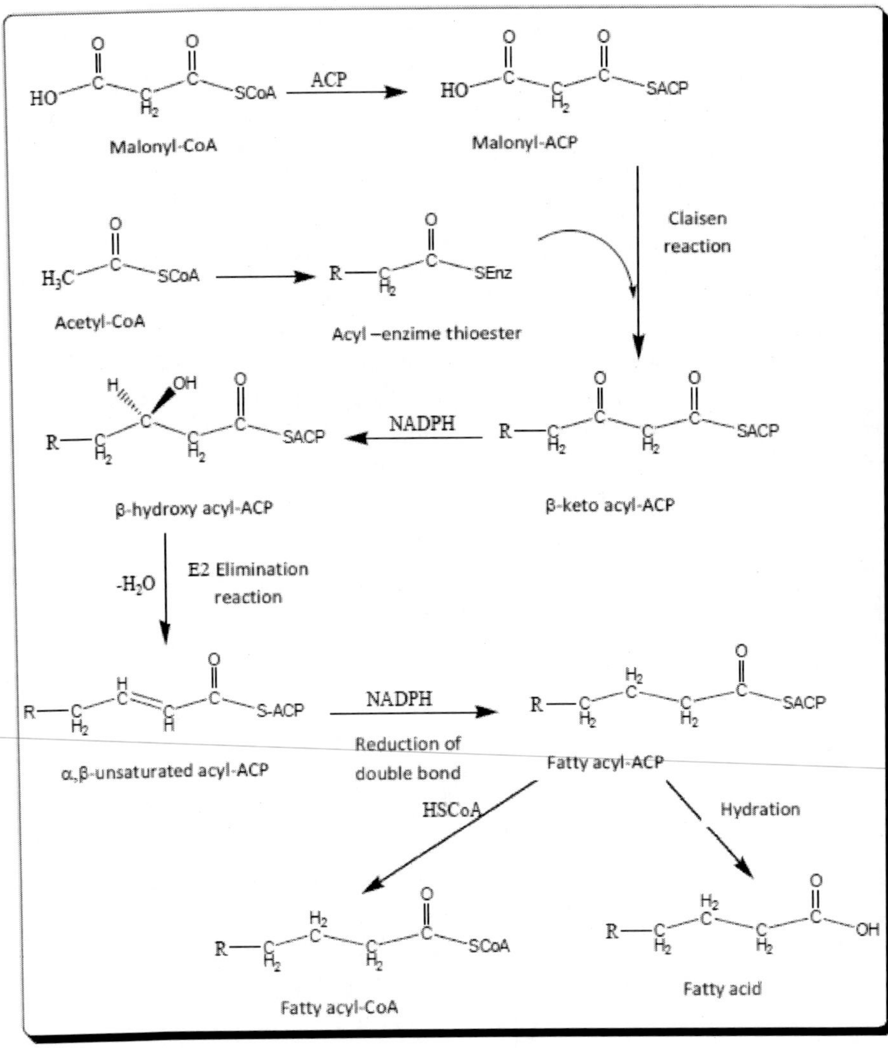

Dewick, 2002.

Figure 3. Acetate-malonate metabolic pathway for the formation of fatty acids.

Table 2. Isolated compounds from *J. curcas* biosynthesized by the acetate-malonate pathway

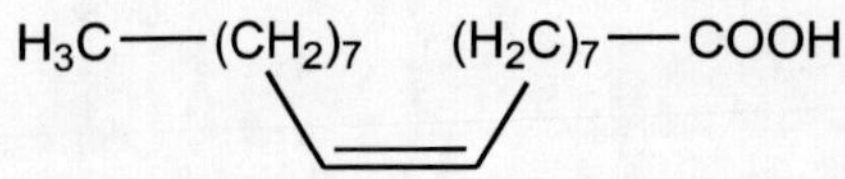

Oleic acid

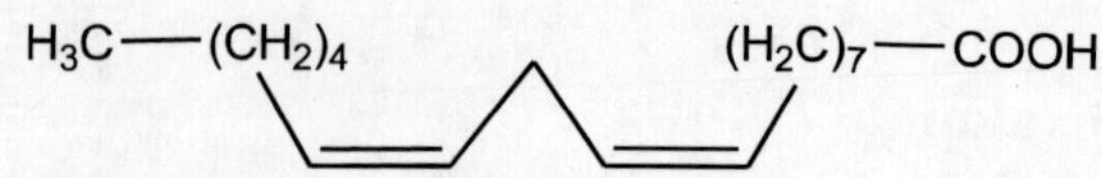

Linoleic acid

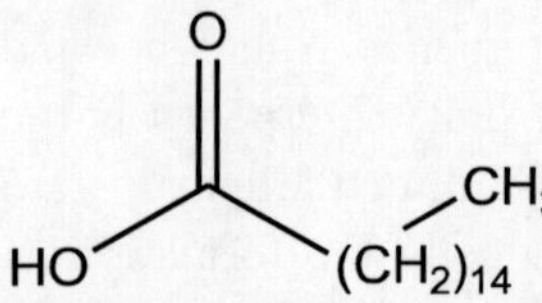

Palmitic acid

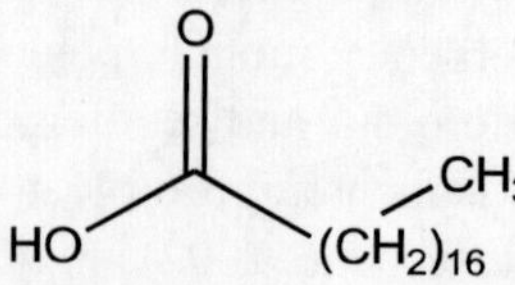

Estearic acid

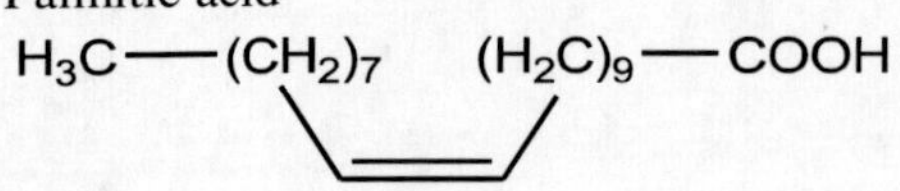

cis-11-eicosenoico acid

cis-11,14-eicosadienoic acid

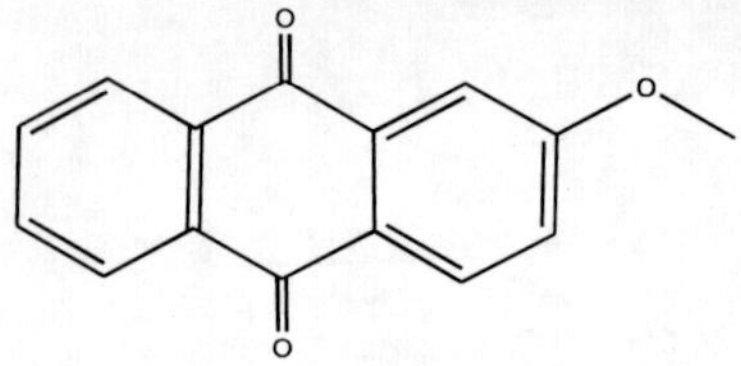

2-Metoxy-anthraquinone

Jatropha curcas is known to synthesize large quantities of oil; however, this oil is extracted from seeds and does not belong to the essential oil classification since it is not volatile and chemical composition contains a number of long chain hydrocarbons instead of monoterpenes and sesquiterpenes. However, diterpenes have been isolated from *Jatropha* genus. Diterpenes are the largest group of terpenes in nature. They are formed by two metabolic pathways, through the route of pyruvic acid (which is limited to certain species of plants such as *Gingko Biloba*) and mevalonate route, via formation of GGPP (geranyl geranyl pyrophosphate).

These are classified according to their biogenetic origin: acyclic (phytanes), bicyclic (labdanes, halimanes, clerodanes), tricyclics (pimaranes, abietanes, cassanes, rosanes, vouacapanes, podocarpanes), tetracyclic (trachlobanes, kaurenes, afidicolanes, estemodanes, estemaranes, bayeranes, atisanes, gibberellanes), diterpenes macrocyclic (taxanes, Cembranes, dafnanes, tiglianes, ingenanes) and mixed compounds (according to the amount and cyclization patterns). Diterpenes isolated from *Jatropha curcas* belong to rhamnofolane, dafnane, latirane, tigliane, dinorditerpene, and pimarane deoxy preussomerin nuclei (Devappa et al., 2011). Examples of these are shown in Table 3 (Adolf et al., 1984; Hirota et al., 1988; Gubitz et al., 1997; Hanson, 1998; Ravindranath et al., 2004a; Ravindranath et al., 2004b; Jing, 2005; Sabandar, 2010; Xu et al., 2011; Méndez and Rojas, 2012).

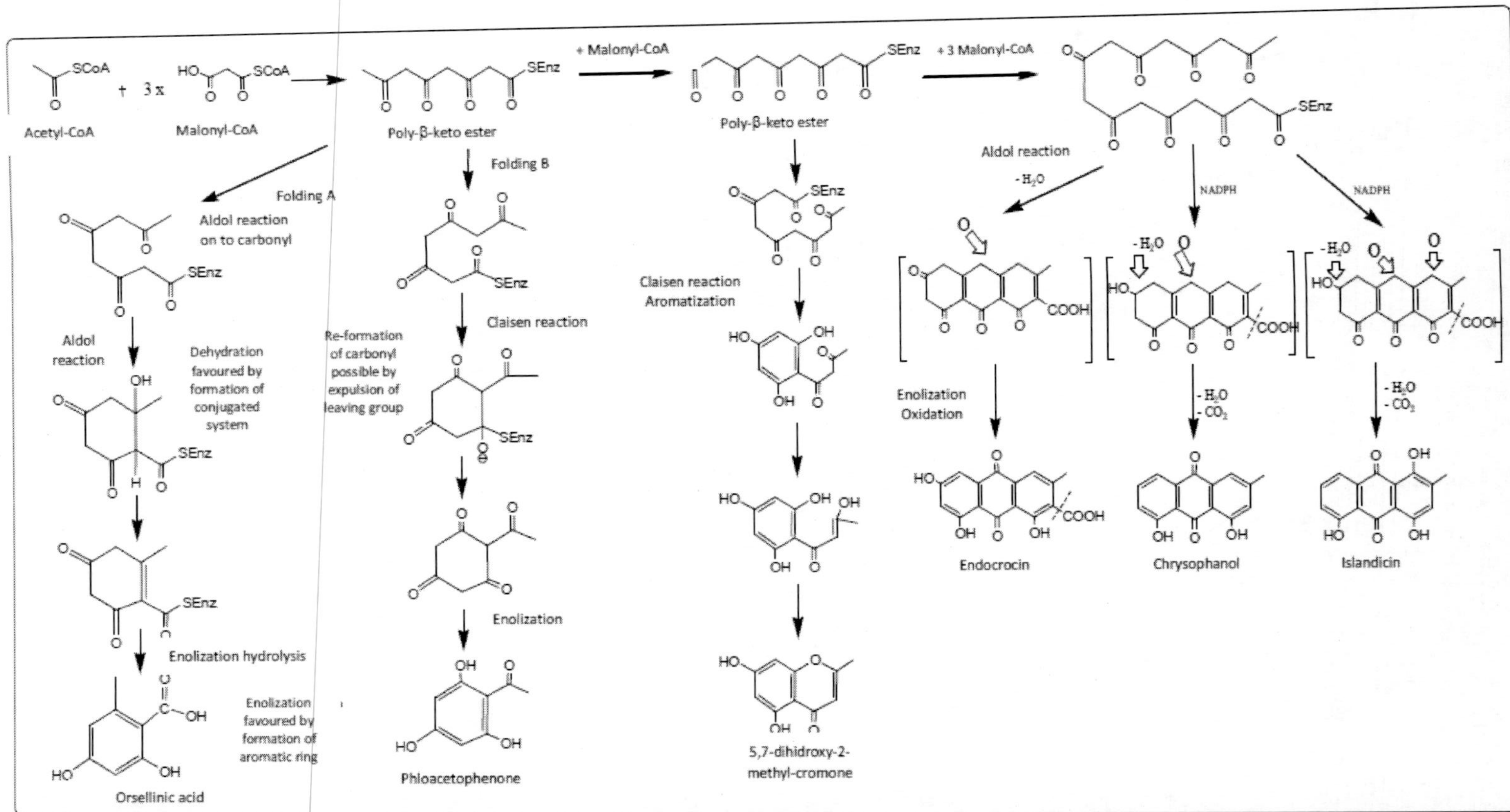

Dewick, 2002.

Figure 4. Acetate-malonate metabolic pathway for the formation of aromatic polyketides.

Table 3. Diterpenes isolated from *Jatropha curcas*

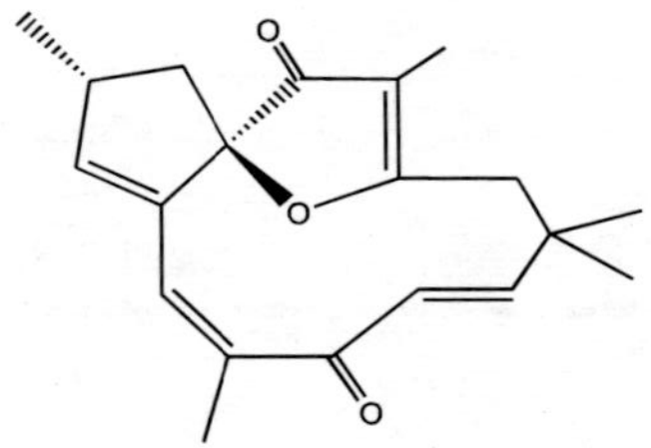

Jatrophone

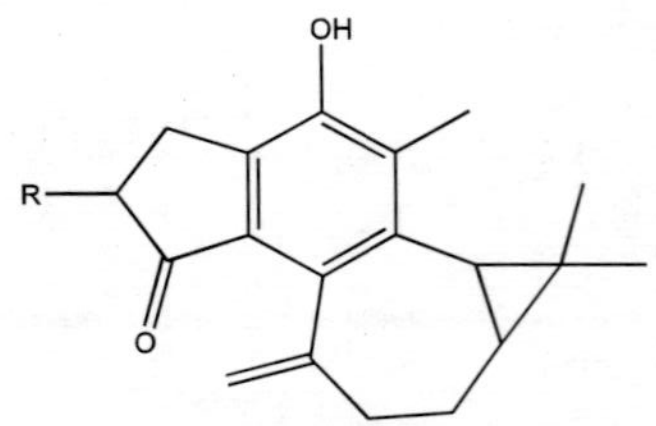

R = 2β-CH_3 Jatropholone A
R = 2α-CH_3 Jatropholone B

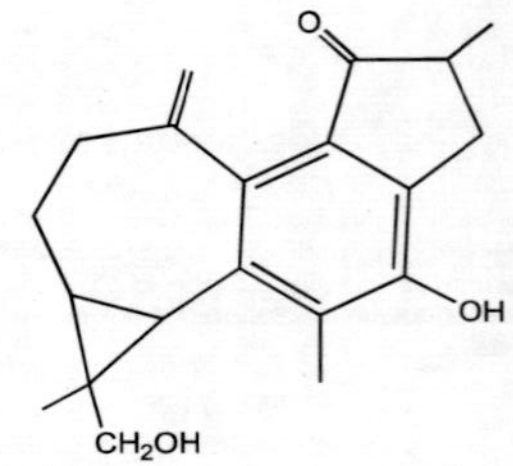

Jatrophol

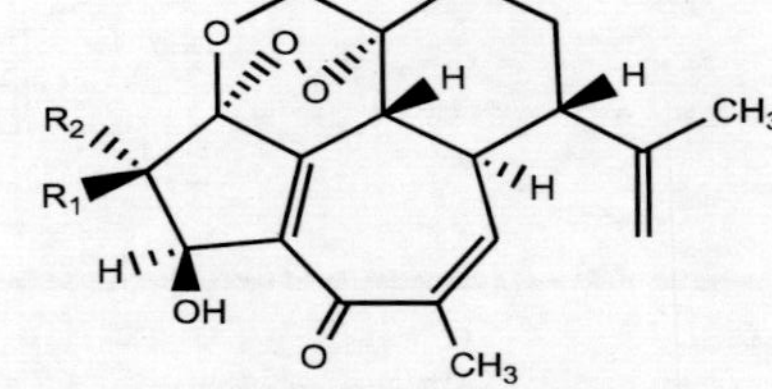

R1 = H and R2 = CH_3 Caniojane
R1 = CH_3 and R2 = H 2-*Epi*-caniojane
Caniojane con 1,11 Bis-*epi* 1,11-Bis-*epi*-Caniojane

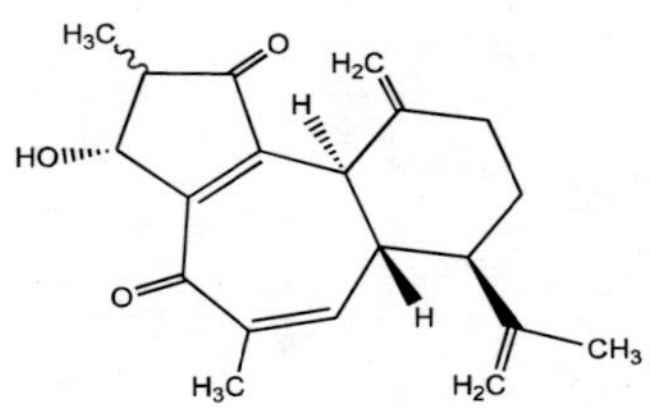

α-CH_3 Jatrogrossidione
β-CH_3 2-*Epi*-Jatrogrossidione

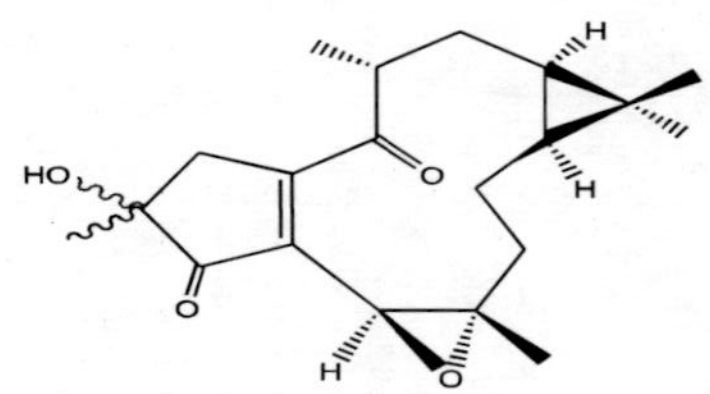

α-OH 2-Hidroxi-*iso*-Jatrogrossidione
β-OH 2-*Epi*-hidroxi-*iso*-Jatrogrossidione

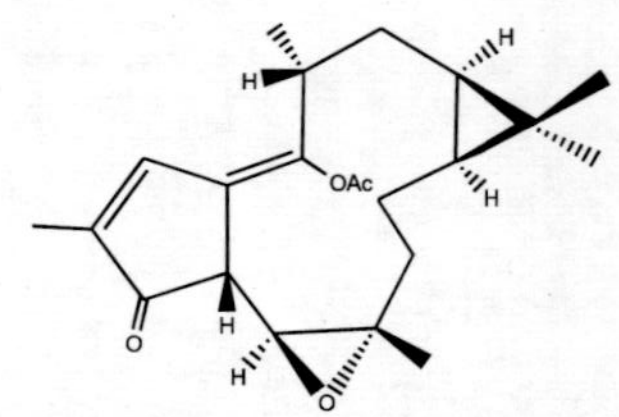

Iso-Jatrogrossidentadione

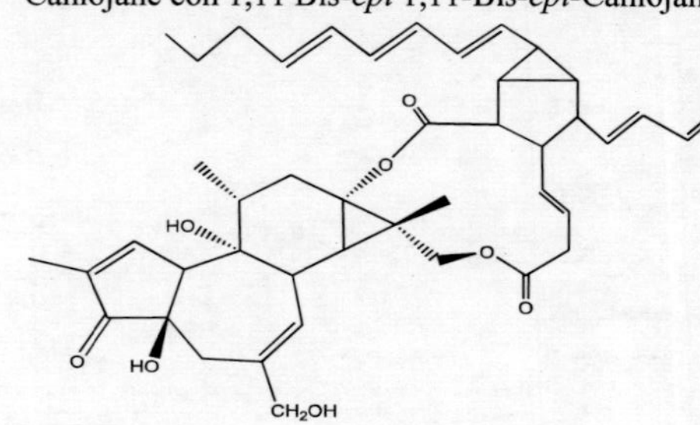

DPHB

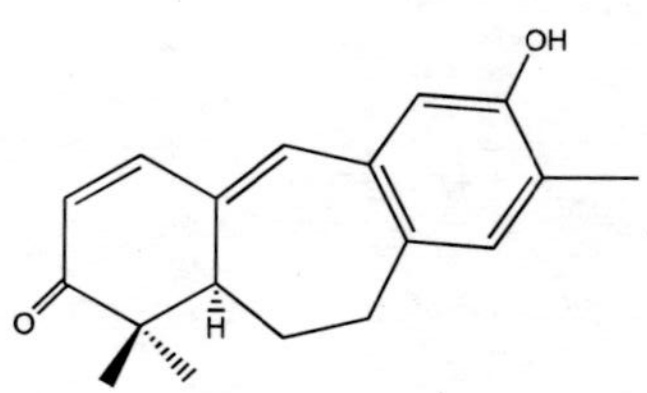

Heudelotinone

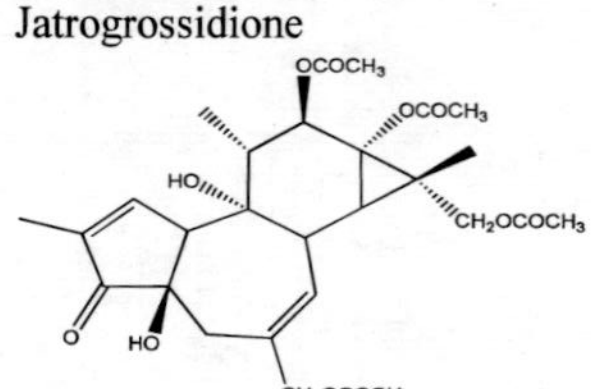

12-Deoxy-16-hidroxiforbol

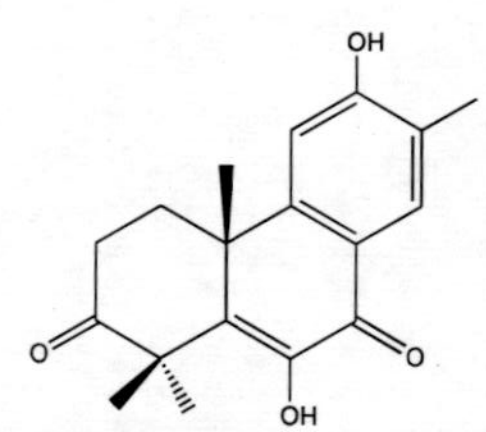

Gossweilone

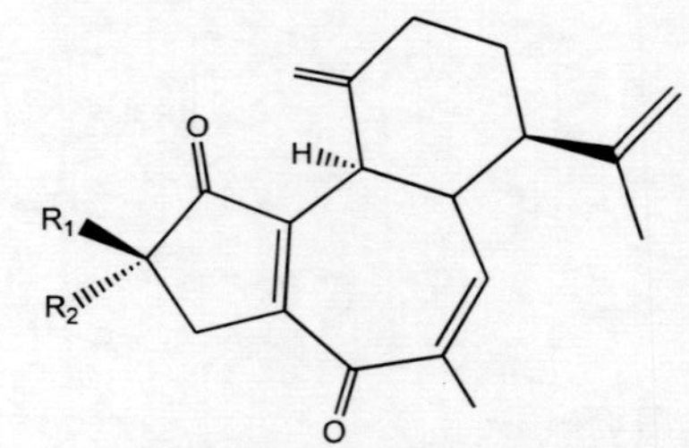

R_1 = CH_3 and R_2 = H Curcusone A
R_1 = H and R_2 = CH_3 Curcusone B
R_1 = CH_3 and R_2 = OH Curcusone C
R_1 = OH and R_2 = CH_3 Curcusone D

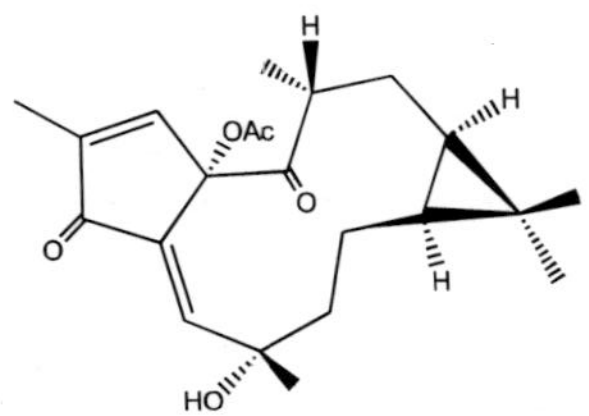
15-*O*-Acetyl-15-*epi*-(4E)-Jatrogrossidentadione

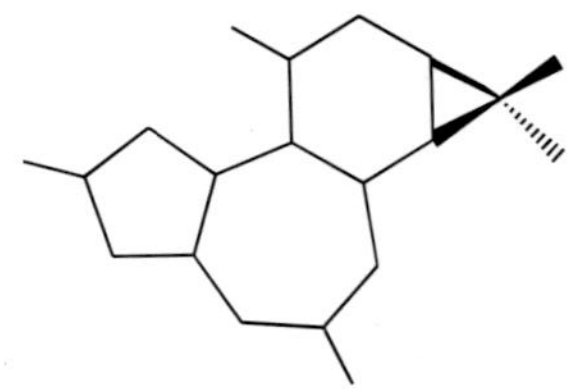
Tigliane

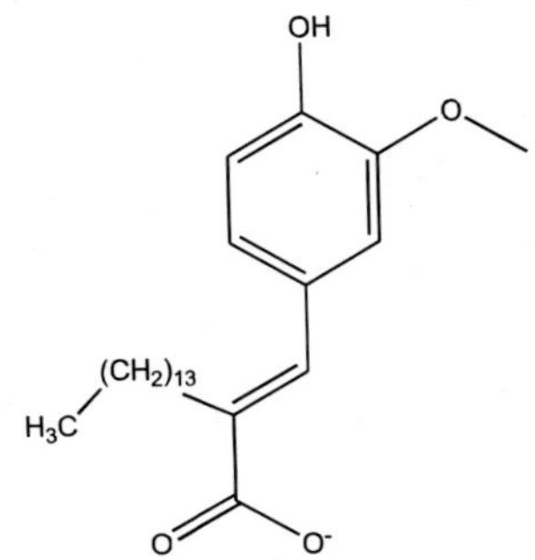
Tetradecil-(E)-ferulate

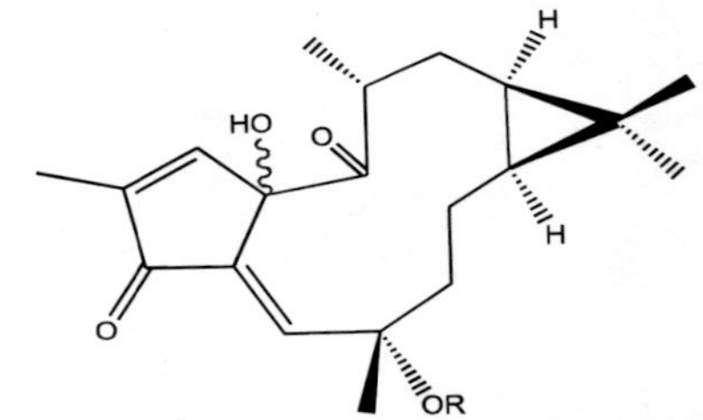
R = H and β-OH (4E)- Jatrogrossidentadione
R = H and α-OH 15-*Epi*-(4E)-Jatrogrossidentadione
R = Ac and β-OH (4E)-Jatrogrossidentadione Acetate

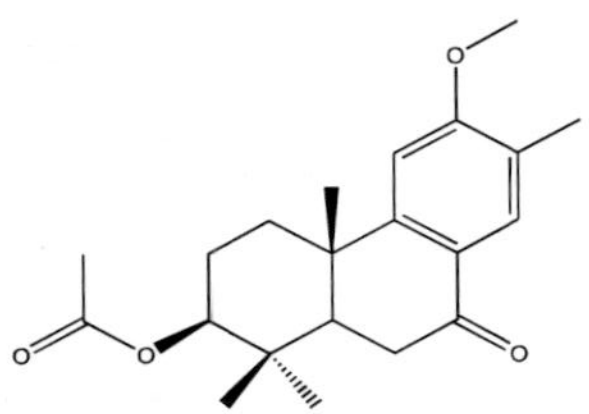
3β-Acetoxy-12-methoxy-13-methyl-Podocarpa-8,11,13-trien-7-one

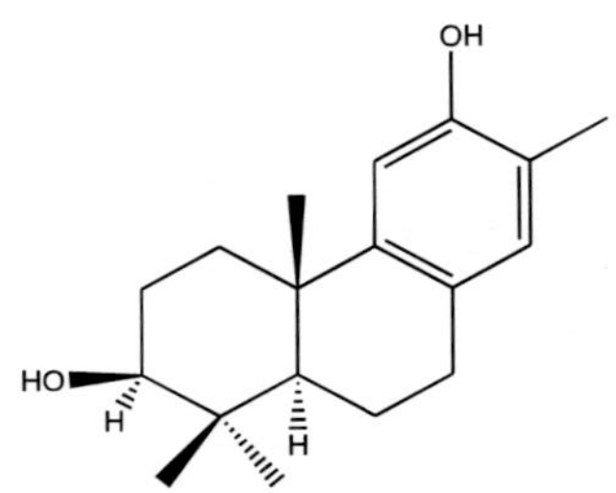
3β-12-dihydroxy-13-methyl-Podocarpane-8,10,13-triene

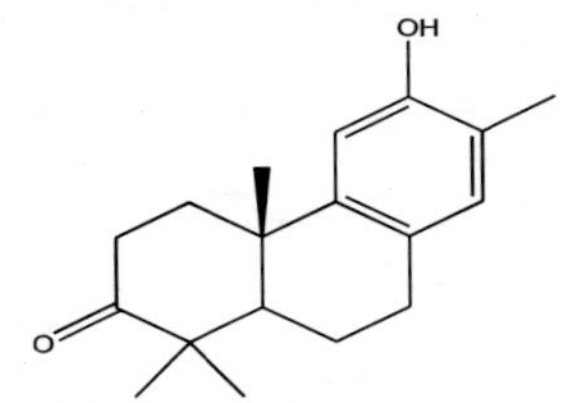
12-Hydroxy-13-methyl-podocarpa- 8,11,13-trien-3-3-one

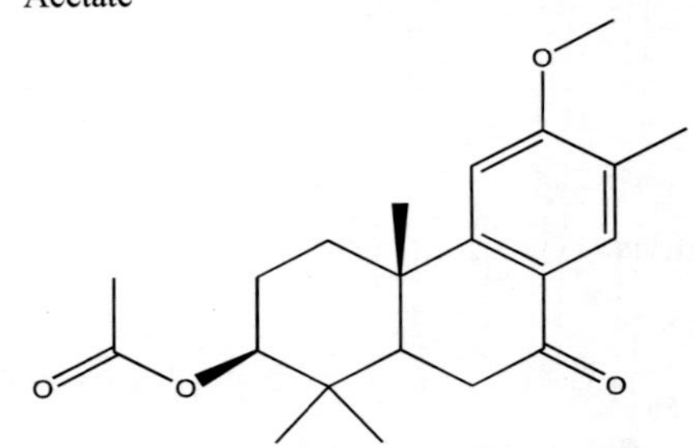
3β-Acetoxy-12-methoxy-13-methyl-Podocarpa-8,11,13-trien-7-one

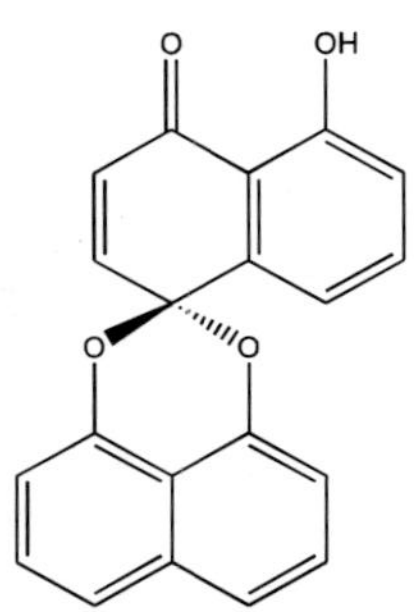
Palmarumicin CP1

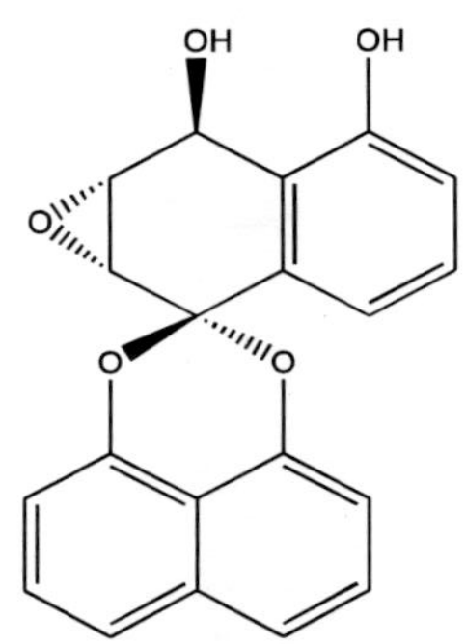
Palmarumicin JC1

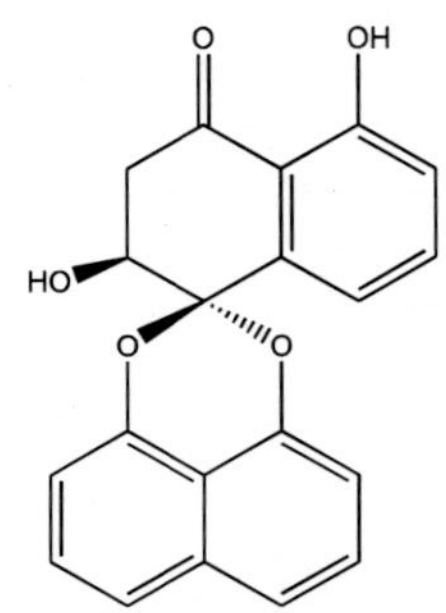
Palmarumicin JC2

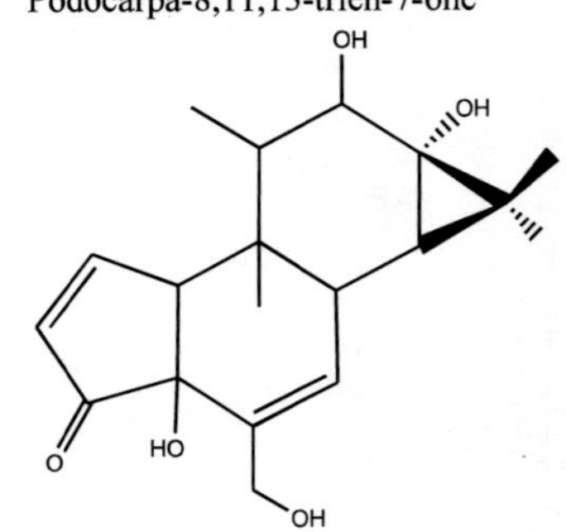
Jatropherol

Table 4. Triterpenes isolated from *Jatropha curcas*

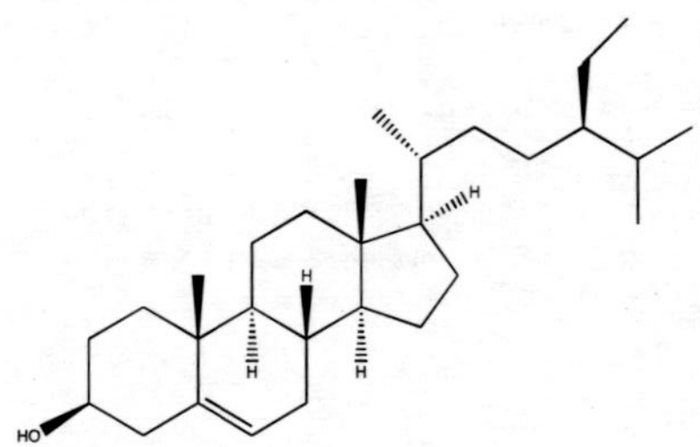

Sitosterol

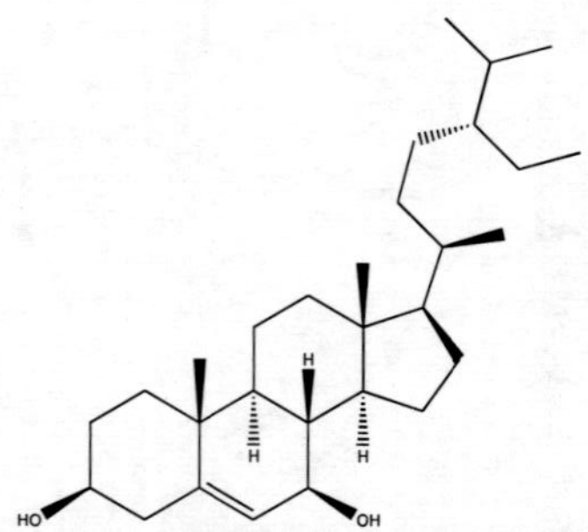

Estigmast-5-en-3β,7(α,β)–diol

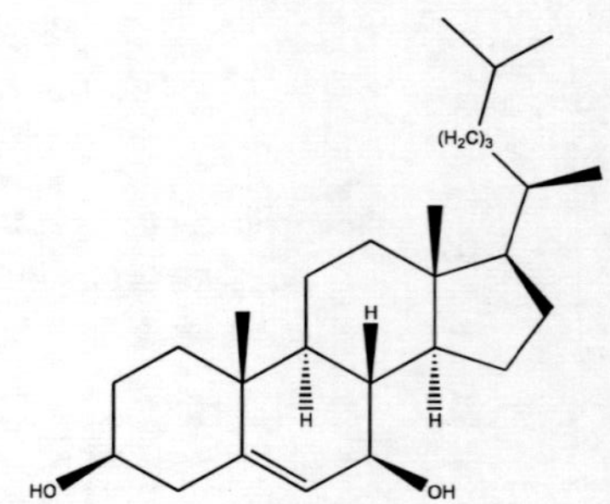

Colest-5-en-3β, 7β-diol

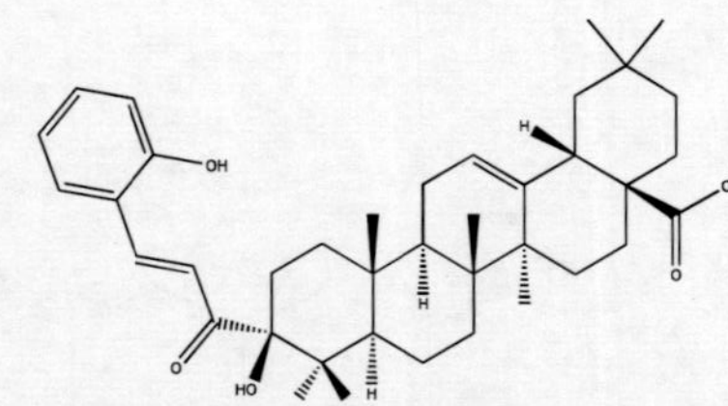

3-*O*-(Z)-coumaroyl oleanolic acid

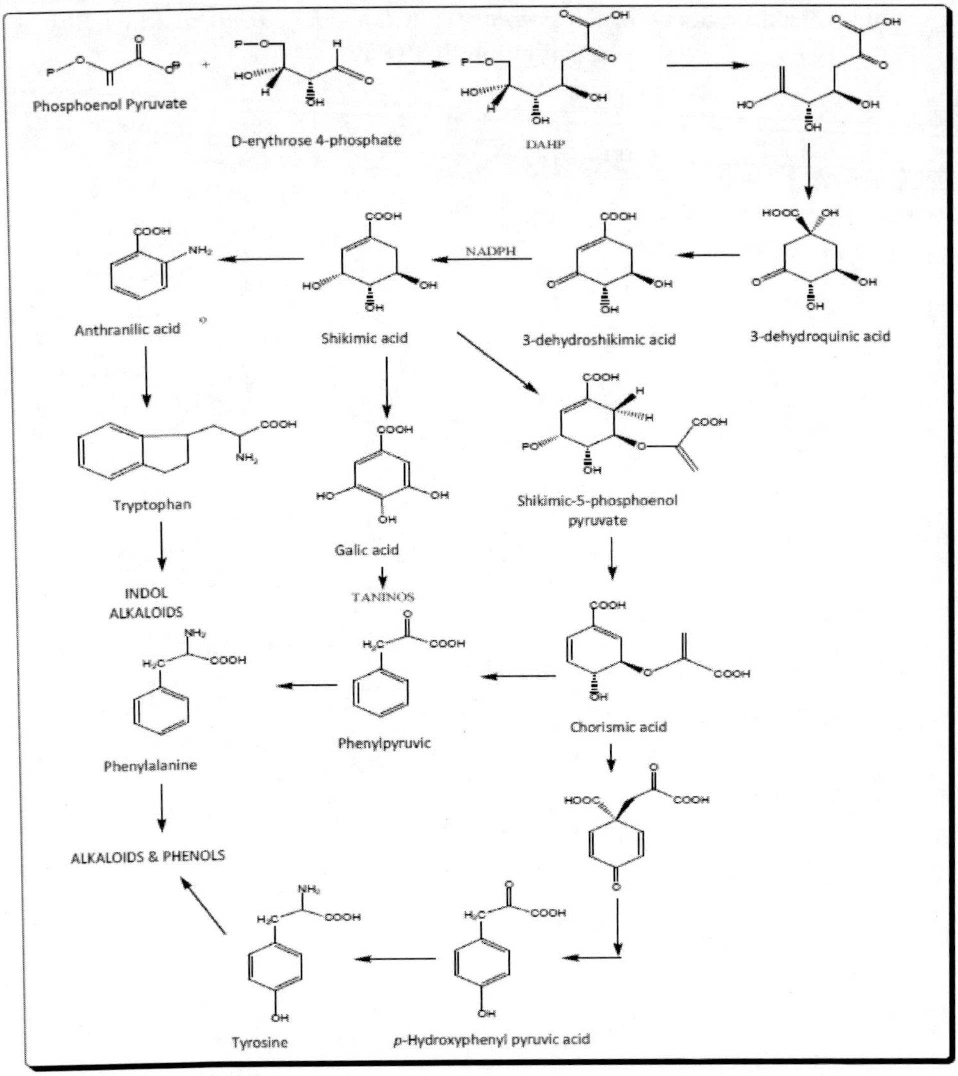

Marcano and Hasegawa 2002.

Figure 5. Shikimate pathway for the formation of alkaloids and phenolic compounds.

Triterpenes found in *Jatropha* species are sitosterol, stigmasterol and its derivatives (Mitra et al., 1970), cholest-5-en-3β, 7β-diol (Khafagy et al., 1977) and acid 3-O-(Z)-cumaroil oleanolic (Ravindranath et al., 2004a). Most of these compounds exhibit mainly different structural functions and are shown in Table 4.

Shikimate Pathway

This pathway produces natural aromatic compounds such as: benzene derivatives, phenylpropanoids and alkaloids (Marcano and Hasegawa 2002). Figure 6 shows the shikimate pathway.

Table 5. Isolated phenolic compounds from *Jatropha curcas* produced by the Shikimate pathway

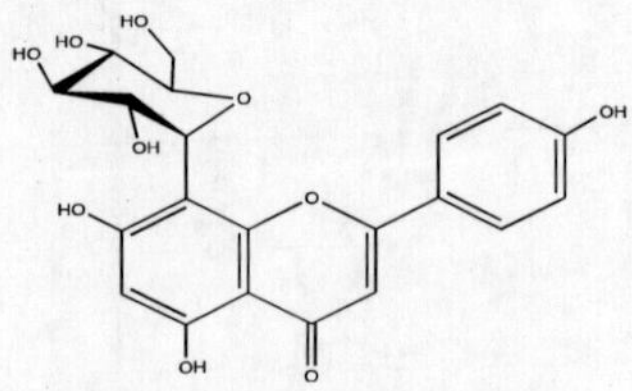
Vitexin

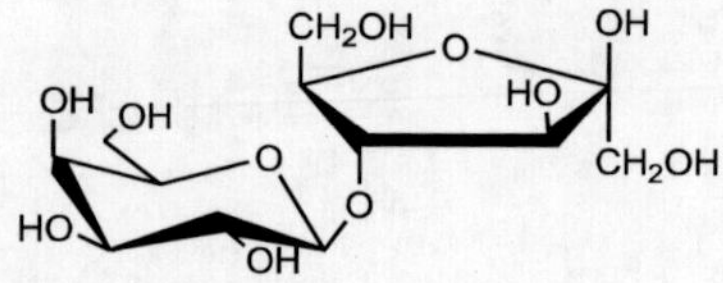

Lactulose

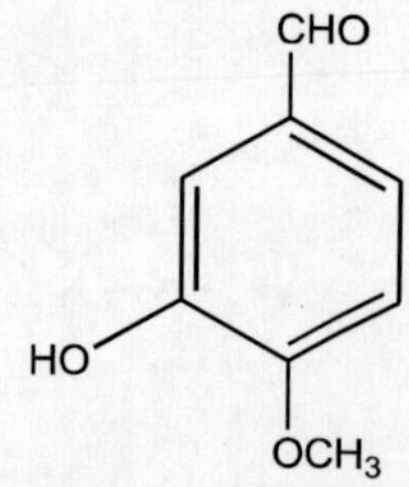

3-Hydroxy-4-methoxy-Benzaldehyde

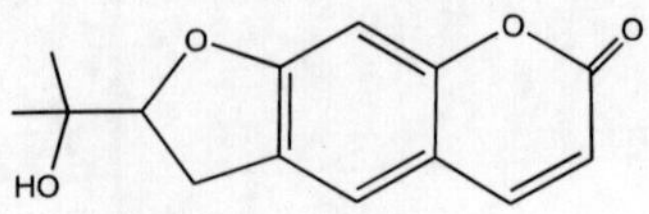

(+)-Marmesin

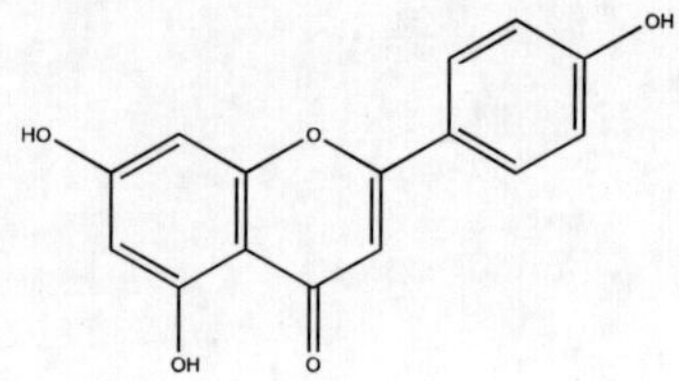
Apigenine

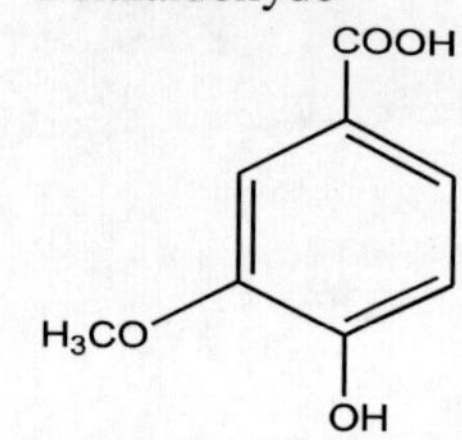

4 hydroxy-3.methoxy-benzoic acid

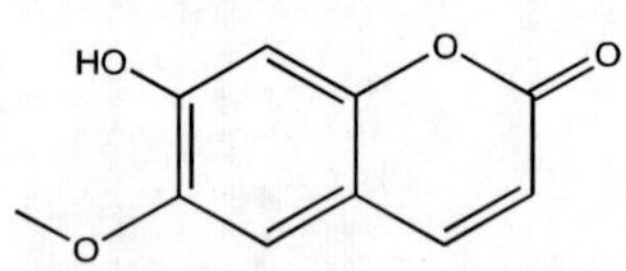

Escopoletin

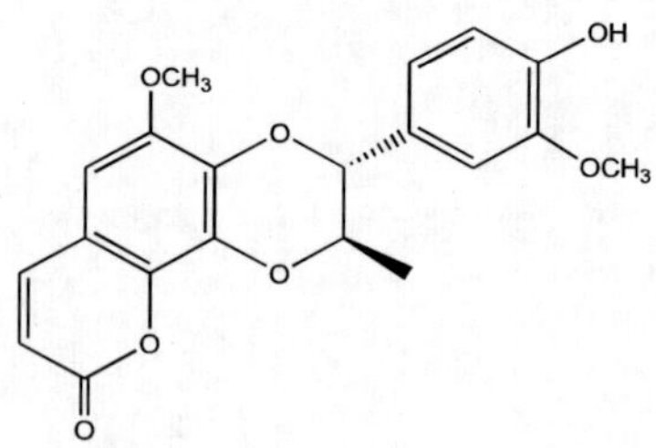
Propacin

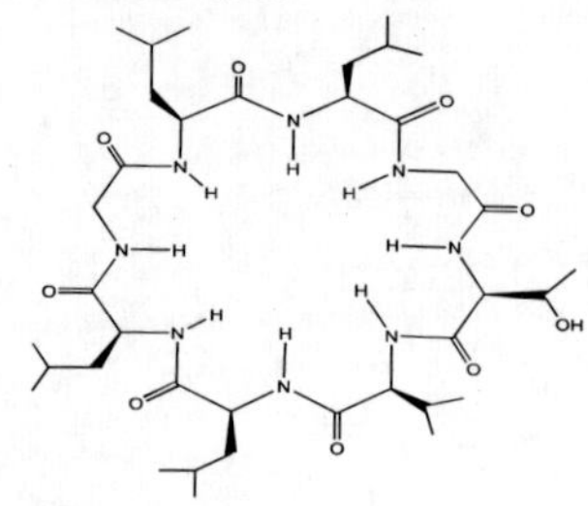
Curcacicline A

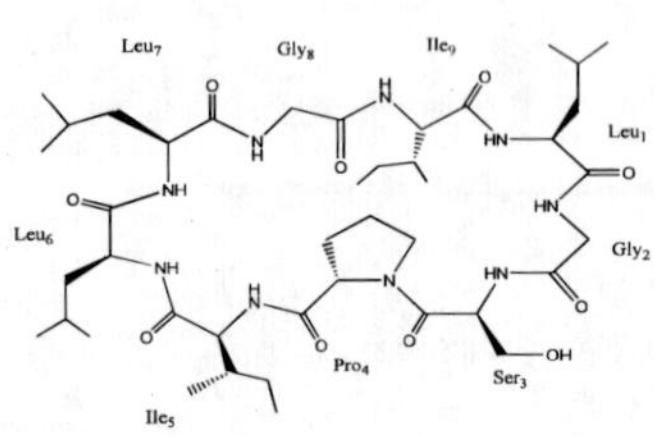
Curcacicline B

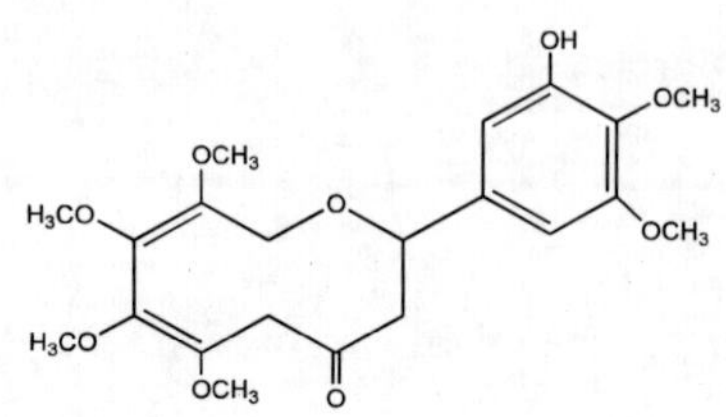
Nobiletin

Phenolic compounds are formed through the shikimate pathway, but also, there are some phenolic derivates which origin is mixed (Shikimate-Acetate) such as lignins, lignans, flavonoids, quinones, tannins, chromenes and coumarins. Some of the metabolized compounds by *J. curcas* through the biosynthetic pathway are shown in Table 5 and are represented by glycosides, flavonoids and phenolic compounds (Shankara et al., 1971; Naengchomnong et al., 1994; Ling-yi et al., 1996; Auvin et al., 1997; Staubmann et al., 1999; Ravindranath et al., 2004a; Sriprang et al., 2010; Das-Gupta et al., 2010).

Table 6. Alkaloids isolated from *Jatropha curcas*

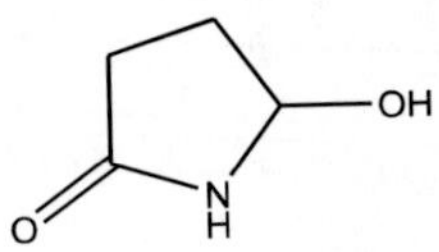

5-Hydroxypirrolidin-2-one

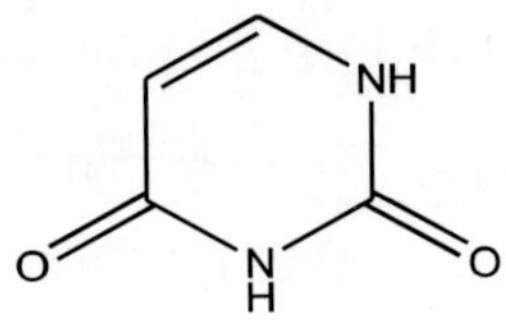

Pirimidin-2,4-dione

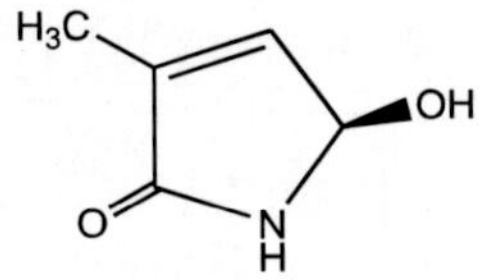

Jatrophine or Jatrofam lilidina

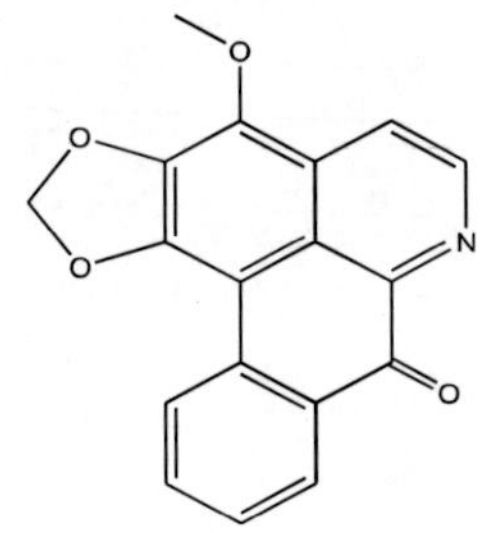

JC-1 (Atheospermidine)

Alkaloids Pathway

Alkaloids are formed by shikimate pathway; through aminoaromatic acids; however, these may also be synthesized through a mixed route, joining with aliphatic structures from the aminoaliphatics acids. These compounds exhibit a great carbon skeleton variety in its structure, thus might be classified as a separate biosynthetic pathway. The nitrogen frequently is present in its structure being formed from aminoacids as precursors (Marcano and Hasegawa, 2002). Isolated alkaloids from *J. curcas* are shown in Table 6 (Naengchomnong et al., 1994; Staubmann et al., 1999; Das-Gupta et al., 2010).

Biological Activities Investigated in *Jatropha curcas*

A number of biological activities have been investigated in *Jatropha curcas* species around the world, such as antibacterial (Aiyelaagbe et al., 2007; Arekemase et al., 2011; Das-Gupta et al., 2010; Igbinosa et al., 2009; Kalimuthu et al., 2010; Luseba et al., 2007; Namuli et al., 2011; Obasi et al., 2011; Oyi et al., 2007; Prashanth et al., 2006; Sangeetha et al., 2009; Sundari and Selvaraj, 2011; Thomas et al., 2008), antifungal (Aiyelaagbe et al., 2007; Arekemase et al., 2011; Córdova-Albores et al., 2011; Das-Gupta et al., 2010; Igbinosa et al., 2009; Kalimuthu et al., 2010; Makun et al., 2011; Obasi et al., 2011; Oyi et al., 2007; Oyi et al., 2007; Rahman et al., 2011; Sundari and Selvaraj, 2011;), antimalarial (Consoli et al.,1989), antitumor (Balaji et al., 2009; De Moraes et al., 1996; Hirota et al., 1988; Lin et al., 2003), anticancer (Balaji et al., 2009; Oskoueian et al., 2011b), anti-inflammatory (Luseba et al., 2007; Mujundar and Misar, 2003; Oskoueian et al., 2011b; Sangeetha et al., 2009; Uche and Aprioku, 2008), rheumatism (Irvine, 1961; Thomas et al., 2008), antiulcer (Kannappan et al., 2008), antihelmintic (Khamer 2008), insecticide (Adebowale and Adedire 2006; Jing, 2005; Jing et al., 2005; Sievers et al., 1949), antipyretic (Martínez, 1959; Bhavesh and Krishnakumar, 2010), antiviral (Matsuse et al., 1999), antidiabetic (Mishra et al., 2010), gastro (Kannappan et al., 2008;), cytotoxic (Das-Gupta et al., 2010; Faria et al., 2006), healing (Esimone et al., 2008; Sachdeva et al., 2011; James et al., 2011), clotting (Osoniyi and Onajobi, 2003), molluscicide (Rug and Ruppel, 2000), abortive (Goonasekera et al., 1995) and antidiarrhoeal (Mujumdar et al., 2000), although it has also been used as animal food resource (Devappa et al., 2010;Gübitz et al., 1997; Makkar et al., 1998; Sonibare et al.,

2008; Ye et al., 2009; Makkar et al., 2010; Murillo et al., 2010; Oseni and Akindahunsi, 2011; Devappa et al., 2011; Valdés et al., 2011). Table 7 shows different biological activities reported for oils, extracts and pure isolated compounds from *Jatropha curcas* species.

ORNAMENTAL, INDUSTRIAL AND MEDICINAL USES

J. curcas species is not commonly used as decorative due to its irritant action, however, the seeds are used as raw material for preparation of candles, soaps, paints, lubricants and biodiesel products (Carels, 2009; Mujumdar et al., 2000; Berchmans and Hirata, 2008; Debnath and Eisen, 2008; Kumar and Sharma, 2008; Parajuli, 2009; Parawira, 2010; Pramanik, 2003; Silva and Ulberich, 2008).

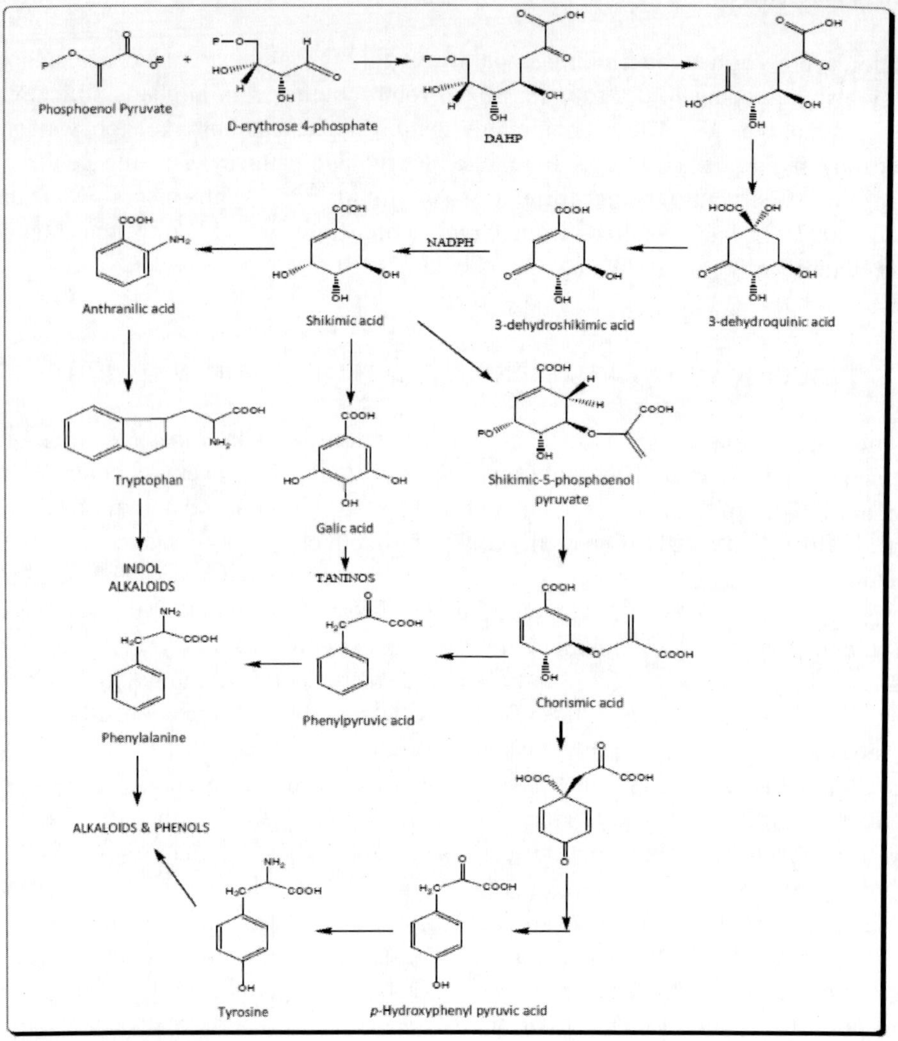

Marcano and Hasegawa 2002.

Figure 6. Shikimate pathway for the formation of alkaloids and phenolic compounds.

Table 7. Biological activities reported for *Jatropha curcas* L.

Structure or name extract	Obtaining source	Biological activity	Observations	References
Fruits	*J. curcas* fruits	Nutritional value	Possible human food by eliminating anti-nutritional compounds	Gübitz et al., 1997
Seeds	*J. curcas* shell seeds	Nutritional value	Food for albino rats on diets with more than 15% of shell seeds	Annongu et al., 2010
Seeds	*J. curcas* and *J. platyphylla* seeds germ	Nutritional value	*J. curcas* has more protein (32.39%), whereas *J. platyphylla* has more fat content (54.53%)	Murillo et al., 2010
Defatted flour	*J. curcas*	Antioxidant and chelanting effect	Antioxidant, hydrolyzing and chelanting peptides that could be used to remove heavy metals in the human body, was found	Gallegos-Tintore et al., 2010
Latex diluted in ethyl acetate, buthyl alcohol and water	*J. curcas* latex	Clotting activity	Dilution in butyl alcohol is more active than dilution in water	Osoniyi and Onajobil 2003
Oil	*J. curcas* seeds	Toxicity	Calf death occurred between 10 and 14 days with a dose of 0.025 g/kg body weight	Ahmed and Adam, 1979
			Mexican varieties of *Jatropha* have fewer phorbol esters that Nigerian varieties	Makkar et al.,1998
			Goats death between 7 and 21 days. They presented hemorrhage, congestion, pulmonary hemorrhage, whole-hepato-nephrotoxicity, and emphysema with doses between 0.25-1.0 g/Kg	Abdel Gadir et al., 2003
		Insecticide	Ovicide activity against *Callosobruchus maculatus*. This oil provides 12 weeks of protection against insects	Adebowale and Adedire 2006
		Nutritional value	The extracted fiber content is between 3.9% and 4.5% dry matter (DM). The main fatty acids found were acids: oleic (41.5-48.8%), linoleic (34.6-44.4%), palmitic (10.5 to 13.0%) and (C20: 1) stearic (2.3 to 2.8%).	Martínez-Herrera et al., 2006
			Seeds are considered as food source with a tolerable anti-nutrients level	Abou-Arab and Abu-Salem 2010

Table 7. (Continued)

Structure or name extract	Obtaining source	Biological activity	Observations	References
Oil	*J. curcas* seeds	Fermentation	Samples treated with *Aspergillus niger* shows increased crude protein content	Belewu and Sam 2010
		Lectin activity	Lectina is active in seeds, using agglutination and hemagglutination latex tests	Aregheore et al., 1998
	J. curcas, *hieronymi* and *macrocarpa* seeds	Bioenergetic potential	Oils possess about 72% of unsaturated fatty acids. Oleic and linoleic acids: acids predominate.	Falasca and Ulberich 2008
	J. curcas seeds	Antibacterial and antifungal	Active against *Pseudomonas aeruginosa, Bacillus subtilis, Escherichia coli, Staphylococcus aureus, Streptococcus pyogenes, Candida albicans and Trichopyton sp.*	Oyi et al., 2007
Oil	*J. curcas* seeds	Antibacterial and antifungal	MIC for *Neisseria gonorrea, Staphylococcus aureus and Candida albicans* = 1.0 mL MIC for *Escherichia coli and Pseudomonas aeruginosa* = 0.5 mL	Arekemase et al., 2011
		Antifungal	Mycelial growth inhibition in three *Fusarium* species	Córdova-Albores et al., 2011
Oily and defatted extract	*J. curcas* seeds	Antifungal	Decreased mycelial growth and reducing the rate of putrefaction in *Fusarium verticilliodes and Aspergillus flavus*	Makun et al., 2011
Petroleum ether, dichloromethane and methanol extracts	*J. curcas* fruits	Abortive	Fetal resorption observed between the sixth and the eighth day of gestation in Sprague-Dawley rats. by oral administration of extracts	Goonasekera et al., 1995
Petroleum ether and methanol extracts	*J. curcas* roots	Antiinflammatory activity	Reducing ear inflammation induced in albino mice	Mujumdar and Misar, 2003
Acetone, ethanol and water extracts	*J. curcas* leaves	Antifungal	100% inhibition against *F. oxysporum*	Falasca and Ulberich 2008
Petroleum ether extract	*J. curcas* stem bark	Antibacterial	Bactericidal against 3 Gram-positive and 5 Gram-negative	Das-Gupta et al., 2010
Hexane, methanol and water extracts	*J. curcas* root wood, seeds and, branches and root bark	Antibacterial	MIC between 1.6-18.8 mg/mL against *Bacillus cereus, B. subtilis, E. coli, K. pneumoniae, Pseudomonas aeruginosa and Staphylococcus aureus.*	Namuli et al., 2011

Structure or name extract	Obtaining source	Biological activity	Observations	References
Hexane and chloroform extracts	*J. curcas* seeds	Antibacterial and Antifungal	MIC between 62.5-500 µg/mL against *S. aureus, E. coli, K pneumoniae, P aeruginosa, S. Typhimurium and V. chlolarae.* MIC between 31.25-500 µg/mL against *C. albicans, A. niger, A. flavus, A. fumigatus, Rhizopus sp and Mucor sp.*	Sundari and Selvaraj 2011
Deffated flour	*J. curcas* seeds meal	Acaricide	J. curcas could be used in the treatment of ectoparasites at level less than 10% in diet.	Abdel-Shafy et al., 2011
Methanol, ethyl acetate and water extracts	*J. curcas* latex	Antibacterial	MIC methanol extract = 0.39 mg/ mL against *Staphylococcus* aureus MIC metanol extract = 6.25 mg/ mL against *Serratia marcescens*	Suhaili et al., 2011
Hexane, ethyl acetate and metanol extracts	*J. curcas* roots *J. curcas* leaves, branches, stem, root bark and roots	Antibacterial and Antifungal	MIC between 0.75 and 12.5 µg/mL against *Gardnerella vaginalis, Neisseria gonorrhoea, Escherichia coli, Klebsiella aerogenes, Proteus mirabilis, Pseudomonas aeruginosa, Staphylococcus aureus, and Candida albicans*	Aiyelaagbe et al., 2007
Methanol and water extracts	*J. curcas* seeds	Antibacterial	Active against *Bacillus cureus, B. subtilis, E. coli, K. pneumoniae, P. aeruginosa, E. aerogenes, Micrococcus luteus and S. aureus.*	Oskoueian et al., 2011a
Ethanol, methanol and water extracts	*J. curcas* stem bark		MIC between 0.5 and 10 mg/mL against S. *aureus, P. aeruginosa, E. coli, S. faecalis, S. epidermidis, S. dysenteriae, M. kristinae, K. pneumoniae, B. cereus, B. subtilis, P. vulgaris and S. marcescens*	Igbinosa et al., 2009
Hexane, dichloromethane and acetone extracts	*J. curcas* seeds	Antibacterial	MIC acetone extract = 30 mg/mL against *E. coli* MIC dichloromethane extract = 50 mg/mL against *E. coli* CIM acetone extract = 50 mg/mL against *S. aureus*	Sriprang et al., 2010
Ethanol extract	*J. curcas* flowers and fruits	Larvicide	Slightly decreases the life of *Aedes fluviatilis*	Consoli et al., 1989
Ethanol extract	*J. curcas* bark and leaves	Antibacterial	Active against *Bacillus subtilis, Escherichia coli, Pseudomonas aeruginosa, Enterobacter aerogenes and Proteus vulgaris*	Sangeetha et al., 2009

Table 7. (Continued)

Structure or name extract	Obtaining source	Biological activity	Observations	References
Ethanol extract	*J. curcas* leaves	Antihyperglycemic	Decreased blood sugar concentration of rats with induced diabetes	Mishra et al., 2010
	J. curcas leaves and roots	Antipyretic and antinociceptive	Significant analgesic activity and decreased induced pyrexia	Bhavesh and Krishnakumar 2010
	J. curcas roots	Antibacterial	MIC against *Neisseria gonorrhea* = 0.5 mL MIC against *Escherichia coli* = 0.1 mL MIC against *Pseudomonas aeruginosa* = 0.1 mL MIC against *Staphylococcus aureus* = 0.1 mL	Arekemase et al., 2011
Ethanol extract	*J. curcas* roots	Antibacterial	MIC against *Aspergillus flavus* = 0.1 mL (Solutions of 5 g of plant in 20 mL of solvent were prepared)	Arekemase et al., 2011
Methanol extract	*J. curcas* sedes oil	Molluscicide	LC_{100} = 25 ppm against *Biomphalaria glabrata* LC_{100} = 1ppm against *Bulinus truncatus and natalensis B*	Rug and Ruppel 2000
	J. curcas leaves	Low cytotoxicity	IC_{50} = 24 μg/mL CC_{50} > 1000 μg/mL using Artemia salina	Matsuse et al., 1999
		Cytotoxicity	IC_{50} = 54.0 ± 0.3 μg/ml ± DP using human cells line HL60	Faria et al., 2006
		Healing	Higher than that obtained with the standard healing. Higher closing rate and decreased time epithelialization	Esimone et al., 2008
		Antiulcerative	Antiulcerative using aspirin for inducing gastric injury in rats Wistar	Kannappan et al., 2008
		Antimetastatic	IC_{50} = 24.8 μg/mL using melanoma cells B16F10	Balaji et al., 2009
Methanol extract	*J. curcas* leaves	Analgesic and anti-inflammatory	Inhibition of inflammation comparable to that obtained using the standard (Piroxicam). Decreased number of contortions, comparable with that obtained using the standard (Paracetamol)	Uche and Aprioku 2008
		Antibacterial and Antifungal	Active against *Staphylococcus aureus, Pseudomonas sp., Penicillium sp. and Mucor sp.*	Kalimuthu et al., 2010

Structure or name extract	Obtaining source	Biological activity	Observations	References
Methanol extract	*J. curcas* leaves	Healing	Healing 91.3% vs. 82.1% with standard treatment	James et al., 2011
		Antioxidant	IC_{50} using DPPH = 9 0.83 μg/mL	James et al., 2011
Methanol extract	*J. curcas* leaves, roots and latex	Cytotoxicity	IC_{50} = 586.79 μg/mL	James et al., 2011
		Antioxidant	IC_{50} latex = 5.9 μg/mL IC_{50} leaves = 6.8 μg/mL IC_{50} roots = 57.9 μg/mL	Oskoueian et al., 2011b
	J. curcas fruits and seeds pulp	Antifungal	% growth inhibition = 78.87% with seeds % growth ihibition = 72.07% with pulp % growth ihibition = 46.42% with fuit	Rahman et al., 2011
	J. curcas shell seeds	Antioxidant	IC_{50} = 1.6 mg/mL using DPPH	Oskoueian et al., 2011b
	J. curcas bark	Baja citotoxicidad	IC_{50} = 24 μg/mL CC_{50} > 1000 μg/mL against MT-4 cells (potent inhibition of HIV-1 cytopathic effect)	Matsuse et al., 1999
	J. curcas bark and leaves	Analgesic	Analgesic effect was observed up to 60 minutes at doses of 300 mg/kg and was similar to aspirin standard (50 mg/kg).	Sangeetha et al., 2009
		Anti-inflammatory	The edema-inhibiting effect was 100% using aqueous leaf extract (300 mg/kg), even after 24 hours and was found effective compared with standard ibuprofen	Sangeetha et al., 2009
	J. curcas roots	Antibacterial	MIC against *Neisseria gonorrhea* = 0.1 mL MIC against *Escherichia coli* = 0.1 mL MIC against *Staphylococcus aureus* = 0.1 mL MIC against *Aspergillus flavus* = 0.1 mL	Arekemase et al., 2011
Phorbol esters	*J. curcas* seeds	Toxicity	LD_{50} = 27.34 mg/Kg of body weight in mice. Toxic to rats, fish, rabbits, sheep, pigs, goats and cattle. The most affected organs are intestine, liver and kidney.	Adam and Magzoub 1975 Gübitz, et al., 1997 Aderibigbe, et al., 1997 Aregheore, et al., 2003 Devappa, et al., 2011

Table 7. (Continued)

Structure or name extract	Obtaining source	Biological activity	Observations	References
Curcusone C and D	*J. curcas* roots	Antibacterial and Antifungal	Antibacterial against *B. subtilis*. Antifungal against *Botrytis cinérea* and *Rhizoctonia solani*	Thippornwong and Tohtong, 2006
DHPB	*J. curcas* seed oil	Tumor promoter	DHPB is a tumor promoter with lower activity than 12-O-tetradecanoylphorbol-13-acetate	Hirota et al., 1988
Jatropherol	*J. curcas* sedes	Insecticide	LC_{50} = 0.22 and 0.83 mg/mL against *Bombyx mori L.* and *Pieris rapae* respectively.	Jing 2005
		Toxicity	Dose oral toxicity 82.2 mg/Kg of body weigth in mice	Jing et al., 2005
JC-1	*J. curcas* stem bark	Antibacterial and antifungal	Active against 5 strains Gram-positive and 7 Gram-negative. Antifungal against 3 different strains	Das-Gupta et al., 2010
JC-2				Das-Gupta et al., 2010
Jatrophine	*J. curcas* latex	Antitumor		Lin et al., 2003
Curcacicline A		Proliferation of human cells	Proliferation of T human cells	Van der Berg et al., 1995
Curcacicline B		Antimalarial	IC_{50} < 10 mM against *Plasmodium falciparum*	Van der Berg et al., 1995 Auvin et al., 1997
Curcaína		Toxicity	LD_{50} = 3 g/Kg orally. LD_{50} = 0.96 g/Kg intraperitoneally	Nath and Dutta 1988, 1989, 1991.

In traditional medicine, *Jatropha curcas* is considered as a major contributor since almost all parts of the plant are used, such as, stem, root, seeds, leaves and latex for the treatment and prevention of many diseases. Many of these uses have been revealed by Tikunas (Indian village located between Brazil, Peru and Colombia), who prepare a bath with crushed leaves. They consider this preparation beneficial for people with fever. For headache treatment, the patient's head is washed with the resulting liquid from filtering a mix of *J. curcas* and *P. alliacea* crushed leaves previously left standing in water. The juice of the petioles is used to treat sore gums in young children. On the other hand, Panare Indians (Venezuela) utilize the root of the same species, crushed and boiled in water for dysentery (Glenboski, 1983; Estrella, 1994; Prasad et al., 2012; Sachdeva et al., 2011).

Other conditions treated are: skin infections, abortion, antiseptic, diuretic, hemostatic, against jaundice, laxative, anti-diarrheal, treatment for rheumatism, snakebite, paralysis sciatica, fever, sexually transmitted diseases such as gonorrhea and syphilis (Dalziel, 1937; Chopra et al., 1956; Martínez, 1959; Adolf et al., 1984; Houghton and Osibogun, 1993; Burkill, 1994, Mujumdar et al., 2000; Osoniyi and Onajobi, 2003; Ferreira et al., 2005). It is also used in some countries to increase the flow of breast milk, to treat liver diseases, as food for anemic people and lactating mothers. The oil extracted from this fruit species is used as emetic and purgative (doses of 10 to 12 drops), against dropsy, chronic pleurisy and parasites.

Latex is also utilized to cure rashes, other skin conditions and conjunctivitis (applying it to the temples for several days); although at high doses produces changes in the gastrointestinal tract, which may lead to death by severe dehydration (Jongschaap et al., 2007).

CONCLUSION

Jatropha curcas L. has been widely studied due to the presence of a variety of secondary metabolites and also for industrial and medical applications. A number of biological activities have been investigated and reported around the world, such as antibacterial, antifungal, antimalarial, antitumor, anticancer, anti-inflammatory, insecticide, antipyretic, diuretic, cytotoxic, molluscicide, herbicide and purgative. However, recent interest for *Jatropha* worldwide cultivation responds to the use of seeds oil to obtain biodiesel, as an alternative for a more ecological fuel. Regarding chemical composition, *Jatropha* genus is composed by a variety of metabolites; alkaloids, diterpenes, triterpenes, tannins, cyanogenic glycosides, glucosinolates, flavonoids, steroids, coumarins, steroidal sapogenins, lignans, cyclic peptides and phenolic glycosides are among the compounds reported. Thus, *Jatropha curcas* is considered an important source of secondary metabolites.

ACKNOWLEDGMENTS

Authors wish to acknowledge Lic. Diego Peña for the translation of present manuscript.

REFERENCES

Abdel Gadir W., Onsa T., Ali W., El Badwi S. and Adam S. (2003). Comparative toxicity of *Croton macrostachys*, *Jatropha curcas* and *Piper abyssinica* seeds in Nubian goats. *Small Ruminant Research*, 48, 61-67.

Abdel-Shafy S., Nasr S., Abdel-Rahman H. and Habeeb S. (2011). Effect of various levels of dietary *Jatropha curcas* seed meal on rabbits infested by the adult ticks of *Hyalomma marginatum marginatum* I. Animal performance, anti-tick feeding and haemogram. *Tropical Animal Health and Production*, 43(2), 347-357.

Abou-Arab A. and Abu-Salem F. (2010). Nutritional quality of *Jatropha curcas* seeds and effect of some physical and chemical treatments on their anti-nutritional factors. *African Journal of Food Science*, 4, 93-103.

Adam S. and Magzoub. M. (1975). Toxic effects of *Jatropha curcas* in goats. *Toxicology*, 4, 347-354.

Adebowale K. and Adedire C. (2006). Chemical composition and insecticidal properties of the underutilized *Jatropha curcas* seed oil. *African Journal of Biotechnology*, 5, 901-906.

Aderibigbe, A., Johnson, C., Makkar, H., Becker, K. and Foidl, N. (1997). Chemical composition and effect of heat on organic matter- and nitrogen degradability and some antinutritional components of *Jatropha* meal. *Animal Feed Science and Technology*, 67, 223-243.

Adolf W., Opferkuch H. and Hecker E. (1984). Irritant phorbol derivatives from four *Jatropha* species. *Phytochemistry*, 23, 129-132.

Ahmed O. and Adam S. (1979). Effects of *Jatropha curcas* on Calves. *Veterinary Pathology*. 16, 476-482.

Aiyelaagbe O., Adeniyi B., Fatunsin O. and Arimah B. (2007). *In vitro* antimicrobial activity and phytochemical analysis of *Jatropha curcas* roots. *International Journal of Pharmacology*, 3, 106-110.

Annongu A., Joseph J., Apata D., Adeyina A., Yousuf M. and Ogunjimi K. (2010). Detoxification of *Jatropha curcas* Seeds for Use in Nutrition of Monogastric Livestock as Alternative Feedstuff. *Pakistan Journal of Nutrition*, 9, 902-904.

Aregheore E., Makkar H. and Becker K. (1998). Assessment of Lectin Activity in a Toxic and a Non-toxic Variety of *Jatropha curcas* using Latex Agglutination and Haemagglutination Methods and Inactivation of Lectin by Heat Treatments. *Journal of the Science of Food and Agriculture*, 77, 349-352.

Aregheore E., Becker, K. and Makkar, H. (2003) Detoxification of a toxic variety of *Jatropha curcas* using heat and chemical treatments, and preliminary nutritional evaluation with rats. *South Pacific Journal of Natural Science*, 21, 50-56.

Arekemase M., Kayode R. M. and Ajiboye A. (2011). Antimicrobial activity and phytochemical analysis of *Jatropha curcas* plant against some selected microorganisms. *International Journal of Biology*, 3, 52-59.

Auvin C., Baraguey C., Blond A., Lezenven F., Pousset J. and Bodo B. (1997). Curcacycline B, a cyclic nonapeptide from *Jatropha curcas* enhancing rotamase activity of ciclophilin. *Tetrahedron letters*, 38, 2845-2848.

Balaji R., Rekha N., Deecaraman M. and Manikandan L. (2009). Antimetastatic and antiproliferative activity of methanolic fraction of *Jatropha curcas* against B16F10 melanoma induced lung metastasis in C57BL/6 mice. *African Journal of Pharmacy and Pharmacology*, 3, 547-555.

Belewu M. and Sam R. (2010). Solid state fermentation of *Jatropha curcas* kernel cake: Proximate composition and antinutritional components. *Journal of Yeast and Fungal Research*, 1, 44-46.

Berchmans H. and Hirata S. (2008). Biodiesel production from crude *Jatropha curcas* L. seed oil with a high content of free fatty acids. *Bioresource Technology*, 99, 1716-1721.

Berry P., Yatskievych K. and Holst B. (1999). *Flora of the Venezuelan Guayana*. Missouri botanical garden press, St Louis, US, pp 72, 162-163.

Bhavesh S. and Krishnakumar N. (2010). Screening of different parts of *Jatropha curcas* for antinociceptive and antipyretic activity on rats. *Malaysian Journal of Pharmaceutical Sciences*. Penerbit Universiti Sains Malaysia, 8, 23-28.

Burkill, H.M., (1994). The Useful Plants of West Tropical Africa (Families E-J). *Royal Botanic Gardens*, Kew. Second edition. ISBN 0947643648, 9780947643645, pp: 90-94.

Calixto J. and Sant'ana A. (1990). Evidence for the mechanism of the inhibitory action of jatrophone in the isolated rat uterine muscle. *General Pharmacology: The Vascular System*, 21, 117-122.

Carels N. (2009). *Jatropha curcas*: A Review. *Advances in Botanical Research*. 50, 39-86.

Chopra, R.N., Nayar S.L. and Chopra I.C. (1956). *Glossary of Indian Medicinal Plants*. Publications and Information Directorate, CSIR. New Delhi, India, pp 45.

Consoli R., Mendes N., Pereira J., Santos B. and Lamounier M. (1989). Influence of several plant extracts on the oviposition behaviour of *Aedes fluviatilis* (Lutz) (Diptera: Culicidae) in the laboratory. *Memórias do Instituto Oswaldo Cruz*, 84, 47-51.

Córdova-Albores L., Hernández-Lopez M., Martinez-Herrera J., Cruz-Hernández A., Barrera-Necha L. and Bautista-Banos S. (2011). Actividad antifúngica del aceite de *Jatropha curcas* sobre el desarrollo de tres especies del género *Fusarium* (antifungal activity of *Jatropha curcas* oil on the development of three species of Fusarium). *Revista Latinoamericana de Química*, 38, 60.

Croteau R., Kutchan T. and Lewis N. (2000). *Natural production (secondary metabolites)*. In: Buchanan B, Gruisssem W, Jones R (Eds.) *Biochemistry and molecular biology of plants*. American Society of Plant Physiologists. Rockville, Estados Unidos, pp 1250-1254, 1262-1267.

Dalziel, J. (1937). *The useful plants of W. Tropical Africa, Crown agents for the colonies*, London. England, pp 18-22.

Das-Gupta D., Haque E., Islam N., Mondal S. and Alam B. (2010). Antimicrobial and Cytotoxic activities of *Jatropha curcas* (Euphorbiaceae). *Dhaka University Journal of Pharmaceutical Sciences*, 9, 139-142.

De Moraes V., Rumjanek V. and Calixto J. (1996). Jatrophone and 12-O-tetradecanoyl phorbol-13-acetate antagonism of stimulation of natural killer activity and lymphocyte proliferation. *European Journal of Pharmacology*, 312, 333-339.

Debnath M. and Bisen PS. (2008) *Jatropha curcas* L., a multipurpose stress resistant plant with a potential for ethnomedicine and renewable energy. *Current Pharmaceutical Biotechnology*. 9(4), 288-306.

Dehgan B. and Webster G. (1979). *Morphology and Infrageneric relationships of the genus Jatropha (Euphorbiaceae)*. University of California publications in botany. Volume 74. London, England, pp 9-34, 46.

Devappa R., Makkar H. and Becker K. (2010). Nutritional, Biochemical, and Pharmaceutical Potential of Proteins and Peptides from *Jatropha*: Review. *Journal of Agricultural and Food Chemistry*, 58, 6543-7108.

Devappa R., Makkar H. and Becker K. (2011). *Jatropha* Diterpenes: a Review. *Journal of the American Oil Chemists' Society*, 88, 301-322.

Dewick P. (2002). *Medicinal Natural Products*. John Wiley and Sons Ltd. London, England, pp 36-37, 61-64.

Esimone C., Nworu C. and Jackson C. (2008). Cutaneous wound healing activity of an herbal ointment containing the leaf extract of *Jatropha curcas L.* (Euphorbiaceae). *International Journal of Applied Research in Natural Products*, 1, 1-4.

Estrella E. (1994). *Plantas medicinales Amazónicas. Realidad y perspectivas* (Amazonian medicinal plants. Reality and perspectives). Tratado de cooperación Amazónica (Amazon Cooperation Treaty). Lima, Perú, pp 280-282.

Falasca S., and Ulberich A. (2008). Southamerican bioenergy potential of forest plantations with *Jatropha* sp (*J. curcas, hieronymi and macrocarpa*). *REDESMA Journal*, Bolivian Center of Multidisciplinary Higher Studies. La Paz, Bolivia, pp 101-115.

Faria M., Carvalho T., Rabenhorst S., Sidrim, J. and Moraes-Filho M. (2006). Cytotoxic and antifungal properties of medicinal plants from Ceara, Brazil. *Brazilian Journal of Biology*, 66, 1133-1135.

Ferreira M., De Sousa J., Feitosa A., Caño M., Goulart A., Genet J., Marquez B., Neuville L. and Moreau N. (2005). Anti-bacterial activity of some Brazilian medicinal plants. *Journal of ethnopharmacology*, 105, 137-147.

Gallegos-Tintoré S., Torres-Fuentes C., Martínez-Ayala A., Solorza-Feria J., Alaiz M., Girón-Calle J. and Vioque J. (Del 25 al 29 de Octubre de 2010). Antioxidant and chelating activity of *Jatropha curcas* L. protein hydrolysates. *International conference on food innovation*. Universidad Politécnica de Valencia, España. pp 1-4.

Glenboski, L. (1983). *The ethnobotany of the Tukuna Indians of Amazonas*. Universidad Nacional de Colombia (National university of Colombia). Bogotá, Colombia.

Goonasekera, M., Gunawardana, V., Jayasena, K., Mohammed, S. and Balasubramaniam, S. (1995). Pregnancy terminating effect of *Jatropha curcas* in rats. *Journal of ethnopharmacology*, 47, 117-123.

Guerrero J. (S/A). Descriptores morfológicos y agronómicos con aplicación en piñón (morphological and agronomic descriptors application sprocket) (*Jatropha curcas* L.). Available in: http://www.ediciona.com/portafolio/document/0/2/2/3/descriptores_cualitativos_y_cuantitativos_aplicados_en_pinon_1_3220.pdf.

Gupta R. C. (*1985*). Pharmacognostic studies on "Dravanti" part I. *Jatropha curcas* Linn. *Proceedings of the Indian Academy* of *Science*, 94, 65-82.

Hanson J.R. (1998). Diterpenoids. *Natural products reports*, 15, 93-106.

Heller, J. (1996). *Physic nut: Jatropha curcas* L. International Plant Genetic Resources Institute. IPK. Italia, pp 7-13.

Hirota M., Suttajit M., Suguri H., Endo Y., Shudo K., Wongchai V., Hecker E. and Fujiki H. (1988). A new tumor promoter from the seed oil of *Jatropha curcas* L., an intramolecular diester of 12-deoxy-l 6-hydroxyphorbol. *Cancer Research*. 48, 5800-5804.

Hokche, O., Berry, P., and Huber, O. (2008). *Nuevo catálogo de la flora vascular de Venezuela* (New catalog of the vascular flora of Venezuela). Fundación Instituto Botánico de Venezuela Dr. Tobías Lasser (Dr. Lasser Foundation Botanical Institute of Venezuela). Caracas Venezuela, pp 524.

Houghton P. and Osibogun I. (1993). Flowering plants used against snakebite. *Journal of ethnopharmacology*, 39, 1-29.

Igbinosa O., Igbinosa E. and Aiyegoro A. (2009). Antimicrobial activity and phytochemical screening of stem bark extracts from *Jatropha curcas* (Linn). *African Journal of Pharmacy and Pharmacology*, 3, 058-062.

Irvine, F. (1961). *Woody Plants of Ghana with special reference to their uses*. Oxford University Press (Second edition). London, England, pp 233-237.

James O., Unekwojo E. and Ojochenemi A. (2011). Assessment of biological activities: a comparison of *Pergularia daemia* and *Jatropha curcas* leaf extracts. *British Biotechnology Journal*, 1, 85-100.

Jing L. (2005). Study of insecticidal active components in *Jatropha curcas* L. seed on its extraction, isolation and toxicity action mechanism. *PhD dissertation, Sichuan University*.

Jing L., Fang Y., Ying X., Wenxing H., Meng X., Syed M.N. and Fang C. (2005) Toxic impact of ingested jatropherol-I on selected enzymatic activities and the ultra-structure of midgut cells in silkworm, *Bombyx mori* L. *Journal Applied Entomology*, 129, 98-104.

Jongschaap R., Corré W., Bindraban P. and Brandenburg W. (2007). *Claims and facts on Jatropha curcas L.* Plant research International B. V. Report 158. Wageningen, Netherlands, pp 1-4, 14, 22, 28.

Kalimuthu K., Vijayakumar S. and Senthilkumar R. (2010). Antimicrobial activity of the biodiesel plant, *Jatropha curcas* L. *International Journal of Pharma and Bio-Sciences*, 1, 1-5.

Kannappan N., Jaikumar S., Manavalan R. and Muthu A. (2008). Antiulcer activity of methanolic extract of *Jatropha curcas* Linn on aspirin induced gastric lesions in wistar rats. *Pharmacologyonline newsletters*, 1, 279-293.

Khamer F. (2008). *Phytochemical, antioxidant and anthelmintic studies on leaves of Jatropha curcas Linn.* (Thesis). Rajiv Gandhi University of Health Sciences. Bangalore, Karnataka, India.

Khafagy S., Mohamed Y., Abdel N. and Mahmound Z. (1977). Phytochemical study of *Jatropha curcas. Planta médica*, 31, 274-277.

Krishna M. and Sankar D. (2008). Simple spectrophotometric determination of torsemide in bulk drug and in formulations. *Journal of Chemistry*, 5, 473-478.

Kumar A. and Sharma S. (2008). An evaluation of multipurpose oil seed crop for industrial uses (*Jatropha curcas* L.): A review. *Industrial crops and products*, 5087, 1-9.

Lin, J., Yan, F., Tang, L. and Chen, F. (2003). Antitumor effects of curcin from seeds of *Jatropha curcas. Acta of Pharmacology Sinica*, 24, 241-246.

Ling-Yi K., Zhi-Da M., Jian-Xia S. and Rui F. (1996). Chemical constituents from roots of *Jatropha curcas. Acta Botanica Sinica*, 38, 161-166.

Luseba D., Elgorashi E., Ntloedibe D. and Van Staden J. (2007). Antibacterial, anti-inflammatory and mutagenic effects of some medicinal plants used in South Africa for the treatment of wounds and retained placenta in livestock. *South African Journal of Botany*, 73, 378-383.

Gübitz, G. M., Mittelbach, M. and Trabi, M. (Eds.). (1997). *Biofuels and industrial products from Jatropha curcas,* Dbv-Verlag für die Technische Universität Graz, pp. 190-205.

Makkar H., Aderibigbe A. and Becker K. (1998). Comparative evaluation of non-toxic and toxic varieties of *Jatropha curcas* for chemical composition, digestibility, protein degradability and toxic factors. *Food Chemistry*, 62, 207-215.

Makkar H., Kumar V., Oyeleye O., Akinleye A., Angulo-Escalante M. and Becker, K. (2010). Traditional wisdom confirmed by scientific research: *Jatropha* species from Mexico is non-toxic. *Nature Precedings*, 1-21.

Makun H., Anjorin S., Adeniran L., Onakpa M., Muhammad H., Obu O. and Agbofode Y. (2011). Antifungal activities of *Jatropha curcas* and *Ricinus cumunis* seeds on *Fusarium verticilliodes* and *Aspergillus flavus* in yam." *ARPN Journal of Agricultural and Biological Science*, 6, 22-27.

Marcano, D. and Hasegawa, M. (2002). *Fitoquímica organic* (organic phytochemical). Consejo de Desarrollo Científico y Humanístico (Council of Scientific and Humanistic). Universidad Central de Venezuela (Central University of Venezuela). Caracas, Venezuela, pp 60-71, 171, 361-368.

Martínez, M. (1959) *Las plantas medicinales de México* (Medicinal plants in Mexico). Ediciones Botas. (Quinta edición (fifth edition)). Ciudad de México, México, pp 656.

Martínez-Herrera J., Siddhuraju P., Francis G., Dávila-Ortíz G. and Becker K. (2006). Chemical composition, toxic/antimetabolic constituents, and effects of different treatments on their levels, in four provenances of *Jatropha curcas* L. from Mexico. *Food Chemistry*, 96, 80-89.

Matsuse I., Lim Y., Hattori M., Correa M. and Gupta M. (1999). A search for anti-viral properties in Panamanian medicinal plants. The effects on HIV and its essential enzymes. *Journal of ethnopharmacology*, 64, 15-22.

Méndez L., Rojas J., Izaguirre C., Contreras B. and Gómez R. (2014). *Jatropha curcas* leaves analysis, reveals it as mineral source for low sodium diets. *Food chemistry*, 165, 575-577.

Méndez L. and Rojas J. (2012). *Revisión bibliográfica sobre los estudios fitoquímicos realizados en el género Jatropha* (Euphorbiaceae) (Review of the literature on studies conducted phytochemicals gender Jatropha (Euphorbiaceae)). Editorial Académica Española (Spanish Academic Publishing). (Primera Edición (first edition)). Madrid, España, pp 5-70.

Mishra S., Vijayakumar M., Ojha S. and Verma A. (2010). Antidiabetic effect of *Jatropha curcas* L. leaves extract in normal and alloxan-induced diabetic rats. *International general of pharmaceutical sciences*, 2, 482-487.

Mitra C., Bhatnagar S. and Sinha M. (1970). Chemical examinations of *Jatropha curcas*. *Indian Journal of Chemistry*, 8, 1047.

Morales A. and Negrón E. (1988). Los constituyentes químicos de algunas plantas, en especial de los Andes Venezolanos; su importancia económica (The chemical constituents of some plants, especially the Venezuelan Andes; its economic importance). *Ciencia*, 6, 3-19.

Mujumdar A. and Misar A. (2003). Anti-inflammatory activity of *Jatropha curcas* roots in mice and rats. *Journal of ethnopharmacology*, 90, 11-15.

Mujumdar A., Upadhye A. and Misar A. (2000). Studies on antidiarrhoeal activity of *Jatropha curcus* root extract in albino mice. *Journal of ethnopharmacology*, 70, 183-187.

Murillo I., Angulo M., Salazar E., Verdugo A., Pérez V., Sánchez E. and Rubio W. (Del 11 al 13 de Octubre del 2010). Valor nutrimental de semillas silvestres de *Jatropha curcas* y *Jatropha platyphylla* del estado de Sinaloa (nutritional value of wild Jatropha seeds and *Jatropha curcas* platyphylla state of Sinaloa). 7mo Encuentro Nacional de Biotecnología del Instituto Politécnico Nacional en Mazatlán, Sinaloa (The 7th National Meeting on Biotechnology of the National Polytechnic Institute in Mazatlan, Sinaloa).

Naengchomnong W., Tarnchompoo B. and Thebtaranonth Y. (1994). (+)-Jatrophol, (+)-marmesin, propacin and jatrophin from the roots of *Jatropha curcas* (Euphorbiaceae). *Journal of the Science Society of Thailand*, 20, 73-83.

Namuli A., Abdullah N., Sieo C. C., Zuhainis S. W. and Oskoueian E. (2011). Phytochemical compounds and antibacterial activity of *Jatropha curcas* Linn. Extracts. *Journal of Medicinal Plants Research*, 5, 3982-3990.

Nath, L. and Dutta, S. (1988). Extraction and study of certain physicochemical properties of a new proteolytic enzyme from the latex of *Jatropha curcas* Linn. *Indian Journal of Pharmaceutical Sciences*, 50, 125-127.

Nath, L. and Dutta, S. (1989). Kinetic studies on curcain, a protease from the latex of *Jatropha curcas* Linn. *Indian Journal of Pharmaceutical Sciences*, 50, 43-47.

Nath L. and Dutta S. (1991). Extraction and purification of Curcain, a protease from the latex of *Jatropha curcas* Linn. *Journal of pharmacy and pharmacology*, 43, 111-114.

Obasi L., Ejikeme P. and Egbuonu A. (2011). Antimicrobial and phytochemical activity of methanolic extract and its fractions of *Jatropha curcas* Linn. (Eurphorbiaceae) stem bark. *African Journal of Pure and Applied Chemistry*, 5, 92-96.

Oseni O. and Akindahunsi A. (2011). Some phytochemical properties and effect of fermentation on the seed of *Jatropha curcas* L. *American Journal of Food Technology*, 6, 158-165.

Oskoueian E., Abdullah N., Ahmad S., Saad W., Omar A. and Ho Y. (2011a). Bioactive compounds and biological activities of *Jatropha curcas* L. Kernel Meal Extract. *International Journal of Molecular Sciences*, 12, 5955-5970.

Oskoueian E., Abdullah N., Saad W., Omar A., Ahmad S., Bin W., Zolkifli N., Hendra R. and Ho Y. (2011b). Antioxidant, anti-inflammatory and anticancer activities of methanolic extracts from *Jatropha curcas* Linn. *Journal of Medicinal Plants Research*, 5, 49-57.

Osoniyi O. and Onajobi F. (2003). Coagulant and anticoagulant activities in *Jatropha curcas* latex. *Journal of Ethnopharmacology*, 89, 101-105.

Oyi, A., Onaolapo J., Haruna A. and Morah C. (2007). Antimicrobial screening and stability studies of the crude extract of *Jatropha curcas* Linn latex (Euphorbiaceae). *Nigerian Journal of Pharmaceutical Sciences*, 6, 14-20.

Parajuli R. (2009). *Jatropha curcas* and Its Potential Applications; A Compilation Paper on Plantation and Application of *Jatropha Curcas.*

Parawira W. (2010). Biodiesel production from *Jatropha curcas*: A review. *Scientific Research and Essays*, 5, 1796-1808.

Pramanik K. (2003). Properties and use of *Jatropha curcas* oil and diesel fuel blends in compression ignition engine. *Renewable energy*, 28, 239-248.

Prasad D., Izam A. and Rahman Khan M. (2012). *Jatropha curcas*: Plant of medical benefits. *Journal of Medicinal Plants Research*, 6, 2691-2699.

Prashanth V., Chauhan N., Padh H. and Rajani M. (2006). Search for antibacterial and antifungal agents from selected Indian medicinal plants. *Journal of Ethnopharmacology*, 107, 182-188.

Rahman M., Ahmad S., Mohamed M., Zaki M. and Rahman A. (2011). Extraction of *Jatropha curcas* fruits for antifungal activity against anthracnose (*Colletotrichum gloeosporioides*) of papaya. *African Journal of Biotechnology*, 10, 9796-9799.

Ravindranath N., Ravinder Reddy M., Mahender G., Ramu R., Ravi K. and Das B. (2004a). Deoxypreussomerins from *Jatropha curcas*: are they also plant metabolites?. *Phytochemistry*, 65, 2387-2390.

Ravindranath N., Reddy M., Ramesh C., Ramu R., Prabhaker A., Jagadeesh B. and Das B. (2004b). New Lathyrane and Podocarpane Diterpenoids from *Jatropha curcas. Chemical and pharmaceutical Bulletin*, 52, 608-611.

Rojanapo, W., Pimbua, J. and Glinsukon, T. (1988). Failure of diterpenes from *Jatropha curcas* to induce mutation in *Salmonella typhimurium* TA98 and TA100. *Journal of Ethnopharmacology*, 23, 345.

Rug M. and Ruppel A. (2000). Toxic activities of the plant *Jatropha curcas* against intermediate snail hosts and larvae of schistosomes. *Tropical Medicine and International Health*, 5, 423-30.

Sabandar C. (2010). Secondary metabolite compounds from *Jatropha* species. Available in: http://carlasabandar.wordpress.com/2010/06/29/secondary-metabolite-compounds-from-jatropha-species/.

Sachdeva K., Singhal M. and Srivastava B. (2011). A review on chemical and medicobiological applications of *Jatropha curcas. International Research Journal of Pharmacy*, 2, 61-66.

Sangeetha J., Divya K., Prashanth M., Vamsikrishna A. and Leela R. (2009). Anti-inflammatory and antibacterial activity of *Jatropha curcas* Linn. *Journal of Pharmaceutical Research and Health Care*, 2, 258-262.

Sievers A., Andrewarcher W., Morre R., and Mcgovran, E. (1949). The phorbol ester fraction from *Jatropha curcas* seed oil: potential and limits for crop protection against insect pests. *Journal of Economic Entomology*, 42, 549-551.

Silva L. and Ulberich A. (2008). Potencialidad bioenergética sudamericana a partir de forestaciones con *Jatropha sp.* (*J. curcas*, *hieronymi* y *macrocarpa*) (American bioenergy potential from afforestation with Jatropha sp. (J. curcas, hieronymi and macrocarpa). *Revista Virtual REDESMA*. Centro Boliviano de Estudios Multidisciplinarios, 2, 102-115.

Sonibare O., Sonibare M. and Akharame E. (2008). "Comparative chemical composition of lipids in some *Jatropha* species." *European Journal of Scientific Research*, 21, 209-211.

Sriprang, S., Sriprang, N., Sumpradit, T. and Shimbhu, D. (2010). Antibacterial activities of crude extracts from physic nut (*Jatropha curcas*) seed residues. *Science Asia*, 36, 346-348.

Staubmann R., Schubert-Zsilavecz M., Hiermann A. and Kartnig T. (1999). A complex of 5-hidroxypyrrolidin-2-one and pyrimidie-2,4-dione isolated from *Jatropha curcas*. *Phytochemistry*, 50, 337-338.

Suhaili, Z., Yeo, C., Yasin, H., Badaludin, N. and Zakaria, Z. (2011). Antibacterial profile of *Jatropha curcas* latex extracts against selected human pathogenic bacteria. *African Journal of Microbiology Research*, 5, 5147-5154.

Sundari, J. and Selvaraj, R. (2011). Antibacterial and antifungal activity of seed extract from *Jatropha curcas* Linn. *International Journal of Current Research*, 33, 084-087.

Thippornwong, M. and Tohtong, R. (2006). Effect of curcusone B on KKU-100 cell invasion. *Proceedings of the American Association for Cancer Research*, 3, A112.

Thomas R., Sah N. and Sharma P. (2008). Therapeutic biology of *Jatropha curcas*: a mini review. *Current Pharmaceutical Biotechnology*, 9, 315-324.

Uche F. and Aprioku J. (2008). The phytochemical constituents, analgesic and anti-inflammatory effects of methanol extract of *Jatropha curcas* leaves in Mice and Wister albino rats. *Journal of Applied Sciences and Environmental Management*, 12, 99-102.

Valdés O., García R., Sánchez O. and Pérez A. (2011). Isolation and pathogenicity of a possible *Pythium aphanidermatum* in *Jatropha curcas* L. non toxic. *Tropical and Subtropical Agroecosystems*, 14, 649-660.

Van Der Berg A., Horsten S., Kettenes J., Kroes B., Beukelman C., Leeflang B. and Labadie Ri. (1995). Curcacycline A, a novel cyclic ocatpeptide isolated from the latex of *Jatropha curcas* L. *FEBS Letters*, 358, 215-218.

Webster, G. L. (1994). Classification of the Euphorbiaceae. *Annals of the Missouri Botanical Garden*, 81, 3-32.

Xu J., Fan J., Zeng G. and Tan N. (2011). A New Tetracyclic Diterpene from *Jatropha curcas*. *Helvetica Chimica Acta*, 94, 842-846.

Ye M., Li C., Francis G. and Makkar H. (2009). Current situation and prospects of *Jatropha curcas* as a multipurpose tree in China. *Agroforestry System Impact factor*, 76, 487-497.

In: The Promising Future of *Jatropha curcas*
Editors: Maira R. Segura-Campos et al.
ISBN: 978-1-63484-989-0
© 2016 Nova Science Publishers, Inc.

Chapter 3

JATROPHA CURCAS L. BIOMASS RAW MATERIALS: THE CURRENT STATE AND FUTURE POTENTIALITIES

N. A. Sri Aprilia[1,*], H. P. S. Abdul Khalil[2], M. Dani Supardan[1], Adisalamun[1], Amri Amin[3], M. K. Mohamad Haafiz[2], M. I. Syakir[2] and M. R. Nurul Fazita[2]

[1]Department of Chemical Engineering, Engineering Faculty, University of Syiah Kuala, Banda Aceh, Indonesia
[2]School Industrial Technology, Universiti Sains Malaysia, Penang, Malaysia
[3]Engineering Faculty, University of Abulyatama, Aceh Besar, Indonesia

ABSTRACT

Due to the increased in scarcity of non-renewable resources and the lack of natural resources as well as in attempt to reduce the environmental problems created from many industries. Researchers have concerned and shifted to the utilization of crops for many applications. *Jatropha curcas L.* is one of the crops which the different part of this crops plant shows a potential to be used in various industrial application. Due to its excellent chemical composition and unique properties from stem, bark, leaves, part of fruit, and latex, made this crop has a great material for product development. Beside that this plant is an excellent raw material for the future because of non-edible plant as compared with another bioresources. Lignocellulose content parts from this crops has potential to be applied in biocomposites for the future this will widen the applications of *Jatropha curcas L.* Therefore this crop not only help in reducing the environmental pollution issue but also provide an opportunity in developing renewable and sustainable material to be used in various advanced applications in the future. This chapter deals to discuss different part of *Jatropha curcas L.* biomass with its present applications and future potentialities.

Keywords: *Jatropha curcas L.*, biomass, biocomposites, and renewable material

---

* Corresponding author: N. A. Sri Aprilia. Tel.: + 62 651 23991, fax: + 62 651 52222, e-mail address: sriaprilia@ unsyiah.ac.id.

INTRODUCTION

During the last two decades the world are concerned on the use of renewable resources or also known as bioresources. Bioresources are the new materials that used for many applications to replace the dependency to of non-renewable resources. One of bioresources abundantly in the world (tropical and sub-tropical region) is *Jatropha curcas Linnaeus* also recognized as purging nut or physic nut. It is a tropical plants that has spermatophyta division, dicotyledonae, and spurge *(Euphorbiaceae)* family (Matsumoto et al. 2014; Makkar et al. 2011; Dhillon et al. 2011), species plantarum and has approximately 170 known species (Kumar and Sharma, 2008). This plant can be grown in low to high rainfall areas (Abdul Kalil et al. 2013), in marginal land (Valdes-Rodriguez et al. 2013), and in gradually sandy saline soil (Lateef et al. 2014) and as a wild-growing hardy plant well adapted to various soil and climate condition (Myers et al. 2013). As biomass *Jatropha* plant is a multipurpose species that has lignocellulosic content can be considerable potential (Sri Aprilia et al. 2012) for many purpose. This plant however, commonly used as; to reclaim land, hedge to protect fields, oil-soap production, pharmaceutical and biodiesel feed stock (Matsutomo et al. 2014). This plant also claimed as renewable energy resources because of can produce biodiesel.

Jatropha curcas L. as multipurpose plant that also known as a robust energy plant which in addition to seed oil, also produces wood, fruit shells, seed husk and press cake (Abdul Khalil, 2013). On the other hand, all part of *Jatropha* plant such as stem, bark, leaves, part of fruit, and latex have a potential for a variety needs and application in the future. They have different chemical and physical properties in different parts of *Jatropha* make them could potential in many application so that a zero-waste biomass system in *Jatropha*. Development of new materials, especially as a composite material and innovative products that will enrich new material in the future.

Nowadays, a large number of interesting applications of *Jatropha* plant such as biocomposites, pulp and paper, construction, automotive, medical, packaging, aerospace, pharmaceutical and biomass energy production are emerging from this biomass plant. This due to recent progress in technological advances, biomass material based on *Jatropha* plant development, genetic engineering, and composite science technology that offer significant opportunities for an exploration and development of improved materials from renewable resources based on *Jatropha.*

POTENTIALLY COMPONENT OF *JATROPHA CURCAS L.*

The *Jatropha curcas L.* plant has high agro-industrial potential due to its various potentially beneficial products, such as potential to diversify rural incomes, reclaim unproductive lands, reduce importation of fossil fuels, and consequently accumulation of greenhouse gases in the atmosphere. *Jatropha curcas* grows almost anywhere, make it easy to cultivate (Cai et al. 2012). It is a great potential biodiesel plant (Xu et al. 2011), a non-edible, oil bearing and draught hardy shrub with ecological advantages (Vyas and Singh, 2007), drought resistant, perennial (Dhillon et al. 2011). It is a glabrous soft wooded deciduous shrub or erect tree of 3-5 m height, open pollinated and has an annual yield of up 5 ton per hectare (Makkar et al. 1998). For favourable conditions, the tree can grow up to 10 m (Jingura et al.

2010). The plant is monoecious and the terminal inflorescence contains unisexual flowers (Achten et al. 2008). The plant itself is believed to prevent/control soil erosion and can be used as a biological fence for reducing carbon dioxide. Every part of *Jatropha curcas* such as stem, bark, leaves, part of fruits and latex has potential for many products such as for fuel, pharmacy, candle and soap, cosmetic, and so on. Part of *Jatropha* plant show in Figure 1.

Stem, Bark and Roots

The stem of *Jatropha* plant is very important parts, the stems of plants for the body can be likened to axis of the body. The branches and stems are hollow and the soft wood is of little value (Brittaine and Lutaladio, 2010). *Jatropha curcas* is a small tree or large shrub with smooth gray bark, which exudates a whitish colored watery latex, upon cut (Verma and Gaur, 2009). Normally, five roots are formed from seeds, one central (tap root) and four peripheral.

Leaves

Jatropha curcas has large green to pale green leaves, alternate to sub-opposite, three to seven lobed with a spirally phylotaxis, hypostomatic and stomata are of paracytic type (Verma and Gaur, 2009; Kumar and Sharma, 2008). The leaves, arranged alternately on the stem, are shallowly lobed and vary from 6 to 15 cm in length and wide. The leaves shed during the winter months form mulch around the base of the plant. The organic matter from shed leaves enhance earthworm activity in the soil around the root-zone of the plants, which improves the fertility of the soil. The leaf size and shape can differ from one variety to another. It has green leaves with a length and width of 6 to 15 cm, with 5 to 7 shallow lobes (Gunaseelan, 2009). The leaves are arranged alternately.

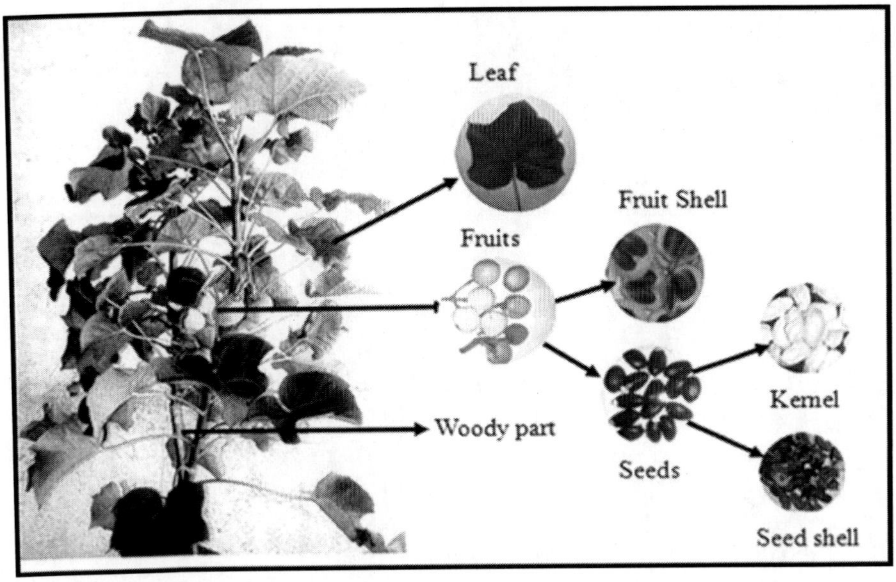

Figure 1. Different part of *jatropha curcas L.*

Part of Fruits

The shape of *Jatropha* fruits are ellipsoidal, green and fleshy, turning yellow and then brown depend on they age. The fruits will mature and ready to be harvested around 90 days after flowering. Each fruit contains two or three black seeds, around 2 cm x 1 cm in size (Brittaine and Lutaladio, 2010). The fruits from in bunches, they have around ten and initially olive green color. One fruit content three seed, every seed composed a kernel and a shell as shown in Figure 2.

Fruiting is continuous throughout the year, in wetter climates, and the harvest could be confined until two months in semi-arid regions. The fruits will ripe in different time, it is require weekly picking and making the harvest labour intensive. The yellow and brown fruits are harvested by beating the branches with sticks to knock them to the ground, or by hand picking. The fruits are dried and the seeds removed from the fruit shells by hand, by crushing with a wooden board or by using a mechanical decorticator. For best oil yields, the seeds should be harvested at maturity. Seeds are mature if the color of the fruits has changed from green to yellow-brown (Achten et al. 2008).

All part of the fruits such as shell, seed and kernel has potential use for many products because of rich in oil contents. The seeds are shade dried for sowing but dried in the sun for oil production to reduce moisture content to around 6-10 percent. The seed yield can be obtained up to 40-50 years from the day of plantation. Singh et al., (2008) stated that dry *Jatropha curcas* fruit contain about 37.5% shell and 62.5% seed. The seed contains about 42% hull or seed shell and 58% kernel. *Jatropha* seeds contain an outer layer of hard and black shell and the white kernel inside.

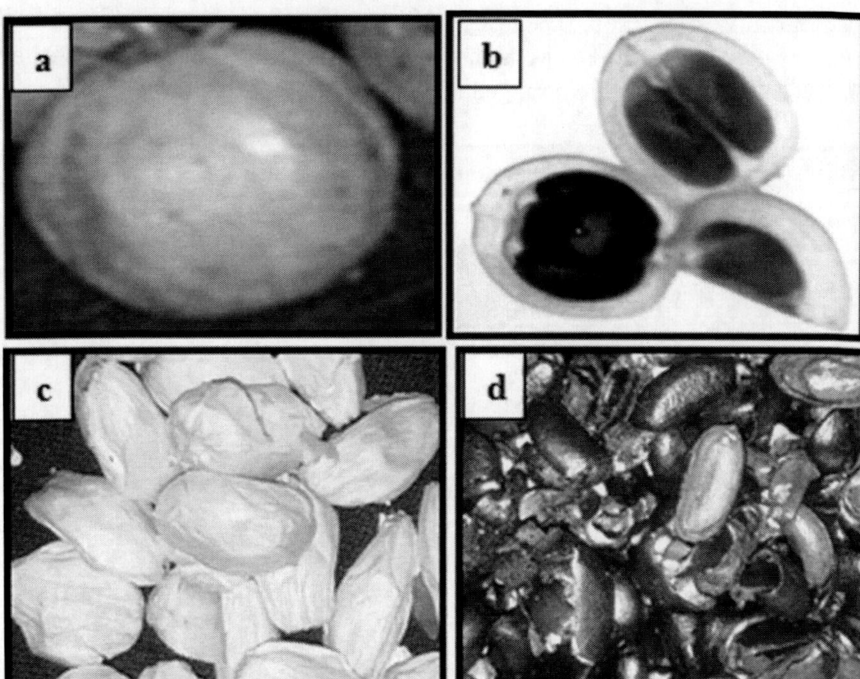

Figure 2. Part of *Jatropha* fruits: a. Fruit; b. Seed and seed shell; c. Kernel; d. Seed shell.

Latex

Jatropha plant extracts have many uses in traditional societies. The branches normally five and contain latex. The latex has a widespread reputation for healing wounds and stopping bleeding, and for curing various skin problems. It is used against pain and the stings of bees and wasps. The fresh stems are used as chewing sticks to strengthen gums and treat gum disease. The branches contain whitish latex, which causes brown stains, and very difficult to be remove.

PROPERTIES OF *JATROPHA* PARTS

Chemical Component

As multipurpose application, *Jatropha curcas* has chemical component in each part. The bark of *Jatropha* contains tannin, the flowers attract bees and thus the plant has a honey production potential (Behera et al. 2010; Openshaw, 2000). Parts of wood in *Jatropha* have lignocellulose compounds. From phytochemical screening, extraction of *Jatropha* stem bark produce secondary metabolites such as saponins, steroids, tannins, glycosides, alkaloids, flavonoids and also yields dark blue dye (Prasad et al. 2012).

As reported early by other researchers, *Jatropha curcas* well known as energetic for energy fuel. *Jatropha curcas* seed content viscous oil, so that suitable for bio-diesel, very good for candles and soap production (Cai et al. 2012). *Jatropha* oil content oleic and linoleic acid derived from unsaturated fatty acid (Pecina-Quentiro et al. 2011). The composition of fatty acid ethyl ester depend on origin of the crop (Rodrı´Guez et al. 2011). The fatty acid profile made up of palmitic acid (10.5-13.0%), stearic acid (2.3-2.8%), oleic acid (41.5-48.8%), and linoleic acid (34.6-44.4%) (Coize-Valladares et al. 2012; Chen et al. 2012). *Jatropha* seeds contain high levels of tocochromanols, mainly in the gamma-tocotrienol form (Coize-Valladares et al. 2012). Tocochromanols are minor plant compounds that exhibit vitamin E activity. In addition, it is a little alkaloids contents in the seeds (Chen et al. 2012).

The seed cakes obtained after oil extraction from the seed can be used for animal food after detoxification (remove porbhol esters (crucin, saponin, etc.)). Because of protein and carbohydrate content, the seed cakes (Abdul Khalil et al. 2013; Liang et al., 2010) suitable for animal. Leaves, stems and fruits have toxic matter and after remove toxic component will be useful for animal. *Jatropha* seed cake also production of methane (CH_4) after an-aerobic digestion (Rahenman and Mondal, 2012; Behera et al. 2010). The fruit is rich in carbohydrates and could be applied for bioethanol production by fermentation (Manurung et al. 2009). The product from the fruit (fruit shell and seed cake) rich with nitrogen, phosphorous, and potassium, these compounds are suitable for fertilizer (Openshaw, 2000). Fruit shell contains low nutrient, but it is abundant in cellulose, hemicellulose and lignin (Tongpoothorn et al. 2011). Moreover, *Jatropha* seed cakes contain anti nutritional factors such as saponin, phytate, trypsin inhibitor, glucosinolates, amylase inhibitors, cyanogenic glucosides, curcin, lectin, and phorbol ester. Glycerin come from by product of bio-diesel (Kumar and Sharma, 2008). *Jatropha* kernel rich protein and lipid, and viewer of moisture and ash content (Makkar et al. 1998; Aderibigbe et al. 1997). The main chemical compositions of the seed shell are fiber and lignin and very little protein (Makkar et al. 1998). The kernel mainly contains crude fat and protein which the high concentration of phorbol ester contain (Achten et al. 2008; Martinez-Herera et al. 2006).

Jatropha leaves has protein content which rich with amino acid. The major components in *Jatropha* leaf are lignin and protein beside polysaccharide. The main constituent of latex are curcin, curcacycline A and curcacycline B (Hudlikar et al. 2012; Joglekar et al. 2011; Bar et al. 2009). The latex contains a protease of curcain product after isolation (Osoniyi and Onajobi, 2003), tannin, saponin, wax and resin (Prasad et al. 2012).

Physical and Mechanical Properties

Physical and mechanical properties of *Jatropha* fruit, shell and kernel are important to design equipment for dehulling, nut shelling, drying and oil extraction, and also in other processes like, transportation and storage (Abduk Khalil et al. 2013; Karaj et al. 2008). The data of size and shape of fruit, shell and kernel are very important to determine the lower size limit of the conveyors such as belt conveyor, bucket elevator and screw conveyor (Sirisombon and Kitchaiya, 2009).

Fruit of *Jatropha* is a drupe ovoid capsule with the dimension range between 1.5 to 3.0 cm in diameter. The seed shape was elliptical with length 17.0 ± 0.96 mm, width 8.4 ± 0.50 mm (Makkar et al. 2011). Ruzbarsky et al. 2014 were determined of the seed size: the seed thickness 11.25 ± 0.43 mm, the seed length 18.27 ± 0.77 mm and the seed height 8.94 ± 0.50 mm. An average seed mass (calculated from the masses of 200 seeds) amounted 0.6476 ± 0.1117 g at the proportion of the kernel mass to the coat mass 1:0.62. The kernel mass was 0.4041 ± 0.0909 g.

The seed and kernel of *Jatropha curcas* have a geometrical shape which could be approximated by prolate ellipsoids. The relations between length, width, breadth, arithmetic mean, geometrical mean, volume, sphericity, bulk and solid density, porosity, surface and specific surface follow a strict pattern, which enables the reliable correlation of geometrical dimensions to the unit mass (Karaj and Muler, 2010).

The moisture content of the fruit are about 7.50-23.33% in dry basis (Yahya et al., 2013; Salawu et al. 2013) and the shell about 88.95% and the kernel 51.87% in wet basis (Sirisombon et al. 2007). The bulk and true densities of seeds are important parameters in determining the storage capacity. The mean solid density of the fruit is 0.9735 gr/cm^3 (Yahya et al. 2013). The seed has a bulk density of 0.44 g/cm^3 and a solid density of 0.79 g/cm^3, and kernels had a bulk density of 0.48 g/cm^3 and a solid density of 0.99 g/cm^3 (Karaj and Muller, 2010). The solid density of the dried kernel (951-990 kg m^3) (Sirisombon and Kitchaiya, 2009). Karaj and Muller, (2010) imply that in general bulk and solid density were higher for kernels than seeds. Since kernel size was smaller than seed size, more unit of kernels were entering into the container volume compared with unit of seeds resulting to a higher bulk and solid density. The mean value of porosity of the kernel is 53.91% and the seed is 34.64%. The surface area of the seed approximately 532.89 mm^2 and kernel 339.59 mm^2.

The physical properties of the kernel oil: in room temperature are liquid, density at 0.903 g/ml, kinematic viscosity at 42.88 cp, iodine value was determined at 103.62 g I_2/100 g, peroxide value at 1.93 meg/kg, and saponification value at 193.55 (Akbar et al. 2009). The seed oil have physical properties such as: density 914 kg/m^3, kinematic viscosity at 31.2

mm^2/s, iodine value at 100 g $I_2/100$ g (Karaj et al. 2008). The energy for the pressing amounted of the *Jatropha* seed at 439,450 J and the *Jatropha* kernel at 307,507 J, and the maximum loading force at 55,000 N (Ruzbarsky et al. 2014).

PRODUCT APPLICATION OF *JATROPHA CURCAS* L. BIOMASS

Jatropha curcas is a multipurpose shrub or small tree that mainly grows in the wild and has many useful attributes and numerous application. Traditionally, *Jatropha* is used to prevent and control erosion, to reclaim land or to act as a living fence for excluding farm animals. Many plant parts have potential economic value, such as tannin from the bark for treating leather and honey from its flower. Figure 3 shows the economic value of *Jatropha*. Among various economic values, biodiesel from *Jatropha* plant has been attracting attention for its potential for conventional diesel fuel substitution and favourable effects on the environment. Some product applications of *Jatropha* will be discussed in the next section.

Medicinal Uses

All parts of *Jatropha* plant such as seeds, leaves and bark, have been used in traditional medicine for a long time. *Jatropha* has been used as traditional medicinal plants by native people in the tropical and subtropical countries. The medicinal property, phytochemistry and pharmacology of several *Jatropha* species has been completely review by Sabandar et al. (2013). Some parts of *Jatropha* plants are employed to heal skin-related ailments. The seed oil, latex, leaf, stem bark or root of *Jatropha* plants are pounded and applied on infected skin i.e., eczema, itches, carbuncles, mouth blisters, wounds and swelling. On the other side, the seeds of *Jatropha* species are highly toxic and advised against their usage in herbal medicine. The toxic element in the seed is a toxalbumin named *Jatrophin* which causes agglutination and haemolysis of red cells and also injurious to other cells (Lucas and De Silva, 2006).

Jatropha species are famous for the purgative effect of the seed oil. This purgative effect has been directed to cure digestive system symptoms i.e., diarrhoea, dysentery, vomiting, retching and stomachache (Sabandar et al. 2013). The latex of *Jatropha* contains alkaloids including jatrophine, jatropham and curcain (Joglekar et al. 2011). It can be used as a source of pharmaceutical components.

Jatropha contains some kinds of phorbol esters. Phorbol esters from *Jatropha* species have been reported to possess tumor-promoting activity where their actions stimulate protein kinase C involved in signal transduction and development of most cells and tissues. These actions were found to be responsible for skin irritants and tumor promotion (Goel et al. 2007). *Jatropha* leaves can be used for silkworm feed, antiseptics, or remedy for dermatitis. The leaves and latex are used in healing of wounds, refractory ulcers, and septic gums and as a styptic in cuts and bruises. Curcain is a protein from the seed oil of *Jatropha* was reported to exhibit wound-healing property when tested on mice at two different enzyme concentrations (Villegas et al. 1997).

Figure 3. *Jatropha* economic value.

Investigation of the coagulant activity of the latex of *Jatropha* showed that whole latex significantly reduced the clotting time of human blood. Diluted latex, however, prolonged the clotting time, at high dilutions, the blood did not clot at all (Osoniyi and Onajobi, 2003). In addition, some *Jatropha* species have antibacterial, antifungal and antiviral activities.

Absorbent

Activated carbon is a popular adsorbent widely used in industries for variety of applications. It is widely prepared from variety of biomass precursors and the industry demands availability of cheap source of biomass that produce good quality activated carbons. Activated carbon van be derived from many parts of the *Jatropha* plants, such as cake, husk, and seeds can be considered as a potential low-cost adsorbent. It can be used for the removal of heavy-metal ions such as zinc and cadmium as well as reactive dyes from industrial effluents. Activated carbons can be prepared by physical and chemical activation methods. Chemical activation method has two important advantages compared than physical activation, i.e., (a) lower temperature in which the process is accomplished; and (b) the global yield of the chemical activation tends to be greater since burn-off char can be avoided. Acid activation followed by the thermal treatment increases the adsorption capacity to a large extent due to the increased surface area and pore volume (Yang et al. 2010).

Many studies reported the application of *Jatropha* as natural adsorbent. Mohammad et al. (2013) successfully tested *Jatropha* seed hull as an alternative adsorbent for the removal of cadmium metal ions. Meanwhile, Sathishkumar et al. (2012) reported the the utilization of jatropha pods as an activated carbon for the removal of reactive dye, Remazol Brilliant Blue R (RBBR) from textile effluent. *Jatropha* hull has the potential to be a good precursor for the production of activated carbon due to its high lignin content. *Jatropha* may be used widely because of its exceptional adsorbent properties. It may be giving the solution for cleansing of *Jatropha* wastes and their reutilizations as well as protecting the environment from pollution. Preparation of activated carbon from *Jatropha* seed shell by chemical activation was investigated by Karthikeyan et al. (2008).

Biogas Production

A proper utilization of non-edible seed cake is required to utilize a huge amount amount of oil seed cake from biodiesel production. The residual seeds obtained after the recovery of oil were found to be good substrate for the biogasification to obtain biomethane. It is a sugar-rich material. Some studies have been conducted on anaerobic digestion of *Jatropha* seed cake for biogas production. Raheman and Mondal, (2012) reported that biogas produced from kg of total solid contents *Jatropha* seed cake was 0.170 m^3 as compared to 0.166 m^3 for cow dung. The optimum total solid and C:N ratio of *Jatropha* seed cake slurry should be between 15 and 20% and 22:1 to 27:1, respectively for getting maximum biogas production at 40 HRT. Meanwhile, Singh et al. (2008) reported that the total biogas production from *Jatropha* seed cake of 348 l/kg of TS from the reactor with 10% TS as compared to 241 l/kg when the TS was 15% after the incubation period of 40 days. In both cases, methane content in biogas produced was 66 ± 2%. It was also reported that *Jatropha* seed cake gave 60% more biogas compared to biogas obtained from cattle dung.

Biogas formation using a semicontinous upflow anaerobic sludge blanket reactor was studied by Staubmann et al. (1997). A maximum production rate of 3.5 m^3 m^{-3} day^{-1} was obtained in the anaerobic filter with a loading rate of 13 kg COD m^{-3} day^{-1}.

Adhesive

Adhesives come from either natural or synthetic sources. Due to sustainable and environmental concern, there have been several attempts to replace part of the petroleum-derived phenolic compounds in wood bonding adhesives with phenolic-type compounds obtained from renewable sources. A sustainable alternative of natural adhesives is adhesives based on proteins. The most common technical applications of the industrial proteins, e.g., adhesives, coatings, and surfactants have been reported by Vaz et al. (2003).

Protein from *Jatropha* has used for technical applications an emulsifier, coating, bioplastic, foaming agent or adhesive. Lestari et al. (2011) investigated adhesive properties of the protein extracted from *Jatropha* seed and compared it with adhesive properties of casein, which commercially is being used as labeling adhesives. *Jatropha* seed protein had comparable adhesive properties with casein adhesives. Table 1 shows *Jatropha* protein adhesive properties. Based on their properties, *Jatropha* seed de-oiled press cake protein may have potential uses for paper adhesives. Meanwhile, Hamarneh et al. (2010) were successfully extracted proteins from *Jatropha* seeds using the principle of isoelectric precipitation. The proteins were used at different contents as a reactive component in polyketone-based wood adhesive formulations. Aung et al. (2014) has mentioned to used protein based *Jatropha* oil as adhesive. They were synthesized the hydroxylated *Jatropha* oil-based polyol to generate the polyurethane adhesive. Ong et al. (2012) has studied the *Jatropha* seed meal based on melamine urea formaldehyde extender on the bonding strength of plywood.

Table 1. Properties of different part of *Jatropha* protein as adhesive

Part of *Jatropha*	Methods	pH	Viscosity	Gel Time	Ref.
Seed cake	0.055 M NaOH	8	170 ± 3.5 (cP) 2 days 4338 ± 442 (cP) direct	2.25 ± 0.35 min	Lestari et al. 2011
Seed oil	Hydolysis Alcoholysis	3.5 7.5	75 91	210-290 min	Aung et al. 2014
De-oiled press cake	0.055 M NaOH	8	931 ± 115 (cP) 2 days 383 ± 6.2 (cP) direct	1.25 ± 0.4 min	Lestari et al. 2011
Kernel	0.01 M NaOH	11	0.25 to 10 (Pa.s) within 1-30 days	-	Hamarneh et al. 2010

Biodiesel

Biodiesel is one of potential sources to substitute for conventional diesel fuel. It produces favourable effects on the environment. One of the challenges to widespread adoption of biodiesel is its cost compared with petroleum-based fuel. The high cost of the refined edible oils that are predominant feedstocks for biodiesel production causes the relative high price of biodiesel. Among several non-edible oil seed species, *Jatropha* is a multipurpose species with many attributes and considerable potential as a feedstock for biodiesel production.

Biodiesel is made from virgin or used vegetable oils and animal fats through transesterification. Both edible and non-edible can be used as oil source. The transesterification reaction can be catalysed by a homogeneous catalyst using base or acid catalyst; or a heterogeneous catalyst. Base catalysts are usually preferred over acid catalysts because of the higher reaction rates and lower process temperatures required as compared to the acid-catalysed process. However, the oil or fat used in base transesterification reaction should contain no more than 1% FFA, which is equivalent to 2 mg KOH/g triglyceride. If the FFA level exceeds this threshold, saponification hinders separation of the ester from glycerine and reduces the yield and formation rate of biodiesel.

Berchmans and Hirata, 2008 have been developed two-stage transesterification process to produce biodiesel from jatropha oil with high FFA (free fatty acids) contents. The first stage involved the acid pretreatment process to reduce the FFA level of crude *Jatropha* seed oil to less than 1% and second was the alkali base catalyzed trans-esterification process gave 90% methyl ester yield. In order to reduce the cost of biodiesel fuel production from *Jatropha*, the lipase producing whole cells of *Rhizopus oryzae* immobilized onto biomass support particles was used (Tamalampudi et al. 2008). It was found to be a promising biocatalyst for producing biodiesel.

The conventional methods for biodiesel production from oil seeds of *Jatropha* involves the following steps: oil extraction, purification (degumming, dewaxing, deacidification, dephosphorisation, dehydration, etc.), and subsequent esterification or transesterification. These multiple biodiesel processing steps constitute over 70% of the total cost of biodiesel production if refined oil is used as the feedstock. Biodiesel can be produced by reactive extraction of oil seeds to reduce the cost and processing time associated with conventional methods. This process widely applied for biodiesel production from *Jatropha* seed by many researcher such as Shuit et al. (2010) and Kaul et al. (2010). Reactive extraction differs from the conventional biodiesel production process in that the oil-bearing material is brought into contact with the alcohol directly instead of reacting with pre-extracted oil. In this process, alcohol acts both as a reagent for transesterification and a solvent for extraction. As a consequence, a higher amount of alcohol is required compared to conventional process. On the other hand, reactive extraction eliminates the requirement of two separate processes of oil extraction and transesterification reaction, thus reducing processing time, cost, and the amount of solvent required. Meanwhile, a batch transesterification process is the most common method for producing biodiesel. It has suffered several disadvantages compared to continuous processes such as requiring larger reactor volumes, resulting in higher capital investment. A continuous process has been developed to reduce higher procurement costs and also to enhance mixing of the reactants in order to improve the reaction rates (Azhari et al. 2008).

The properties of diesel, *Jatropha* oil, methanol and methyl ester of *Jatropha* oil are summarized in Table 2. Sarin et al. (2007) have been examined the blends of *Jatropha* and palm biodiesel for their physico-chemical properties and to get optimum mix of them to achieve better low temperature and improved oxidation stability needed for South Asian and South East-Asian countries.

Currently, it is making a global transition from a research and demonstration concept to commercial production of biodiesel. The by-products in *Jatropha* biodiesel processing industry such as shell and cake of *Jatropha* must be utilized to produce valuable products such as adsorbent, organic fertilizers, adhesive and soap. It will make the biodiesel production from *Jatropha* is very profitable. In addition, high profitable and environmentally friendly processing can help *Jatropha* biodiesel production get more profits.

Table 2. Properties of *Jatropha* oil biodiesel from several researcher

Density ($Kg.m^{-3}$)	Kinematic viscosity (cSt)	Free fatty acid content (%)	Cetane number	Flash point (min) °C	Specific gravity ($g.cm^{-3}$)	Heat of combustion ($MJ.kg^{-1}$)	Ref.
930	50	14.6	-	241	-	37.05	Goyal et al. 2013
901		-	41-51	-	-	39.63	Rodriguez et al. 2011
810		8.14	59.24	175	0.912	-	Kafuku et al. 2010
-	37-54.8	0.18-3.40	38-51	210-240	0.860-0.933	-	Achten et al. 2008
840	4.59	-	45-55	50	-	-	Kumar and Sharma, 2008
-		-	57	110	0.862	-	Vyas, 2007

CURRENT AND FUTURE APPLICATIONS OF *JATROPHA CURCAS* BIOMASS

Scientist, researcher, academia, and industry still concern to study in development of new materials, especially the use of biomass as raw materials for many applications, until now. One of biomass is *Jatropha curcas* plant, it is become sources raw materials especially in energy crop. Global energy demand leads to massive opportunity for biofuel production. *Jatropha* oil is much considered as the best source of biodiesel production among the various biomass based fuel resources (Belewu et al. 2010). There is, globally, massive demand for new sources of oil driven by ever-increasing demands for energy. At the same time, the reserves of fossil fuels are declining which is leading to ever-increasing prices and concerns over security of supply. Thus there are a number of reasons to have a strong to use biomass for many applications, not the least reason for environmental and social-development, but a stand-alone business from an economic perspective. Proved that, there are a variety of potential applications, products and related market available for byproducts.

Presently, some countries are producing *Jatropha* oil to supplement the fuel requirements for lamps, cooking, and small diesel engines. Beside as robust energy plant, *Jatropha curcas* also produces wood, fruit shells, seed, seed shells and press-cake. These by product potentially for technical applications such as glues, film coatings, emulsifier, this is an interesting area of value creation, concurrently avoiding the competition with food production (Gofferje et al. 2014). Besides this attribute, *Jatropha* is also effective for the prevention of soil erosion, ecological restoration, improvement of rural economy. Because of chemical component in different part of *Jatropha* plant, pharmaceutical application and medicinal value growing up until now. This plant also call as doctor plant, because of application for medical benefit. As traditional herbal medicine, different part of *Jatropha* such as leaves, fruit, seed, stem bark, branches, twigs, latex and root has been used (Prasad et al. 2012). From variously chemical component in different part of *Jatropha curcas* such as seed oil and seed kernel has establish as soap production, rub medicine and cosmetic product.

Lignocellulosic material utilization of biomass for a number of very significant value-added products through chemical, physical and biological innovation to create innovative products and competitive as in various fields. One of the *Jatropha* plant is the answer, one of biomass that can be used for reinforcement in biocomposite material in the future. *Jatropha curcas* with their attribute has given more attention to the challenges in the provision as raw material for reinforcement biocomposites in future. The current and the future applications in various field of part *Jatropha* plant show in Figure 4.

FUTURE POTENTIAL OF BIOCOMPOSITE INDUSTRY

The use of biomass as a reinforcement in biocomposites continues to grow, both in the laboratory and industrial scale. Plantation of *Jatropha curcas* plant has developed in several countries, this is show that the biomass from this resources will never run out, if continued planting seriously. Different part of *Jatropha* which have varying chemical content suitable to be used as reinforcement in biocomposites. From the stem bark, woody part, leaves, fruit, seed, kernel, fruit shell, seed shell, seed cake, and oil of *Jatropha* has potentiality as raw material in reinforcement in biocomposites. So that the material biokomposite base on different part of *jatropha* plant can be improved in an industrial scale. Many researchers have conducted research in the field of biocomposite-based different part of *Jatropha*, but still in small scale. Given the success of researchers in isolating, extracting, and analysis various elements in different part of *Jatropha* plant, then the development of biocomposites can be carried forward for industrial scale.

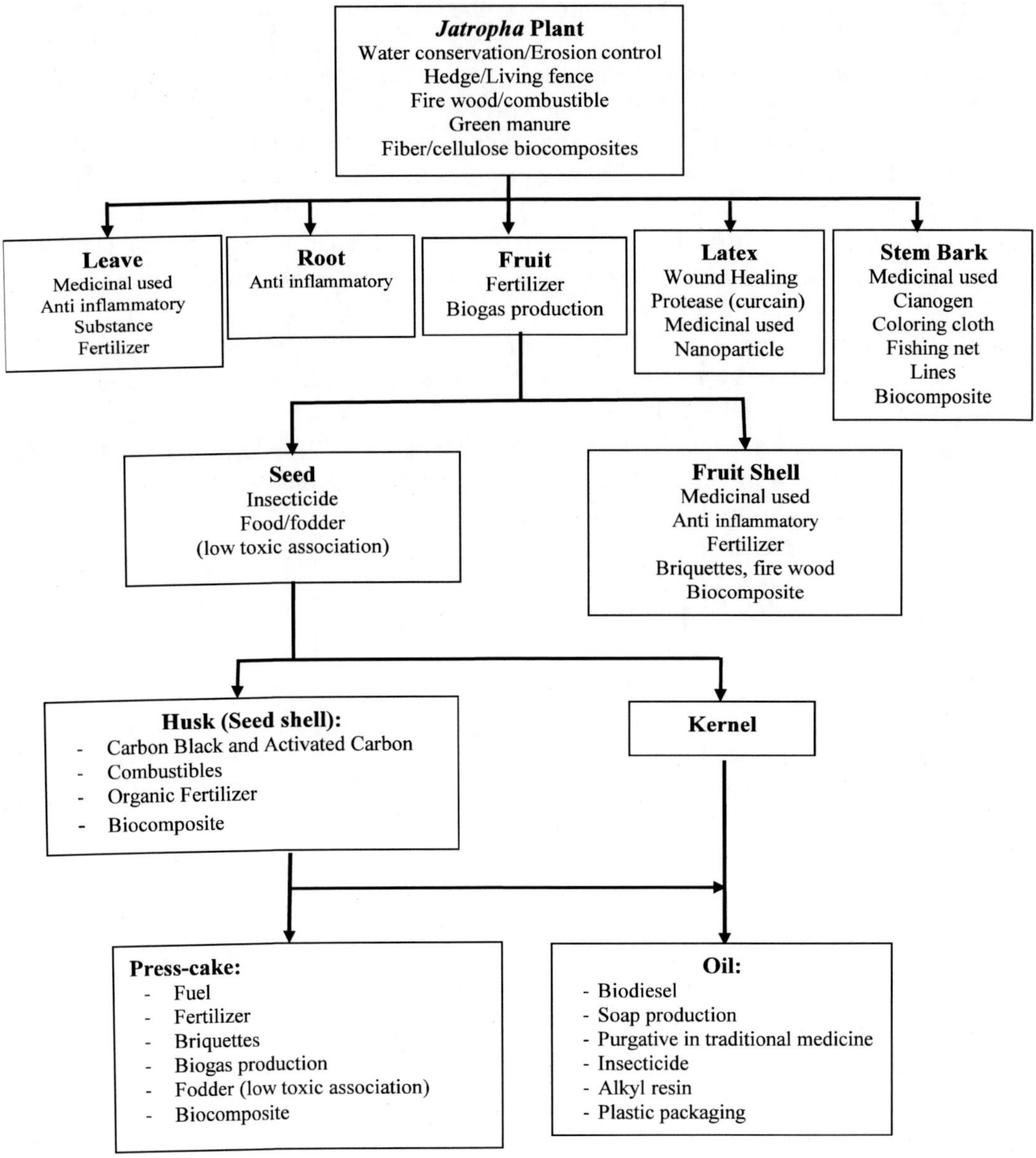

Figure 4. The current and the future application of *Jatropha curcas* plant.

To support industrial biocomposites, many researchers have conducted in develop raw materials based on different part of *Jatropha* curcas plant. Sri Aprilia et al. (2015) has conducted their research in use carbonized *Jatropha* seed shell to reinforced vinyl ester composite. Gofferje et al. (2014) has studied *Jatropha curcas* protein cast film suitable for food packaging applications. Gogoi et al. (2014) has prepared biocomposites based on *Jatropha* oil of alkyl/epoxy resin. Valasek, (2015) has reported epoxy biocomposite based on filler in micron size from seed of *Jatropha*.

The area of nanotechnology encompasses synthesis and characterization of nano scale materials, and the understanding and utilization of their potential applications in biocomposites, nanoparticles has synthesized from latex and seed of *Jatropha curcas* (Hudlikar et al. 2012; Joglekar et al. 2011; Bar et al. 2009).

Several researcher report that carbon black and activated carbon base on fruit shell, seed shell and seed cake of *Jatropha curcas* have chemical, physical and mechanical properties similar to other biomass products (Xin-Hui et al. 2011; Tongpoothorn et al. 2011; Ramakrishnan and Namasivayam, 2011; Raja et al. 2009; Sricharoenchaikul et al. 2007; Ramakrishnan and Namasivayam, 2009). These materials are suitable for dyes, fillers as reinforcement in thermoplastic, thermoset and elastomer to produce various material biocomposites on an industrial scale. Diego-Armando et al. (2012) has characterized the seed shell, from extraction process, they found hemicellulose, cellulose, and lignin, these component not in seed shell, but also in woody part, fruit shell rich this component, so that could used for reinforcement in biocomposites (Makkar et al. 2011). As zero waste of different part of *Jatropha curcas* have potentiality raw material to reinforcement and enhancement in biocomposites industry.

ECONOMIC CAPITAL

Jatropha curcas biomass as non-edible plant become one of the plants economically advantageous in the present and future. *Jatropha curcas* oil based on seed of the fruit is undoubtedly a product of biodiesel because it possesses excellent properties as a source of energy. The economy vibrancy of a country is indicated by its energy consumption. Until now, oil-based products based on *Jatropha* continues to grow. In other parts of this plant such as leaves, roots, stem, etc. are also very promising as economic value added. Not only the main parts of *Jatropha* potential for many applications, but also *Jatropha* waste can be beneficial to a wide range of products. The diversity of technological options and multiplicity of *Jatropha curcas* processing products, provide an opportunity for a zero-waste approach to *Jatropha curcas* utility as an energy crop.

Of course, the *Jatropha* products will increase the added value for the world community to improve the economy in various fields. Factors such as environment reason, renewable sources, various products, green technology, trend, cost and other factor should also take into account when developing a sustainable products. Potentiality of *Jatropha* crop can be attention about the major economic activity providing income and employment opportunities to the rural communities (Verma and Gaur, 2009). In addition, improving the local economy can also be taking advantage of this opportunity for the development of *Jatropha* plant would require developing countries to overcome policy, public investment and regulatory constraints that generally affect their agricultural development (Brittaine and Lutaladio, 2010). A very important step is to utilize the waste *Jatropha* biomass into production of value-added and innovative new products.

CONCLUSION

Jatropha curcas L. plant is multipurpose because different part of this crop has potential to be used in various applications. Not only for medicine but also technical applications have been develop from the different part of this *Jatropha* plant. All of product applications based on *Jatropha* parts, well as large-scale production as well as in household scale, has been carried out in the development that benefit for the needs of people in the world. This plant is not competitive with others because of non-edible plant and could be food application after toxic compound remove from its source. Because it can grow in various kinds of climates, *Jatropha* plant becomes an abundant source in the world. In the field of biocomposites, this plant can also be used as a raw material. Although still developments in laboratory scale, *Jatropha* plant will be increasing benefit to develop biocomposites in industries for the future. All of the parts of this plant will be promising, both economically and for the use of the products are widely. *Jatropha curcas* biomass with their attribute as an opportunity to beat their wings very promising industries in the world. Advances in science and technology enable the world's manufacturing industry to manipulate matter at the policy level to improve the overall properties of alternative materials base on *Jatropha* biomass to replace conventional materials. Technology-based research with a focus on *Jatropha* biomass with high potentialities can be grown to be applied in various industries.

REFERENCES

Abdul Khalil, H.P.S., Sri Aprilia, N.A., Bhat, A.H., Jawaid, M., Paridah, M.T., and Rudi, D. (2013). A Jatropha biomass as renewable materials for biocomposites and its applications, *Renewable and Sustainable Energy Reviews*, 22, 667-685.

Achten, W.M.J. Verchot, L. Franken, Y.J. Mathijs, E. Singh, V.P. Aerts, R. and Muys, B. 2008. *Jatropha* bio-diesel productionanduse, *Biomass and bioenergy*, 32, 1063-1084.

Aderibigbe, A.O. Johnson, C.O.L.E. Makkar, H.P.S. Becker, K. and Foid, N. (1997). Chemical composition and effect of heat on organic matter- and nitrogen-degradability and some antinutritional components of *Jatropha* meal, *Animal Feed Science Technology*, 67, 223-243.

Aung, M.M. Yaakob, Z. Kamarudin, S. and Abdullah, L.C. (2014). Synthesis and characterization of Jatropha (*Jatropha curcas* L.) oil-based polyurethane wood adhesive, *Industrial Crops and Products*. 60, 177-185.

Azhari, Ghazi, T.I.M. Resul, M.F.M.G. Yunus, R. and Yaw, T.C.S. (2008). Preliminary design of oscillatory flow biodiesel reactor for continuous biodiesel production from jatropha triglycerides. *Journal of Engineering Science and Technology*, 3, 138-145.

Bar, H. Bhui, D. Kr. Sahoo, G.P. Sarkar, P. Pyne, S. and Misra, A. (2009). Green synthesis of silver nanoparticles using seed extract of *Jatropha curcas*. *Colloids and Surfaces A: Physicochem. Eng. Aspects*, 348, 212-216.

Behera, S.K. Srivastava, P. Tripathi, Singh, R.J.P. and Singh, N. (2010). Evaluation of plant performance of *Jatropha curcas L.* Under different agro-practices for optimizing biomass – A case study, *Biomass and bioenergy*, 34, 30-41.

Berchmans, H.J. and Hirata, S. (2008). Biodiesel production from crude *Jatropha curcas* L. seed oil with a high content of free fatty acids. *Bioresource Technology*, 99, 1716-1721.

Brittaine, R., andLutaladio, N.B. (2010). Jatropha: A Smallholder Bioenergy Crop, The Potential for Pro-Poor Development, *Integrated Crop Management*, 8. Roma, Food and Agriculture Organization of The United Nations.

Cai, F. Yang, C. O. Duan, P. Gao, S. Xu, Y. and Fang, C. (2012). Purification and characterization of a novel thermal stable peroxidase from *Jatropha curcas* leaves, *Journal of Molecular Catalysis B*: *Enzymatic*, 77, 59-66.

Chena, C-R. Chenga, Y-J. Ching, Y-C. Hsiangd, D. and Chang, C-M. J. (2012). Green production of energetic Jatropha oil from de-shelled *Jatropha curcas L.* seeds using supercritical carbon dioxide extraction, *Journal of Supercritical Fluids, Article in Press.*

Corzo-Valladaresa, P.A. Fernández-Martínezb, J.M. and Velasco, L. (2012). Tocochromanol content and composition in *Jatropha curcas* seeds, *Industrial Crops and Products*, 36 304-307.

Dhillon, R.S. Hooda, M.S. Pundeer, J.S. Ahlawat, K.S. and Chopra, I. (2011). Effects of auxins and thiamine on the efficacy of techniques of clonal propagation in *Jatropha curcas* L. *biomass and bioenergy*, 35. 1502-1510.

Diego-Armando, Z. Wever, H.J. Heeres, and Broekhuis, A.A. (2012). Characterization of physic nut (*jatropha curcas L.*) shells, *biomass and bioenergy*, 37, 177-187.

Emil Akbar, E.E. Yaakob, Z. Kartom Kamarudin, S.K. Ismail, M. andSalimon, J. (2009). Characteristic and composition of *jatropha curcas o*il seed from Malaysia and its potential as biodiesel feedstock. *European Journal of Scientific Research*, 29 (3), 396-403.

Goel, G. Makkar, H.P. Francis, G. and Becker, K. (2007). Phorbol esters: structure, biological activity, and toxicity in animals, *Int. J. Toxicol.*, 26, 279-288.

Gofferje, G. Schmid, M. and Stäbler, A. (2014). Characterization of *Jatropha curcas* L. protein cast films with respect to packaging relevant properties, *Hindawi Publishing Corporation International Journal of Polymer Science*, 1-10.

Gogoi, P. Boruah, M. Bora, C. and Dolui, S.K. (2014). *Jatropha curcas* oil based alkyd/ epoxy resin/expanded graphite (EG) reinforced bio-composite: Evaluation of the thermal, mechanical and flame retardancy properties, *Progress in Organic Coatings*, 77, 87-93.

Goyal, P. Sharma, M.P. and Jain, S. (2013). Optimization of transesterification of *jatropha curcas o*il to biodiesel using response surface methodology and its adulteration with kerosene. *J. Mater. Environ. Sci.* 4 (2), 277-284.

Gunaseelan, V.N. (2009). Biomass estimates, characteristics, biochemical methane potential, kinetics and energy flow from *Jatropha curcus* on dry lands, *Biomass and bioenergy*, 33, 589-596.

Hamarneh, A.I. Heeres, H.J. Broekhuis, A.A. and Picchioni, F. (2010). Extraction of *Jatropha curcas* proteins and application in polyketone-based wood adhesives, *International Journal of Adhesion and Adhesives*, 30, 615-625.

Hudlikar, M. Joglekar, S. Dhaygude, M. and Kodam, K. (2012). Green synthesis of TiO2 nanoparticles by using aqueous extract of *Jatropha curcas L.* Latex, *Materials Letters*, 75, 196-199.

Jingura, R.M. Musademba, D. and Matengaifa, R. (2010). An evaluation of utility of *Jatropha curcas* L. as a source of multiple energy, Carriers, *International Journal of Engineering, Science and Technology*, 2 (7), 115-122.

Joglekar, S. Kodam, K. Dhaygude, M. Hudlikar, M. (2011). Novel route for rapid biosynthesis of lead nanoparticles using aqueous extract of *Jatropha curcas L.* Latex, *Materials Letters*, 65, 3170-3172.

Kafuku, G. Lee, K.T. andMbarawa, M. (2010). The use of sulfated tin oxide as solid superacid catalyst for heterogeneous transesterification of *Jatropha curcas* oil. *Chemical Papers*, 64 (6) 734-740.

Karaj, S. and Müller, J. (2008). Determination of physical, mechanical and chemical properties of seeds and kernels of *Jatropha curcas L.*, *Industrial Crops and Products*, 32, 129-138.

Karthikeyan, S. Sivakumar, P. and Palanisamy, P.N. (2008). Novel activated carbons from agricultural wastes and their characterization. *E-Journal of Chemistry*, 5, 409-26.

Kaul, S. Porwal, J. and Garg, M.O. (2010). Parametric study of Jatropha seeds for biodiesel production by reactive extraction. *Journal of the American Oil Chemists' Society*, 87, 903-908.

Kumar, A. and Sharmar, S. (2008). An evaluation of multipurpose oil seed crop for industrial uses (*Jatropha curcas L.*): A review, *Industrial crops and products*, 28, 1-10.

Lateef, F.A. Jere, P. and Yusuf, D. (2014). Harnessing biofuel and economic potentials of *jatropha curcas* for sustainable development in Nigeria: A review, *International Journal of Renewable Energy Technology Research,* 3 (1), 1-25.

Lestari, D. Mulder, J. and Sanders P.M. (2011). Jatropha seed protein functional properties for technical applications. *Biochemical Engineering Journal*, 53, 297-304.

Liang, Y. Siddaramu, T. Yesuf, J. and Sarkany, N. (2010). Fermentable sugar release from Jatropha seed cakes following lime pretreatment and enzymatic hydrolysis, *Bioresource Technology*, 101, 6417-6424.

Lucas, G.N. and De Silva, T.U.N. (2006). *Poisonous Plants of Sri Lanka*, First ed., Sri Lanka College of Paediatricians, Colombo.

Makkar, H.P.S. Kumar, V. Oyeleye, O.O. Akinleye, A.O. Angulo-Escalante, M. A. and Becker, K. (2011). *Jatropha platyphylla*, a new non-toxic *Jatropha* species: Physical properties and chemical constituents including toxic and antinutritional factors of seeds, *Food Chemistry*, 125, 63-71.

Makkar, H.P.S. Aderibigbe A.O. and Becker, K. (1998). Comparative evaluation of non-toxic and toxic varieties of *Jatropha curcas* for chemical composition, digestibility, protein degradability and toxic factors, *Food Chemistry*, 62 (2). 207-215.

Manurung, R. Wever, D.A.Z., Wildschut, J. Venderbosch, R.H. Hidayat, H. van Damc, J.E.G. Leijenhorst E.J. Broekhuisa, A.A. and Heeres, H.J. (2009). Valorisation of *Jatropha curcas* L. plant parts: Nut shell conversion to fast pyrolysis oil, *Food and bioproducts processing*, 87, 187-196.

Martı´nez-Herrera, J. Siddhuraju, P. Francis, G. Da´vila-Ortı´z, G. and Becker, K. (2006). Chemical composition, toxic/antimetabolic constituents, and effects of different treatments on their levels, in four provenances of *Jatropha curcas* L. from Mexico, *Food Chemistry*, 96, 80-89.

Matsumoto, H. Yeasmin, R. Kalemelawa, F. Aranam, M. Inoue, M. and Nishihara, E. (2014). Effect of soil water environment and simulated savanna climate on growth and mineral nutrition in *jatropha curcas* L. *Agricultural Science*, 2 (3), 13-22.

Mohammad, M. Yaakob, Z. and Abdullah, S.R.S. (2013). Carbon Derived from Jatropha Seed Hull as a Potential Green Adsorbent for Cadmium (II) Removal from Wastewater, *Materials*, 6, 4462-4478.

Myers, A. Wagner, A. Vendrame, and Pinares, A. (2013). In vitro growth of *jatropha curcas* L. cell cultures in microgravity, *Proc. Fla. State Hort. Soc.* 126.

Ong, H.R., Prasad, D.M.R., Khan, M.R., and Rao, D.S. (2012). Effect of jatropha seed oil meal and rubber seed oil meal as melamine urea formaldehyde adhesive extender on bonding strength of plywood, *Journal of Applied Science*, 12 (11), 1148-1153.

Openshaw, K. (2000), A review of *Jatropha curcas*: an oil plant of unfilled promise, *Biomass and Bioenergy*, 19, 1-15.

Osoniyi, O. and Onajobi, F. (2003). Coagulant and anticoagulant activities in *Jatropha curcas* latex, *Journal of Ethnopharmacology*, 89, 101-105.

Pecina-Quintero, V, Anaya-Lo´pez, J.L. Colmenero, A.Z. Garcı´a, N.M. Colı´n a, C.A.N. and Bonilla, J.L.S. Aguilar-Rangel, M.R. Langarica, H.R.J. and Bustamante, D.J.M. (2011). Molecular characterisation of *Jatropha curcas L.* Genetic resources from Chiapas, Me´xico through AFLP markers, *Biomass and bioenergy*, 35, 1897-1905.

Prasad, D.M.R. Izam, A. and Khan, Md.M.R. (2012). *Jatropha curcas*: Plant of medical benefits, *Journal of Medicinal Plants Research*, 6(14), 2691-2699.

Raheman, H. and Mondal, S. (2012). Biogas production potential of *jatropha* seed cake, *Biomass and bioenergy*, 37, 25-30.

Raja, S.A. Kennedy, Z.R. and Pillai, B.C. (2009). Flash pyrolysis of jatropha oil cake in fluidized bed reactor, *eNREE*, 6 (4), 1-16.

Ramakrishnan K. and Namasivayam, C. (2011). Zinc chloride-activated jatropha husk carbon for removal of phenol from water by adsorption: equilibrium and kinetic studies, *Toxicological and Environmental Chemistry*, 93 (6), 1111-1122.

Ramakrishnan, K. and Namasivayam, C. (2009). Development and characteristics of activated carbons from jatropha husk, an agro industrial solid waste, by chemical activation methods, *J. Environ. Eng. Manage.*, 19 (3), 173-178.

Rodrı´guez, R.P. Perez, L.G. Alfonso, M. Duarte, M. Caro, R. Galle, J. Sierens, R. and Verhelst, S. (2011). Characterization of *Jatropha curcas* oils and their derived fatty acid ethyl esters obtained from two different plantations in Cuba, *Biomass and bioenergy*, 35, 4092-4099.

Ružbarský, J. M. Müller, M. and HrabČ, P. (2014). Analysis of physical and mechanical properties and of gross calorific value of *Jatropha curcas* seeds and waste from pressing process, *Agronomy Research*, 12(2), 603-610.

Sabandar, C.W., Ahmat, N., Jaafar, F.M., Sahidin, I. (2013). Medicinal property, phytochemistry and pharmacology of several Jatropha species (*Euphorbiaceae*): A review, *Phytochemistry*, 85, 7-29.

Salawu, A.T. Suleiman, M.L. and Isiaka, M. (2013). Physical Properties of *Jatropha curcas* Seed, *Eloctronic Journal of Polish Agricultural Universities*, 16 (4).

Sarin, R., Sharma, M., Sinharay, S., Malhotra, R.K., (2007). Jatropha–Palm biodiesel blends: an optimum mix for Asia. *Fuel*, 86, 1365-1371.

Sathishkumar, P., Arulkumar, M., Palvannan, T. (2012). Utilization of agro-industrial waste *Jatropha curcas* pods as an activated carbon for the adsorption of reactive dye Remazol Brilliant Blue R (RBBR), *Journal of Cleaner Production*, 22, 67-75.

Shuit, S.H., Lee, K.T., Kamaruddin, A.H.; and Yusup, S. (2010). Reactive extraction of *Jatropha curcas* L. seed for production of biodiesel: Process optimization study. *Environmental Science and Technology*, 44, 4361-4367.

Singh, R.N. Vyas, D.K. Srivastava, N.S.L. and Narra, M. (2008). SPRERI experience on holistic approach to utilize all parts of *Jatropha curcas* fruit for energy, *Renewable Energy*, 33, 1868-1873.

Sirisomboon, P. Kitchaiya, P. Pholpho, T. and Mahuttanyavanitch, W. (2007). Physical and mechanical properties of *Jatropha curcas L.* fruits, nuts and kernels. *Biosyst. Eng*. 97, 201-207.

Sirisomboon, P. Kitchaiya, P. (2009). Physical properties of *Jatropha curcas* L. kernels after heat treatments, *Biosystems engineering*, 102, 244-250.

Sricharoenchaikul, V. Pechyen, C. Aht-ong, D. and Atong, D. (2007). Preparation and Characterization of activated carbon from the pyrolysis of physic nut (*Jatropha curcas* L.) Waste, *American Chemical Society Published.*

Sri Aprilia, N.A. Abdul Khalil, H.P.S. Bhat, A.H. Rudi, D. and Hossain, M.d. S. (2014). Exploring material properties of vinyl ester biocomposites filled carbonized *jatropha* seed shell, *Bioresources*, 9 (3), 4888-4898.

Staubmann, R., Foidl, G., Foidl, N., Gübitz, G.M., Lafferty, R.M., Arbizu, V.M.V., Steiner, W. Biogas production from *Jatropha curcas* press-cake, *Applied Biochemistry and Biotechnology*, 63-65, 457-67.

Tamalampudi, S., Talukder, M.R., Hama, S., Numata, T., Kondo, A., Fukuda, H., (2008). Enzymatic production of biodiesel from Jatropha oil: a comparative study of immobilized-whole cell and commercial lipases as a biocatalyst. *Biochemical Engineering Journal*, 39, 185-189.

Tongpoothorn, W. Sriuttha, M. Homchan, P. Chanthai, S. and Ruangviriyachai, C. (2011). Preparation of activated carbon derived from *jatropha curcas* fruit shell by simple thermo-chemical activation and characterization of their physico-chemical properties, *Chemical engineering research and design,* 89, 335-340.

Vaz, C.M., Graaf, L.A.D., Mulder, W.J. (2003). *Adhesives coatings and bioplastics from protein sources*. In: Fahnestock SR, Steinbuchel A, editors. *Polyamides and complex proteinaceous materials II*. New York: Wiley-VCH, p. 383-404.

Valdés-Rodríguez, O.A. Sánchez-Sánchez, O. Pérez-Vázquez, A. Joshua S. Caplan, J.S. and Danjon, F. (2013), *Research Article: Jatropha curcas* L. Root Structure and Growth in Diverse Soils, *The Scientific World Journal,* 13, 1-9.

Verma, K.C. and Gaur, A.K. (2009). *Jatropha curcas L.*: Substitute for Conventional Energy, *World Journal of Agricultural Sciences*, 5 (5): 552-556.

Villegas, L.F., Fernández, I.D., Maldonado, H., Torres, R., Zavaleta, A., Vaisberg, A.J., Hammond, G.B., (1997). Evaluation of the wound-healing activity of selected traditional medicinal plants from Perú. *Journal of Ethnopharmacology*, 55, 193-200.

Vyas, D.K. and Singh, R.N. (2007). Feasibility study of *jatropha* seed husk as an open core gasifier feedstock, *Renewable Energy*, 32, 512-517.

Xin-hui, D. Srinivasakannan, C. Jin-hui, P. Li-bo, Z. and Zheng-yong, Z. (2011). Preparation of activated carbon from *Jatropha* hull with microwave heating: Optimization using response surface methodology, *Fuel Processing Technology*, 92, 394-400.

Xu, R. Wanga, R. and Liu, A. (2011). Expression profiles of genes involved in fatty acid and triacylglycerol synthesis in developing seeds of *jatropha* (*Jatropha curcas L.*), *Biomass and Bioenergy*, 35, 1683-1692.

Yang, K., Jinhui, P., Srinivasakannan, C., Libo, Z., Hongying, X., Xinhui, D. (2010). Preparation of high surface area activated carbon from coconut shells using microwave heating. *Bioresource Technology*, 101, 6163-9.

In: The Promising Future of *Jatropha curcas*
Editors: Maira R. Segura-Campos et al.
ISBN: 978-1-63484-989-0
© 2016 Nova Science Publishers, Inc.

Chapter 4

PHYTOCHEMICAL ASPECTS OF *JATROPHA CURCAS* AND ITS APPLICATION IN HEALTH

Ludy C. Pabón B.[1,*] ***and Patricia Hernández-Rodríguez***[2,1]
[1]Professor-Researcher, Department of Basic Sciences,
Research Group Molecular Biology and Immunogenetics (BIOMIGEN),
Center Research CISVI, Universidad de La Salle, Bogotá, Colombia
[2]Professor-Researcher, Department of Basic Sciences,
Director Research Group Molecular Biology and Immunogenetics (BIOMIGEN),
Center Research CISVI, Universidad de La Salle, Bogotá, Colombia

ABSTRACT

Jatropha curcas (Euphorbiaceae) is a shrub or small tree native to America but widely cultivated in Asia and Africa; it is recognized as an excellent crop because of its easy adaptation in arid, semi-arid and high rainfall areas and for resistance to pathogens. This species is the most significant of its kind and has been reported for multiple purposes as an alternative source for biodiesel, food and feed, fertilizer production and traditional medicine.

The development of research directed towards knowledge and usefulness of these metabolites in *J. curcas*, have enabled the production of pharmaceuticals, agricultural products, cosmetics and food. In this way, the production of *J. curcas* is an important resource in the economy of many countries, especially those in the developing economies; because of their cultivation are obtained different marketable products. The pharmaceutical industry has sought to validate the knowledge we have of this plant, as a laxative, anti-inflammatory, antimicrobial, healing, contraceptive, antiviral, and coagulant, among others.

---

* Ludy C. Pabón B.: Professor-Researcher, Department of Basic Sciences. Research Group Molecular Biology and Immunogenetics (BIOMIGEN). Center Research CISVI. Universidad de La Salle. Bogotá, Colombia, South America. E-mail: lupabon@unisalle.edu.co.

1 Patricia Hernández-Rodríguez: Professor-Researcher, Department of Basic Sciences. Director Research Group Molecular Biology and Immunogenetics (BIOMIGEN). Center Research CISVI. Universidad de La Salle. Bogotá, Colombia, South America. E-mail: phernandez@unisalle.edu.co.

In this chapter the phytochemical studies in order to know the main metabolites are presented; as well as some functional and biological aspects of *J. curcas*, which allow to value this plant for its high medicinal potential. This work will reveal the usefulness of this species, as an example of the chemical diversity offered by plants for the production of active principles. Multiple studies that relate the composition and chemical analysis of each parts of this plant will be analyzed. Also the wide range of biological activities of this species will be discussed, allowing to consolidate it as a promising medicinal plant, but on which we should promote research to continue with the chemical, biological and genetic study, but even more important to determine if these activities are related with toxicity. Therefore, it seeks to make evident the need for systematic studies for toxicity and develop strategies to minimize the toxic effects that can to optimize and take advantage of biological and medicinal properties of *Jatropha curcas*.

Keywords: *Jatropha curcas*, health, active principles, phytochemical

INTRODUCTION

Jatropha curcas is a fast growing and easily spread with high oil content in its seeds which has been characterized as one of the most promising biofuel production plants; This property has led to the development of research aimed at obtaining new and better varieties with outstanding agronomic characteristics through genotypic changes as an approach to the genetic improvement of the species (Dhakshanamoorthy et al. 2015; Sarmiento et al. 2011; Oliveira et al. 2009). Another outstanding feature of *Jatropha curcas* is the root structure that provides soil resistance to erosion by water and wind generating capacity of the plant adaptation to arid, semi-arid and high rainfall which facilitate the ability to recover nutrients, restore and rehabilitate soils affected by erosion; additionally they will have few natural enemies associated which favors their growth and development (Valdés-Rodríguez et al. 2013; Openshaw, 2000; King et al. 2009; Misra et al. 2010; Aderibigbe et al. 1997).

The roots, leaves and seeds of *Jatropha curcas* have been widely used in traditional medicine in many parts of West Africa, Central and South America because it exhibits bioactive activities for fever, infections, jaundice, sores caused by Guinea worm, diarrhea among other due to their antimicrobial activity, mainly against *Staphylococcus aureus* and *Escherichia coli* (Thomas et al. 2008; Saetae et al. 2010) and antioxidant activity (Bent and Ko, 2004; Oliver-Bever, 1986; Fagbenro-Beyioku et al. 1998; Matsuse et al. 1998; Mujumdar et al. 2001; Igbinosa et al. 2009; Igbinosa et al. 2011; Balaji et al. 2009; Igbinosa et al. 2013). Despite all the beneficial effects of *J. curcas,* has also been shown to exhibit toxicity, this allows its use in pest control for its properties as insecticide and fungicide (Makkar et al. 1997; Igbinosa et al. 2013).

Jatropha curcas has identified several chemical compounds including flavonoids stand, di- and triterpenes, sterols, sapogenins, coumarins, organic acids, iridioides, saponins and tannins that provide the plant with economic, industrial, agricultural and pharmaceutical potential (Villegas et al. 1997, Gubitz et al. 1999; Staubmann et al. 1999; Parawira, 2010; Ravindranath et al. 2004; Bourgaud et al. 2001; Garcia and Pérez, 2009; Pabón and Hernández-Rodríguez, 2012).

The plants have historically been used as a source for obtaining drugs contribute significantly to human health when used as phyto way against many diseases. Additionally, the plant extracts are still offered with coverage of over 80% of the world's population, especially in developing countries (WHO, 2002).

Jatropha species are widely used in traditional medicine, mainly in Africa, Asia and Latin America (Burkill, 1994) to treat various ailments. It has been reported antibacterial and antiparasitic activity of the sap and crushed leaves of *J. curcas*. The water extract of the branches also strongly inhibits HIV-induced cytopathic effects with low cytotoxicity (Matsuse et al. 1999; Aiyela-Agbe, 2001). However, the information available on the medicinal properties of *J. curcas* basically limited to extracts of leaves of the plant. Studies on stem and other structures should be expanded; as well as studies validating the toxicity of this plant on animal and human species as has been observed with different aqueous extracts evidence from different toxic symptoms depend on the dose, mode of administration, and the sensitivity of the animals that were tested (Mariz et al. 2008; Chomchai et al. 2011). In this sense, and based on reports of adverse effects associated with hepatotoxicity and nephrotoxicity of medicinal plants (Seeff, 2007; Tang et al. 2008), have raised concerns about the lack of scientific studies focused on quality control the efficacy and safety of medicinal plants (Rousseaux and Schachter, 2003; Firenzuoli and Gori, 2007).

It is also known that medicinal plants contain several pharmacologically active compounds which can act individually, additively or synergistically to improve health; the problem is that many of these features depend on factors such as concentration, the type of administration and the species to evaluate what is necessary to encourage the development of research that scientifically prove these properties (Igbinosa et al. 2011; Azaizeh et al. 2003; Gurib-Fakim, 2006; Igbinosa et al. 2013).

Vegetable oils obtained from *J. curcas* is currently used for food, raw materials for the chemical industry and has increased the demand for these oils for biofuels due to rising costs and unstable and finite supplies of fossil fuels; also because the use of oils derived from *J. curcas* you have a renewable source for the production of sustainable and affordable biofuels that can reduce the greenhouse effect caused by gases from traditional fuel (Igbinosa et al. 2011; Igbinosa et al. 2013).

Despite the recent interest in *J. curcas* it is little that is known about breeding programs in this species. Currently it is treated as an agricultural crop that has led to an imbalance in the rapid implementation of large plantations, while the essential questions on growth, management and crop production of *J. curcas* have not been addressed adequately carried inappropriate yields seed and the oil. This requires strengthening research on varieties of the species to a range of climates and agro-ecosystems to ensure good profitability and a reduction in emissions that allow environmental protection.

Knowing seed traits, yield and oil composition, behavior flowering tree morphology and disease resistance is important for it is necessary to encourage genetic knowledge of *J. curcas*, a promising species for industry, the economy, agriculture and human and animal health. Therefore, the purpose of this chapter is to make evident the phytochemical studies in order to know the main metabolites in the plant; as well as some functional and biological aspects of *J. curcas*; also, determine whether these activities are related to toxicity. Multiple studies about composition and chemical analysis of each of the parts of this plant will be analyzed. The wide range of biological activities of this specie will be discussed, allowing consolidate as a promising medicinal plant but on which research should be encouraged to continue with chemical, biological and genetic study that can optimize and exploit biological and medicinal properties *Jatropha curcas*.

STRUCTURE AND FUNCTION OF MAJOR METABOLITES STUDIED IN *JATROPHA CURCAS*

Plants synthesize various compounds known as primary and secondary metabolites. The first type of compounds (amino acids, carbohydrates, lipids, glycerides, nucleic acids) are directly involved in the survival, growth and reproduction on plants. While secondary metabolites (terpenes, phenols, glycosides and alkaloids) are involved in ecological interactions with the environment and between functions is to act as a defense mechanism against predators and pathogens, allelopathic agents, attractant pollinators and animals that contribute to seed dispersal. The knowledge, study and utility of these metabolites has favored the development of research with natural products such as phytochemical, through the incorporation of new techniques of extraction, purification, and identification, synthesis and processing of biologically active molecules; directed towards obtaining raw materials for the manufacture of pharmaceuticals, cosmetics, nutraceutical and agricultural products (Bourgaud et al. 2001; Garcia and Pérez 2009; Pabón and Hernández-Rodriguez, 2012).

Jatropha curcas is a species with economic potential due to the diversity of compounds among which are flavonoids, diterpenes, sterols, triterpenes, sapogenins, coumarins, deoxipreussomerinas, organic acids, iridioides, saponins and tannins Villegas et al. 1997; Gubitz et al. 1999; Staubmann et al. 1999 a,b, Parawira, 2010; Ravindranath et al. 2004). In this sense, Agbor et al. 2015 established the content of the main metabolites present in extracts obtained from leaves, roots and bark using solvents such as water, methanol, ethanol and ethyl acetate. In this study, no significant differences in the relative distribution of metabolites with different solvents were found, but there is a higher content when methanol is used. Figure 1 shows the relative distribution of the main metabolites of methanol extracts of the three parts evaluated.

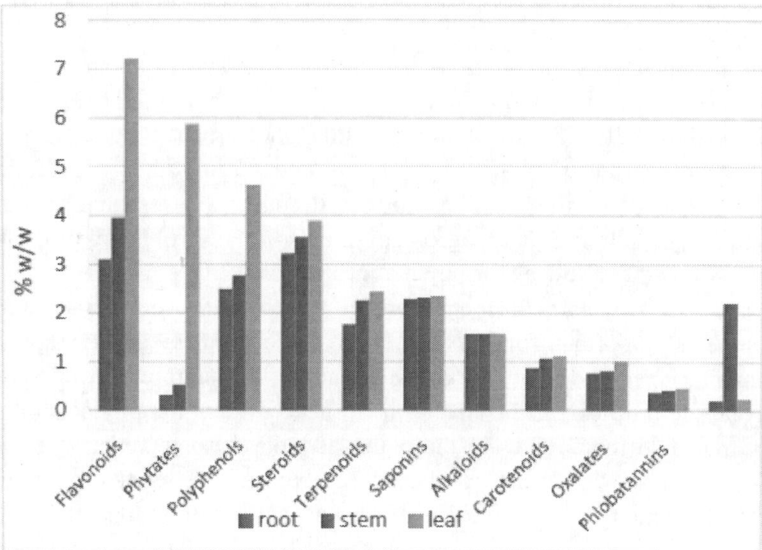

Figure 1. Relative distribution of the main metabolites found in the methanol extracts of leaf and root bark of *J. curcas*.

The roots, bark and leaves of *J. curcas* have similar content of terpenoids, alkaloids, saponins, carotenoids, oxalates and phlobatannins. However, the leaves are those that have the highest content of metabolites, especially flavonoids, phytates and polyphenols, while the bark is the one with the highest content of triterpenoid saponins. The results of this study related to the chemical composition allow you to set the high medicinal potential of this species as a therapeutic agent for the treatment of a wide range of diseases and infections. For example, the high content of polyphenols and flavonoids explains its antioxidant, antiallergic, anti-inflammatory, platelet aggregation, antimicrobial, antiviral and antitumor properties. In addition to contain saponins and steroids in significant but safe amounts (<10%), it represents a potential as agents of fertility; and having alkaloids can be used as medicinal agents due to its analgesic, antispasmodic and antibacterial (Agbor et al. 2015).

Among the studies that have been done to know which are the metabolites present in this species it has been found that the diterpenoids are secondary metabolites characteristic of the genus. Of these compounds it has been found which can have seven different skeletons, includingziglianes, casbenes daphnanes, podocarpanes, lathyreanes, jatrophanes and rhamnofolanes (Figure 2) (Abdelgadir and Staden, 2013; Liu et al. 2013; Liu et al. 2015).

The largest number of diterpenes have been isolated from roots, such as: caniojane, 3-deshidroxi-2-epi-caniojane, curculathyranes A and B, curcusecons A-E, curcusones A-J, 4-epi-curcusone E, jatrocurcasenones A–E, 4Z-jatrogrossidentadion, 15-epi-4Z-jatrogrossidentadion, 4E-jatrogrossidentadion, 15-epi-4E-jatrogrossidentadion, (14E)-14-O-acetyl-5,6-epijatrogrossidentadion, jatrogrossidion, isojatrogrossidion, 2-epi-isojatro-grossidion, 2-epi-hydroxyisojatrogrossidion, 2-hydroxyisojatrogrossidion, jatrophadiketone, jatrophalactam, jatrophalactone, jatrophaldehyde, epi-jatrophaldehyde jatrophalone, jatrophene, jatrophodiones B–E, jatrophol, epi-jatrophol, jatropholone A and B, acetoxyjatropholone, jatrowediol, lathyranlactone, riolozatrione and spirocurcasone (Villegas et al. 1997; Gubitz et al. 1999; Rakshit, et al. 2001; Sabandar, 2010; Liu et al. 2012; Zhang et al. 2012; Abdelgadir and Staden, 2013; Sabandar et al. 2013; Liu et al. 2013; Liu et al. 2015).

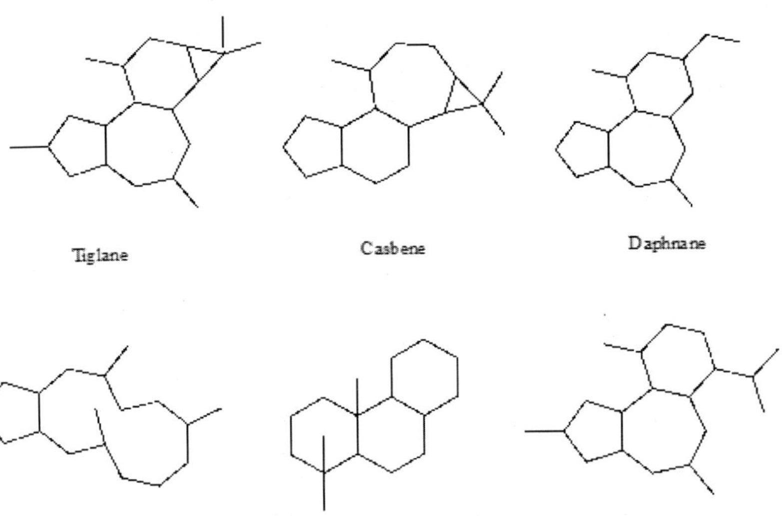

Figure 2. Diterpene skeletons of *Jatropha curcas*.

Likewise heudolotinona have been isolated from leaves (Rakshit et al. 2001; Ravindranath et al. 2004); palmarumycin CP1, JC1and JC2 from stem (Ravindranath et al. 2004; Rakshit et al. 2001; Sabandar et al. 2013); (14E)-14-O-5,6-epoxyjatrogrossidentadione, 15-O-acetyl-15-epi-(4E)-jatrogrossidentadione, isojatrogrossidentadione, 3-β-acetoxy-12-methoxy-13-methyl-8,11,13-trienepodocarpa-7-one, 3β, and 12-hydroxy-13-methyl-8,10,13-trienepodocarpa from the aerial part; and the factors C1-C6 (diesters tiglano core 12-deoxy-16-hidroxiforbol) from seeds (Haas et al. 2002; Roach et al. 2012; Sabandar et al. 2013). The latter group are commonly known as phorbol esters and found spread over most parts of the plant such as roots, outer bark, inner bark, wood, stems, leaves, buds, flowers and kernels containing concentrations 0.55, 0.39, 3.08, 0.09, 0.99, 2.75, 2.10, 1.83 and 6 (mg/g dry matter), respectively. However, they are not found in the latex (Devappa et al. 2013).

The second group of metabolites are more abundant sterols, where from leaves, roots and seeds has been isolated ß-sitosterol and β-D-glucoside; and only in the roots was found daucosterol (Gubitz et al. 1999; Sabandar, 2010; Sabandar et al. 2013) and leaves has been reported stigmasterol, stigmast-5-en-3β,7β-diol, stigmast- 5-en-3β, 7α-diol, cholest-5-en-3β, 7β-diol, cholest-5-en-3β, 7α-diol, campesterol and 7-keto-beta-sitosterol (Gubitz et al. 1999; Staubmann et al. 1999a,b). Regarding the triterpenes have been isolated from the roots and bark is taraxerol (Villegas et al. 1997; Gubitz et al. 1999; Staubmann et al. 1999), of the leaves is α-amyrin, (Gubitz et al. 1999; Staubmann et al. 1999 a,b), of the seed is diploptene and bark additionally reported β-amyrin, epi-friedelinol, friedelin and jatrocurin (Villegas et al. 1997; Gubitz et al. 1999).

As flavonoids have been isolated flavone type, such as apigenin, vitexin and isovitexin, from sheets (Villegas et al. 1997; Gubitz et al. 1999; Staubmann et al. 1999 a,b) and nobieletina from the roots (Sabandar, 2010; Sabandar et al. 2013). Among the isolated coumarins it has been reported the presence of methyl ester scopoletin from roots (Villegas et al. 1997; Gubitz et al. 1999) and tomentin, marmesin, propacin, jatrophin, 5-hydroxy-6,7-dimethoxicoumarin, 6-methoxy-7-hidroxicoumarine from root (Gubitz et al. 1999; Sabandar, 2010; Abdelgadir and Staden, 2013). It also has been reported the presence of lignans and isoprincepin isoamericanin in sedes (Abdelgadir and Staden, 2013; Yao et al. 2012).

Respect to the peptides have been reported for the latex, the presence of three cyclic peptide curcacycline A and B, jatrophidin (Berg et al. 1995; Gubitz et al. 1999; Altei et al. 2014) which compared with linear considered to have greater potential as therapeutic agents because they have better pharmacodynamic properties, chemical and enzymatic stability and selectivity with the recepter (Altei et al. 2014). Likewise, for this part of the plant it has been reported the isolation of enzyme curcain (Villegas et al. 1997; Gubitz et al. 1999) and in seeds of esterase proteins JEA, JEB, lipase JL, curcin and curcin L (Staubmann et al. 1999; Roach et al. 2012; Rodrigues et al. 2013a; Abdelgadir and Staden, 2013) and protein JCGZ 12502 (Jucá et al. 2015).

Phenolic compounds have been reported for roots is 3-methoxy-4-hydroxybenzoate acid and 3-hydroxy-4-methoxybenzaldehide; caffeoylaldehyde in seeds, isoferulaldehyde, 2,4-di-tert-butylphenol, methyl-3- (3,5-di-tert-butyl-4-hydroxyphenyl) propionate and linoleic acid and the cake seed syringaldehyde and caffeoylaldehyde (Abdelgadir and Staden, 2013; Tongpoothorn et al. 2012; Yao et al. 2012; Sabandar et al. 2013). Also, metabolites simpler leaves as the alcohols 1-triacontanol (Gubitz et al. 1999; Staubmann et al. 1999 a,b), seeds have been isolated the β-ethyl-D-glucopyranoside dipalmitin and 2 glycerol monooleate (Yao et al. 2012; Tongpoothorn et al. 2012). Among the amines has been reported for leaves the presence of 5-hydroxypyrrolidin-2-one, uracil, and pyrimidine-2,4-dione and curcamide for roots (Staubmann et al. 1999 b; Yao et al. 2012). The aerial parts have been reported organic acids as o-cumaric, p-cumáric, p-methylbenzoic, p-hidroxibenzoic, protocatechiuc, resorsilic (Gubitz et al. 1999).

Bromatological trials to the leaves of this species have established a moisture content of 86.5% and an ash content, proteins, lipids, carbohydrates and crude fiber 23.4, 28.0, 3.2, 45.4, 3.0% (as dry basis) respectively. These results indicate that *J. curcas* possess similar to those of an edible plant such as spinach but has three times the amount of Fe and Ca (0.3% and 19.3%, respectively) and have a low amount of K and absence of Na, can be consumed by hypertensive patients or require low sodium diet (Mendez et al. 2014).

J. curcas is considered an oleaginous plant for its high oil content in the seeds, and contain between 30 and 32% protein and 60-66% fat, therefore it has a high potential as a renewable energy source and as a substrate for the production of animal feed. This has allowed conducting research to improve the percentage of extraction and determine the chemical composition of these oils (Martinez et al. 2006; Kumar and Sharma, 2008). Among the methods that have been established for obtaining oil from the seeds is mechanical and chemical extraction (Fu et al. 2014), where for the second type was compared Soxhlet method with supercritical fluid, yielding yields of 49.21 and 43.51%, respectively (Heng et al. 2009; Salinas, 2010). Likewise, the physicochemical characteristics of this oil as viscosity, iodine and saponification have established their potential application in the manufacture of biodiesel, varnishes, polishes and soaps (Rakshit et al. 2001). Regarding the chemical composition of the oil, it is reported that this is mainly composed of 64% triacylglycerols, 12% of hydrocarbon compounds and 9% of free fatty acids, among which are the oleic, linoleic, palmitic, stearic and palmitoleic with percentages of 40, 38, 16, 8, and 0.8%, respectively, thus predominantly unsaturated fatty acids content of 81% and saturated fatty acids in lower proportion (Akintayo, 2004; Oliveira et al. 2006; Sharma and Kumar, 2008; Rodriguez et al. 2013; Nakao et al. 2015; Sousa et al. 2015). Likewise, we have determined a high total content of tocochromanols (507.4 mg/kg) and tocopherols (92.04 mg/kg), whose main components were identified as γ-tocoferol, γ-tocotrienol and Δ-tocotrienol, with a composition of 15.4, 83.8 and 0.8%, respectively. This content represents an added value for the oil to have resistance against oxidation, because these lipids are natural antioxidants phenolic by donating hydrogen to free radicals, delaying the process of lipid peroxidation (Rodriguez et al. 2013). The composition of this oil is similar to that of many edible oils has not been possible marketing by toxic effects attributed to the presence of phorbol ester (Rakshit et al. 2001).

To this oil it was identified the presence of fifteen sterols, being among the major sitosterol, stigmasterol and campesterol, with percentages of 70, 9.90 and 8.32% respectively. Among the sterols showed lower percentages than 6% were reported sitostanol, cholesterol, brassicasterol, 24-methilen-cholesterol, campestanol, clerosterol, Δ-7-campesterol, Δ-7-stigmasterol, Δ-7-avenasterol, Δ-5-avenasterol, Δ-5-23-stigmastadienol and Δ5-24-stigmastadienol (Corzo et al. 2012; Rodríguez et al. 2013). For oil it has been reported the isolation of three sesquiterpenoids called jatrophaeudesmenes A-C, three germacranolides identified as pyrethrosin, 8β-acetoxyanhydroverlotorin, chamissarin and two eudesmanolides as 6α-tigloyloxy-1β-hydroxy-4(15),11-eudesmadien-8α,12-olide and chrysanin (Yang et al. 2013).

Additionally, during the chemical extraction for obtaining biodiesel only the seeds are used, leaving the shell (corresponding to 42%) as a byproduct of the process. Although this residue can be used as fuel, green manure, or for the production of biogas or charcoal, it has not yet exploited the potential of this part of the plant by its bioactivity. The high phenolic content having the extracts of this plant part with solvents such ethyl acetate, ethanol and water (262.91 ± 8.35, 311.91 ± 6.07 and 263.81 ± 6.26 mg GAE/g extract), are higher than those obtained in the extracts obtained from seeds where the concentration does not exceed 4.49 mg GAE/g extract and this additional associated the role of this type of compounds in protecting seeds, makes this part of the plant can be used as a source of new natural antioxidants (Fu et al. 2014).

IMPORTANCE AND STUDIES TO MINIMIZE THE TOXIC EFFECTS OF *J. CURCAS*

The toxicity that have many of the plants used in traditional medicine, make unsafe because its effects can go from mild to lethal, depending on the concentration, the type of active ingredient, usage time and geographical location of the plant (Pabon and Hernandez-Rodriguez, 2012). Often these toxic principles can be used to control pests and pathogens, by obtaining products that benefit agricultural production, human and animal health, but even though it is intended that these products are safe to environment, there are cases that pollute water resources (Kongmany et al. 2014). This is why it is important to conduct systematic and controlled studies to determine the degree of toxicity, its influence on the pharmacological applications in living things, and seek ways to effectively degrade these compounds.

The toxic effect of *J. curcas* has been reported for most plant structures, being seed the most studied for its effects on living organisms ranging from microorganisms to animals (Yan et al. 2010; Lin et al. 2010). Most trials that have evaluated the toxic effect of the seeds in animals have shown common clinical symptoms such as abdominal pain, diarrhea, bleeding, loss appetite, nervous imbalance and sunken eyes. In addition, histopathological findings reported gastrointestinal inflammation, necrosis of the liver, heart and kidneys (Gubitz et al. 1999; Abdel et al. 2003; Rakshit et al. 2008). Toxicity studies have identified differences in the toxic effect, relate to: the dose, treatment given to seeds, mode of administration and the animal that is used in the bioassay (Yan et al. 2010). Regarding the dose has been reported that when given to goats orally daily concentration of 1 g/kg death occurs between days 7 and 11, if the dose is reduced to 0.25 g/kg, is raised death between 18 and 21 d (Abdel et al. 2003). The treatment performed to seeds also has an impact on the toxicity, finding that in rats fed raw seeds death occurs 2 to 3 d, when used meal defatted death occurs after 6 to 8 d and if fed with roasted or boiled seeds is extended from 14 to 16 d (Rakshit et al., 2008). The mode of administration also causes differences in the toxic effect of the substance to be evaluated; oral administration causes death in most animals, whereas if it is used topically occurs to a local disturbance that can range from a rash to a necrotic lesion (Yan et al. 2010). The type of animal also influence the toxicological assessment of a substance, there have been considerable differences in the minimum lethal concentrations in animals such as sheep, goats and calves with values of 7.4 g/kg, 1.5 g/kg and 3, 0 g/kg, respectively (Gubitz et al. 1999).

This toxic effect has also been observed in humans in 1854 in Birmingham, where 30 children were poisoned seeds of *J. curcas*. These toxic effects are attributed to the presence of anti-nutritional factors that has led to the search of antidotes as orellana and *Peltrophorum africanum*, order to counteract poisoning in the population of Mexico and South Africa (Gubitz et al. 1999; Martínez et al. 2006; Kumar and Sharma, 2008; Rakshit et al. 2008; Hamarneh et al. 2010).

In the interest of knowing what anti-nutritional factor should be the major toxic effects, several studies have been developed with some of them, coming to the conclusion that some saponins, proteins, peptides, diterpenes, saponins, trypsin inhibitors, phytic acid, and phorbol esters, are the major metabolites of toxicity presented by this species (Guedes et al. 2014; Martínez et al. 2015). For example, curcin protein has an acute oral toxicity LD_{50} 104.737 mg/kg and a binder effect in rabbit erythrocytes when used at a concentration of 7.8 mg/L; their toxic effect associated with participation in protein inhibition *in vitro* (Gubitz et al. 1999; Lin et al. 2010). However, this compound is found in many foods such as grains and vegetables, so their presence is not a barrier to the use of this plant (Joon et al. 2011). With respect to curcaine toxicity in mice was determined to be toxic when orally administered to LD_{50} value of 960 mg/kg, but according to histopathological studies in mice, this compound has superior healing properties to conventional ointments as nitrofurazone or propamidine isothionate used for the treatment of infectious diseases of the skin (Gubitz et al. 1999).

In addition, phorbol esters are molecules having an LD_{50} 27.34 mg/kg and its toxicity has been associated with several activities (inflammatory, molluscicide, etc.). These compounds also cause gastrointestinal bleeding and microscopic changes in kidneys, lungs, heart, liver and spleen (Rakshit et al. 2001). Additionally, it has been reported for this group of substances participating in cell proliferation, activation of blood platelets, lymphocyte mitogenesis, skin erythema, stimulation of prostaglandin production and degranulation of neutrophils (Saetae et al. 2011; Altei et al. 2014; Baldini et al. 2014). Likewise, have determined their participation in signal transduction to be considered analogs diacyl glycerol (DAG) and co-carcinogens to be activators of protein kinase C, which means that these substances do not induce tumor formation but if they promote growth and increase the risk of appearance when exposed simultaneously to a chemical carcinogen (Berg et al. 1995; Gubitz et al. 1999; Haas et al. 2002; Osoyini and Onajobi, 2003; Abdel et al. 2006; Acda, 2009; Yan et al. 2010; Joon et al. 2011; Abdelgadir and Staden, 2013; Guedes et al. 2014).

These toxic effects have led to propose studies to decrease the toxicity of *J. curcas,* taking into account the potential of this species for animal feed due to high crude protein content of 48-64% and containing all the amino acids essential, except for lysine whose higher values than those reported by the FAO for children 5 years old. Among the detoxification methods that have been implemented are heat treatment, solvent extraction or short chain bases, extraction with ionizing radiation, isoelectric precipitation, fermentation, plasma treatment, among other methods (Kumar and Sharma, 2008; Joon et al. 2011; Phengnuam and Suntornsuk, 2013; Kongmany et al. 2014; Hidayat et al. 2014; Nakao et al. 2015; Widiyastuti et al. 2015).

The heat treatment is considered to be an ineffective method to decrease the concentration of antinutritional factors, so it had to additionally incorporate the use of strong bases, yielding a reduction of 86% of these compounds, in addition to improving the organoleptic properties of the food, increasing the amount of food consumed, reducing the loss of body mass and delaying mortality in rats (Kumar and Sharma, 2008; Rakshit et al. 2008). The alkaline extraction was then added soelectric precipitation, for obtain protein flour with small amounts of phorbol esters, phytic acid, and saponin and any lectin (Widiyastuti et al. 2015).

Another method is evaluated ethanol extraction, with good results to completely remove phorbol esters and lectin (99%), while partially phytic acid, trypsin inhibitor and saponins (Saetae et al. 2011; Nakao et al. 2015). When using this method to obtain flour to feed pigs, lower weight gain and daily intake was found when was used flour mainly detoxified from seeds of *Jatropha*, that when soybean meal is used, the characteristics are attributed to undesirable characteristics in terms of taste, smell and texture, due to the presence of phorbol esters (Berenchtein et al. 2014; Li et al. 2015). Likewise, fed flour *Jatropha* pigs exhibited liver damage was repaired gradually with the growth of pigs and therefore it is suggested that this flour can only replace a maximum percentage of 45% to soybean meal (Li et al. 2015). Another method is Soxhlet in which is having as variables: extraction time, the solvent and solute relationship proportion of the mixture methanol: ethanol. When comparing against an untreated flour, it was found that none of the heat treatments had an impact on the moisture content, ash and dietary fiber, whereas if reduced lipid content and increase the protein. Also it has been found that all treatments eliminates phorbol esters with higher percentages to 83% and lower than the maximum values percentages to be considered as non-toxic (0.11 mg/g) (Guedes et al. 2014).

Thus it has been found that different chemical treatments which allow detoxification of flour, such as extractions with a mixture methanol: ethanol 50:50, 92% methanol, 90% ethanol followed by 0.07% NaHCO3; this content decreases to 0.10, 0.09, 0.08% respectively if one takes into account that phorbol esters in a flour without any treatment is 3.60 mg/g. Processes such as alkaline extraction with methanol, enzymatic hydrolysis followed by washing with alcohols, or γ ray radiation, completely eliminate phorbol esters, while the extraction with ethanol or methanol for 2h only can reduce toxicity indices without arriving reach the limits to be considered as non-toxic (Guedes et al. 2014; Kongmany et al. 2014).

In addition to chemical methods, it has also been evaluated biological methods to reduce the toxicity of the plant. In this regard Phengnuam and Suntornsuk 2013, employed two types of fermentation (submerged and solid state) with *Bacillus subtilis* and *B. licheniformis*, for degradation of the main antinutritional factors, such as phorbol esters, phytic acid, trypsin inhibitors, and lectins. The submerged fermentation with *B. licheniformis* was the most efficient method in reducing the concentration of phorbol ester and trypsin inhibitors with a percentage of 60 and 75% respectively, while for phytic acid was solid-state fermentation with the same microorganism to obtain a reduction ratio of 50%. Likewise it was determined that none of the methods was effective to degrade lectin. This study demonstrates that there is no single methodology to efficiently degrade all anti-nutritional compounds present in the seeds of this species. Another study for the degradation of these toxic compounds in seeds, has been performed with Pleurotus ostreatus, finding that has the capacity to degrade antinutritional factors, specifically phytic acid (95%) Tannin (85%), besides is having a high productivity for producing fungi with high nutritional value (Rodrigues et al. 2013b).

Another microorganism has been evaluated to test the efficacy in degradation of phorbol esters is the bacterium *Pseudomonas aeruginosa*, with which is obtained the complete removal during solid-state fermentation (Joshi et al. 2011). Also other microorganisms were tested as *Aspergillus oryzae* and *Lactobacillus acidophilus*, with a reduction of phorbol esters 38 and 41% respectively were obtained (Widiyastuti et al. 2015). Likewise, there have been trials for degradation by enzymatic activity of phorbol esters, due to the presence of ester bonds between other phorbol and a fatty acid, being analogs diacyl glycerol and therefore can be hydrolysed by lipases, resulting percentages reduction 99% (Hidayat et al. 2014).

Thus it has been found that inhibitors of trypsin and lectin present in seed cake can be inactivated by heat treatment, while it is not possible to eliminate phorbol esters because they are heat stable and therefore to reduce its concentration in the seed cake should be used chemical and biological treatments (Guedes et al. 2014). Therefore, you must continue to perform additional studies to complement the toxic effects they have on different organs of living things, establish a robust and reliable method for the identification and quantification of phorbol esters, in addition to developing a method of economic detoxification and efficient, which can be efficiently led to large-scale, for the use of the seeds in the agricultural industry or in human and veterinary medicine (Kumar and Sharma, 2008; Parawira, 2010; Baldini et al. 2014; Li et al. 2015).

A viable option for the use of this plant without being affected by the toxic effects is the implementation of seed other places as Mexico or Central America, which there are no reports of toxic effects. For example, seeds of species from Mexico are used to feed chickens and pigs and even after being boiled or roasted are used to prepare traditional dishes of this country. It has also been reported weight gain in mice fed seeds of this species from Veracruz and Puebla, Mexico (Gubitz et al. 1999). This has been confirmed by the study by Baldino et al. 2014; Bernal et al. 2004 where from the analysis performed by HPLC-UV and LC-MS/MS with seeds from Morelos and Sinaloa (Mexico) respectively, was determined the absence of phorbol ester and a decrease of 92% of trypsin inhibitors when heat treatment is applied. Additionally, it was found that lambs fed with seeds *J. curcas* from Sinaloa (Mexico) showed no alteration of health, and serum levels of liver and pancreatic function were within normal ranges. Genetic distinction of edible varieties compared to toxic varieties, has been confirmed by the analysis of molecular markers for germplasm, which has led to the conclusion genetic existence of two types of seeds based on the content of phorbol ester, where toxic variety it is having a concentration 1000 times higher, strongly suggesting that the phorbol ester content is controlled by a single genetic locus (Francis et al. 2013).

Considering the above and reports like the one made by SWANSEA (Sustainable Way for Alternative Fuels and Energy for Aviation) of the European Commission, the global economy *Jatropha* could be boosted if the nontoxic variety was implemented (Francis et al. 2013). This situation has not been valued and controversial in Mexico, has increased the number of hectares to cultivate imported seed, although it is known that the species that grows wild not present any toxicity to living beings, as if it occurs with imported seeds. This represents a disadvantage in medical, biological and economic terms, as these species adapted is presenting a health risk by toxicity and with low genetic variability have caused extensive crop losses associated with various insect pests (Martinez et al. 2015). Despite this, the European Union have "JATROPT" project (www.jatropt.eu), which seeks to promote the implementation of non-toxic varieties of seed; instead of promoting the development of decontamination and detoxification processes (Francis et al. 2013).

BIOLOGICAL ACTIVITY AND GENETIC STUDIES IN *JATROPHA CURCAS*

The current interest in the evaluation of natural products that reveal the chemical and biological characteristics has been encouraged due to the potential use of active substances with pharmacological activity (Villegas et al. 1997; Soares et al. 2015). Therefore, studies have been conducted to assess the biological potential of plants as alternative treatment for diseases such as malaria pose problems for public health worldwide; in this case being the leading cause of mortality and morbidity in many developing countries. Additionally, the spread of multi-drug resistant strains and the adverse side effects of drugs have necessitated the search for new sources of efficient drugs, found in plants an alternative solution (Liamah and Idu, 2015).

The natural advantage offered by plants in the production of bioactive substances that act as defense mechanisms against predators has been used to assess the biological activity as antibacterial, antiparasitic and antifungal has been able to corroborate from the design of various trials *in vivo* and *in vitro* (Aiyela-Agbe et al. 2007; Sarmento et al. 2011; Saetae et al. 2010; Ravindranath et al. 2004).

On leaves and *J. curcas* oil molluscicidal properties have been identified, fungicides and insecticides; this has been demonstrated by studying the biological activity of extracts of the leaves and oil against molluscs, fungi and insects (Saetae and Worapot, 2010). *J. curcas* extracts are capable of inhibiting the mycelial growth of *Colletotrichum musae* causes anthracnose disease affecting banana cultivation. Extracts from the sheet were effective in controlling fungal pathogen *Sclerotium* sp., Causing illness in Azolla fern that makes symbiosis with cyanobacteria and is used effectively to supply nutrients to crops and improve the substrates (Saetae and Worapot, 2010; Saetae and Worapot, 2011; Parawira, 2010; Rakshit et al. 2011; Liu et al. 1997; Rug et al. 2000).

Diptera larvae that have phytosanitary issues and constitute a public health problem have been evaluated with extracts obtained from *Jatropha curcas* showing positive effects to the control. Kovendan et al. 2011 evaluated different concentrations of plant extracts on larvae of Culex quinquefasciatus, vector of filaria; concluding the potential of extracts of this species as a control vector of lymphatic filariasis to be used as an ideal ecological control against insect. The insecticide effect has also been reported against *Aedes aegypti* larval stages; identifying which extracts petroleum ether shows a higher mortality of larvae of this insect. The LC_{50} value of petroleum ether extracts of *J. curcas*, *P. tithymaloides*, *P. amarus*, *E. tirucalli hirtay* against *A. aegypti* was 8.79, 55.26, 90.92, 272.36 and 4.25 ppm, respectively. Relationships of various tests, extracts of petroleum ether *J. curcas* showed some effectiveness; therefore it concluded that extracts of this species can be applied as an ideal larvicide against *A. aegypti.* Its use is a safe method for the environment and represents a control of vectors of dengue and filariasis (Rahuman et al. 2008). Extracts obtained from *J. curcas* as n-hexane, ethyl acetate and methanol extracts showed effectiveness as an insecticide against whitefly larvae (Bemisia tabaci) and other mosquitoes as *Ochlerotatus triseriatus*, *Helicoverpa zea* and *Heliothis virescens*. 100% mortality larval and adult stages were obtained at a concentration of 250 mg/mL of extract (Georges et al. 2008).

From the knowledge about the properties of *J. curcas* seed meal plant it has been used in the diet of infested rabbits to assess acaricidal *Hyalomma* potential in order to make control against ticks. Abdel-Shafy et al. 2011 used an experimental model in rabbits to evaluate the performance in feeding, the effect on weight and playback ticks present in rabbits. The researchers found that the feed intake and body weight gain decreased with increasing level of flour *J. curcas* seed in the diet, they concluded that the flour should be used in a percentage lower than 10% for the treatment of ectoparasites (Abdel-Shafy et al. 2011).

In a study in which the extracts obtained from the leaves and sap of *J. curcas* against Staphylococcus, Bacillus and Micrococcus were evaluated; and Anopheles mosquito larvae and eggs of nematodes antibacterial and insecticidal activity of the extracts was demonstrated. The sap was extracted germicidal action and inhibited bacterial growth; also it exhibited a potent inhibitory effect on the normal growth of mosquito larvae; therefore, this study concluded that extracts of *J. curcas* would be an excellent and economical disinfectant and insecticide that can be used as biological control (Fagbenro-Beyioku et al. 1998).

These properties evaluated in the postulate *Jatropha curcas* as a promising plant species, with utility in controlling pests through the use of active ingredients obtained from different parts of the plant that can efficiently control fungi, parasites and other organisms that affect growth and production of economically important crops and affect the health of human and animal population. In this sense, replace synthetic pesticides plant substances represents a viable alternative because they are inexpensive, decompose rapidly and despite being toxic do not have a long residual effect; however, it is necessary to use them with the same caution as chemical pesticides. The basis of plant-insect reactions has been studied and identified different metabolites that act as insecticides, followed by vegetable substances that control nematodes and others may fight mites, slugs and rats. By contrast, plant-fungus reactions have been less studied; however, it is known that the presence of an antifungal substance induces lignification of cell walls, act as a defense mechanism against the presence of aggressors (Gubitz et al. 1999; Liu et al. 1997; Rug and Ruppel, 2000).

Another important in the study of *J. curcas* aspect is its importance in the energy industry; with population growth, industrialization, pollution produced during the extraction and use of fossil fuels and the growing demand for energy sources has prompted the search for new energy alternatives that do not cause pollution, which are renewable and cheaper. Therefore, it is necessary to develop alternative generation biofuels exercising less pressure on the soil and water supply, ensuring the conservation of ecosystems forms. To achieve these purposes has encouraged the production of biodiesel feedstock using vegetable oils. *J. curcas* has been presented as an excellent alternative for the development of these goals; however, not much is known about the genetic diversity in *Jatropha curcas* and this makes the selection of varieties with higher value in obtaining raw material for the production of biodiesel and better adaptation to different biotic and abiotic factors may affect cultivation. This genetic knowledge *J. curcas* would identify the most promising for guiding the crop varieties and maximize use of varieties with genetic features that allow optimization of raw material that provides higher yields and with properties that facilitate growth and development of the plant against the presence of pathogens or adverse environmental conditions (Gu et al. 2015; Costa et al. 2014).

To evaluate the properties that offer varieties of *J. curcas* an *in vitro* study in which an unconventional bioprocess was applied to determine the biomass and oil production from *J. curcas* was raised. To this seed tissue was used in callus induction was isolated. Once obtained callus cell cultures were established; cell viability them, fatty acid profiles and oil content was evaluated. According to these characteristics the cell line with more and better results are selected, identifying it as the most promising (Correa and Atehortúa, 2012).

Polymorphisms are DNA sequences have utility in identifying genetic variations that show diversity characteristics expressed in the phenotype allowing species exhibit properties that favor growth and in the case of *J. curcas* the best profiles express fatty acids and is necessary to improve the content of oils to produce biodiesel. In this regard, Tang et al. 2011 six polymorphisms amplified AFLP (Amplified fragment length polymorphism) for analyzing genetic variations of *J. curcas* species originating in Asia and China; finding a lower species diversity in China. Similarly *J. curcas* species originating in Africa and South America, especially Brazil analyzed DNA polymorphisms possible to establish that the most biodiverse species found in Central and South America for what is considered that *Jatropha* is native to America and the low diversity I found in Africa and Asia is because few genotypes were brought from America to these territories. However, report the degree of genetic variation of this plant species in these continents simultaneously (Ovando-Medina et al. 2011; Basha et al. 2009). High genetic variation of *J. curcas* was of material from Chiapas Mexico, border region with Guatemala, where a strong genetic variation was also found (Pecina-Quintero et al. 2011; Ovando-Medina et al. 2011).

Molecular studies focused on identifying characteristics related to bioenergy production are limited; however, use such biomarkers for selection of varieties of *J. curcas* would improve the conditions of production of biodiesel in selecting genetic variation to be used in breeding programs *Jatropha* cultivation. In this regard it is important to promote molecular analyzes to determine the variation in seed traits (oil content and fatty acid composition), and characteristics of plant growth (Pecina-Quintero et al. 2011; Ovando-Medina et al. 2011).

Although there are genetic mapping and genome data *Jatropha* assisted selection by molecular markers it has not yet been implemented due to the lack of studies of genotype-phenotype association. For the identification of useful alleles for breeding a quantitative analysis was performed using QTL (quantitative trait locus) in two-parent populations where QTL mapping were analyzed for each of the following features: oil production, plant height, diameter stem, branch, total seeds per plant, weight of 100 seeds, oil content and seed fatty acid composition. QTLs were detected for each of these features, finding a mapping populations co-localization of QTLs for stem diameter and height of the plant, both overdominant, suggesting a possible effect locus with pleotropic heterosis. By integrating data candidates and physical genome maps *Jatropha* genes was also possible to determine a large number of genes involved in lipid biosynthesis in the genetic map of *Jatropha*. Therefore, these data will serve as a starting point for the selection of *J. curcas* varieties with better agronomic performance of the seed and oil increased productivity, allowing more control and there is an introduction of desirable alleles favoring management and crop production (Salinas, 2010; Gu et al. 2015).

MEDICINAL IMPORTANCE OF ACTIVE PRINCIPLES IN *JATROPHA CURCAS*

Various natural sources such as microorganisms, animals and plants have been evaluated to obtain new compounds with pharmacological potential; however, studies to assess the medicinal importance have been conducted mainly in plants because these have traditionally been used to treat various diseases (Janovka et al. 2003). Plants are a good source of antibiotics safe and biodegradable, they offer a pharmacological potential against pathogens associated with bacterial, fungal and parasitic infections. They have also been identified active ingredients for the treatment of viral and chronic diseases (Moshi et al. 2006; Runyoro et al. 2006).

Toxoplasma gondii is a parasite of medical and veterinary importance worldwide distribution and causes toxoplasmosis. There are few available treatments for toxoplasmosis and search for plant extracts and compounds with anti-Toxoplasma are important for identification of new active drugs. Therefore, we investigated the action of a protein extract of *J. curcas* seed versus Vero cells infected with *T. gondii.* The protein extract (JCCE) was used in *in vitro* assays for evaluating the anti-Toxoplasma activities at concentrations of 0.01, 0.1, 0.5, 1.5, 3.0 and 5.0 mg/mL for 24 hours. The results showed that the protein extract JCCE *J. curcas* interferes with intracellular growth of *T. gondii*; JCCE was observed that decreased the percentage of infection and the number of intracellular parasites. The effect on the morphology of the Vero cells appeared at a concentration of 3 mg/mL. These results suggest the use of this protein extract as an active agent derived from plants for use as phyto against toxoplasmosis (Soares et al. 2015).

The anti-inflammatory activity and use as treatment of rheumatism has been demonstrated using the plant leaves the affected area. The antiinflammatory activity of *J. curcas* has also been evaluated in albino mice and rats, to this root extracts of the plant, obtained by ether and ethanol were used, to control inflammation in the ear of animals. The antiinflammatory activity was evaluated by measuring the volume of the exudate, reduction morphological granulation tissue and the migration of leukocytes (Mujumdar and Misar, 2004).

The root has been used to treat pneumonia, syphilis and as an abortifacient, and local anti-inflammatory (Gubitz et al., 1999; Mujumdar et al., 2004). Seeds are the foundation of many drugs used for ascites, gout and disease of the skin and the latex has been used as astringent against bruises and in promoting healing wounds and ulcers (Thomas et al. 2008; Villegas et al. 1997; Osoyini and Onajobi, 2003). Peruvian researchers used *J. curcas* extracts as agents for the recovery of tissues associated with surface and internal (gastric ulcers) wound healing activity showing the results of evaluating the healing activity showed effectiveness in the healing process (Villegas et al. 1997).

Extracts obtained from seeds, leaves, bark and *J. curcas* oil have proven effective as a natural laxative action (Gubitz et al. 1999; Abdel et al. 2003). In Egypt *J. curcas* seeds for arthritis, gout and jaundice are used. In Mali, Nigeria, Cameroon, Cambodia, the Philippines and Malaysia the leaves have been used for malaria, jaundice, rheumatism, diarrhea and as a cathartic and healing respectively. In Mexico the latex has shown activity against infections caused by bacteria, fungi, bee and wasp stings. The use of the plant in traditional medicine and veterinary use has been reported in India, Africa and Latin America (Gubitz et al. 1999; Staubmann et al. 1999; Villegas et al. 1997; Akintayo, 2004).

In *J. curcas* metabolites have been identified with antiretroviral activity; among the most interesting compounds the prostatina, phorbol ester not tumorigenic. This metabolite has been shown to have 2 activities; first, it acts as an inhibitor of the replication of human immunodeficiency virus (HIV) internalizing receptors that this used to enter the host cell and secondly, it promotes transcription activating step latency reactivation. These two activities make the Prostratin reactivating a candidate agent in order to remove cellular reservoirs of the virus (Matsuse et al. 1999; Balunas and Kinghorn, 2005; Márquez et al. 2008).

To determine the coagulant and anticoagulant activity latex *J. curcas* an experiment where time to clot formation in latex samples and control samples was recorded was designed. Additionally, the study evaluated the anticoagulant activity in plasma by prothrombin time (PT) and partial thromboplastin time (APTT), showing that *J. curcas* latex has coagulant and anticoagulant effect (Osoyini and Onajobi, 2003).

The use of *J. curcas* for the treatment of cancer was the result of a survey conducted in Mexico on the use of plants with anticancer effect; the anticancer activity of active compound obtained from *J. curcas* has been widely reported (Abdelgadir and Staden, 2013; Alonso-Castro et al. 2011). Diterpenes obtained from different parts of *J. curcas* has shown toxic and cytotoxic activity as tumor inhibitors. The antimetastatic activity in methanolic fractions obtained from the leaves of the plant (Balaji et al. 2009; Devappa et al. 2011) was also evidenced. The root extract was more active compared to the leaf and stem bark suggesting his candidacy as a source of therapeutic anticancer agent (Oskoueian et al. 2011). Studies in four human cancer cell lines showed antimetastatic potential curcusone B, *J. curcas* obtained compound (Mungman et al. 2005). The antiproliferative activity of the compound jatropholone B was tested in cell lines of lung, bladder and blood cancer among others (Theoduloz et al. 2009). Also reported cytotoxic activity jatrophalactone B (Liu et al. 2012).

Diabetes is another chronic disease on which research has been conducted with *J. curcas*. Water extracts were obtained from the bark and the leaves of the plant into tea, boiled or burned to control blood glucose (Abdelgadir and Staden, 2013). In diabetic rats experiments showed that 50% of animals showed an antihyperglycemic effect after oral administration of ethanol extracts of *J. curcas* in doses of 250 and 500 mg mL-1 of body weight. From these dose reduction was achieved in glucose levels in rats of 219.5 to 116.5 and 237 to 98.8, respectively. And further reducing the levels of cholesterol and triglyceride was confirmed. Information on these effects in humans is limited; therefore, it is necessary to propose research leading to check the reduction in sugar levels in this population (Mishra et al. 2010).

CONCLUSION

Jatropha curcas is a species with economic potential due to the diversity of compounds with important medicinal properties such as flavonoids, diterpenes, sterols, triterpenoids, sapogenins, coumarins, deoxipreussomerinas, organic acids, iridioides, tannins and saponins. Also to be an oilseed plant, with high in protein and lipids in the seeds, they demonstrate the high potential as a renewable energy source and as a raw material for the production of animal feed, but it has not been possible marketing by toxic effects primarily attributed to phorbol esters. This has led to the realization of various methodologies for detoxification, finding that there is no single effective methodology to degrade all antinutritional factors which are present in the seeds. Therefore, there is a need to continue conducting studies where toxic effects have on different organs of living beings, establish a robust and reliable method for identifying and quantifying phorbol esters, as well as developing a method of detoxification economical and efficient, that can be carried on a large scale. Another viable option to strengthen the global economy *Jatropha* is to promote the implementation of non-toxic varieties of seeds, thereby reducing costs related to decontamination and detoxification processes.

J. curcas is presented as a promising plant and on which research should be conducted to validate ethnobotanical knowledge and consolidate the results of the medicinal importance of the plant. *J. curcas* derived products have been used for treatment of gastric problems, inflammatory, jaundice, diabetes, dysentery, fever and skin and reproductive diseases. Tests with extracts and metabolites of this plant have been conducted mainly *in vitro* and *in vivo* with animals showing significant biological activities that posit a wide pharmacological potential.

Propose genomic studies in *J. curcas* is critical in identifying genes expressing proteins associated with properties that optimize the raw material for biodiesel production. Also with genetic studies can identify *Jatropha* varieties resistant to adverse environmental effects and pathogens; and, knowing that plant varieties can promote the ability to prevent erosion and improve crop growth in different soil conditions.

REFERENCES

Abdel, W. Onsa, T. Ali, W. Badwi, S. Adam, S. (2003). Comparative toxicity of *Croton macrostachys*, *Jatropha curcas* and *Piper abyssinica* seeds in Nubian goats. *Small Ruminant Research*, 48(1), 61-67.

Abdelgadir, H. Van Staden, J. (2013). Ethnobotany, ethnopharmacology and toxicity of *Jatropha curcas* L. (Euphorbiaceae): A review. *South African Journal of Botany*, 88, 204-218.

Abdel-Shafy S, Nasr SM, Abdel-Rahman HH, Habeeb SM. (2011). Effect of various levels of dietary *Jatropha curcas* seed meal on rabbits infested by the adult ticks of *Hyalomma marginatum* I. Animal performance, anti-tick feeding and haemogram. *Trop Anim Health Prod*, 43(2), 347-57.

Aderibigbe A, Johnson C, Makkar H, Becker K, Foidl N. (1997). Chemical composition and effect of heat on organic matter and nitrogen degradability and some antinutritional components of *Jatropha* meal. *Animal Feed Science Technol*, 67(2-3), 223-43.

Acda M. (2009). Toxicity, tunneling and feeding behavior of the termite, *Coptotermes vastator*, in sand treated with oil of the physic nut, *Jatropha curcas*. *J Insect Science*, 9, 1-8.

Agbor, A. Akpana, M. Dasofunjo, K. James, A. (2015). The biomedical significance of the phytochemical, proximate and mineral compositions of the leaf, stem bark and root. *Asian Pacific Journal of Tropical Biomedicine*, 5(8), 650-657.

Aiyela-Agbe OO, Adeniyi BA, Fatunsin OF, Arimah BD (2007). *In vitro* Antimicrobial activity and photochemical analysis of *Jatropha curcas* roots Intern. *J. Pharmacol*, 3(1), 106-110.

Aiyela-Agbe OO (2001). Antibacterial activity of Jatropha multifida roots. *Fitoterapia*, 72, 544-546.

Akintayo E. (2004). Characteristics and composition of *Parkiabi globbossa* and *Jatropha curcas* oils and cakes. *Bioresource Technol*, 92(3), 307-310.

Altei, W. Picchi, D. Abissi, B. Giesel, G. Flausino O. Reboud-Ravaux, M., Verli, H. Crusca, E. Silveira, E. Cilli, E. Bolzani, V. (2014). Jatrophidin I, a cyclic peptide. *Phytochemistry*, 107, 91-96.

Alonso-Castro, A.J., Villarreal, M.L., Salazar-Olivo, L.A., Gomez-Sanchez, M., Dominguez, F., Garcia-Carranca, A. (2011). Mexican medicinal plants used for cancer treatment: pharmacological, phytochemical and ethnobotanical studies. *Journal of Ethnopharmacology*, 133, 945-972.

Azaizeh H, Fulder S, Khalil K and Said O. (2003). Ethnobotanical knowledge of local Arab practitioners in the Middle Eastern región. *Fitoterapia*, 74(1-2), 98-108.

Balaji R, Suba R, Rekha N and Deecaraman M. (2009). Hepatoprotective activity of methanolic fraction of *Jatropha curcas* on aflatoxin b1 induced hepatic carcinoma, *International Journal of Physical Sciences*, 2, 287-296.

Balunas M, Kinghorn A. (2005). Drug discovery from medicinal plants, *Life Sci*. 78(5), 431-41.

Basha SD, Francis G, Makkar HPS, Becker K, Sujatha M. (2009). A comparative study of biochemical traits and molecular markers for assessment of genetic relationships between *Jatropha curcas* L. germplasm from different countries, *Plant Sci*, 176(6), 812-823.

Baldini, M. Ferfuia, C. Bortolomeazzi, R. Verardo, G. Pasacali, J. Pasentier E. Franceschi L. (2014). Determination of phorbol esters in seeds and leaves of *Jatropha curcas* and in animal tissue. *Industrial Crops and Products*, 59, 268-276.

Bent S and Ko R. (2004). Commonly used herbal medicines in the United States: a review, *The American Journal of Medicine*, 116(7), 478-485.

Berenchtein, B. Abdalla, A. Prado, T. Sbardella, M. Louvandini, H. Filho, A. Dhanasekaran, D. Santos P. (2014). Effects of detoxified *Jatropha curca*s kernel meal in finishing pig diets on their performance, carcass traits, *Livestock Science*, 165, 100-103.

Berg, A. Horsten, S. Kettenes, J. Kroes, B. Beukelman, C. Leeflang, B. et al. (1995). Curcacycline A novel cyclic octapeptide isolated from the latex of *Jatropha curcas* L. *FEBS Letters*, 358(3), 215-218.

Bernal, J. Angulo, M. Estrada, A. Heredia, J. Muy, D. López, M. Barreras, A. Plascencia A. (2014). Feeding value of nontoxic *Jatropha curca*s seed. *Animal Feed Science and Technology*, 198, 107-116.

Bourgaud, F. Gravot, A. Milesi, S. Gontier, E. (2001). Production of plant secondary metabolites: a historical perspective. *Plant Science*, 161(5), 839-851.

Burkill HM (1994). The useful plants of West Tropical Africa (Families EJ), Royal Botanical Gardens Kew, pp. 90-94.

Chomchai C, Kriengsunthornkij W, Sirisamut T, Nimsomboon T, Rungrueng W, and Silpasupagornwong U. (2011). Toxicity from ingestion of *Jatropha curcas* ('saboo dum') seeds in thai children. *Southeast Asian Journal of Tropical Medicine and Public Meath*, 42(4), 946-950.

Correa SM, Atehortúa L. (2012). Lipid profile of *in vitro* oil produced through cell culture of *Jatropha curcas*. *J AOAC Int*, 95(4), 1161-9.

Costa HP, Oliveira JT, Sousa DO, Morais JK, Moreno FB, Monteiro-Moreira AC, Viegas RA, Vasconcelos IM. (2014). JcTI-I: a novel trypsin inhibitor from *Jatropha curcas* seed cake with potential for bacterial infection treatment. *Front Microbiol*, 30(5), 5-12.

Corzo, P. Fernández, J. Velasco, L. (2012). Tocochromanol content and composition. *Industrial Crops and Products*, 36(1), 304-307.

Devappa, R. Roach, J. Makkar, H. Becker, K. (2013). Occular and dermal toxicity. *Ecotoxicology and Environmental Safety*, 94(1), 172-178.

Devappa, R.K., Makkar, H.P.S., Becker, K. (2010). Nutritional, biochemical, and pharmaceutical potential of proteins and peptides from Jatropha: review. *Journal of Agriculture and Food Chemistry*, 58, 6543-6555.

Dhakshanamoorthy D, Selvaraj R, Chidambaram A. (2015). Utility of RAPD marker for genetic diversity analysis in gamma rays and ethyl methane sulphonate (EMS)-treated *Jatropha curcas* plants. *C R Biol*, 338 (2), 75-82.

Fagbenro-Beyioku AF, Oyibo A, and Anuforom B: (1998). Disinfectant/antiparasitic activities of *Jatropha curcas. East African Medical Journal*, 75(9), 508-511.

Firenzuoli F and Gori L. (2007). Herbal medicine today: clinical and research issues. *Evidence-based Complementary and Alternative Medicine*, 4(1), 37-40.

Francis, G. Oliver, J. Sujatha, M. (2013). Non-toxic jatropha plants. *Industrial Crops and Products*, 42, 397-401.

Fu, R. Zhang, Y. Guo, Y. Liu, F. Chen, F. (2014). Determination of phenolic contents and antioxidant. *Industrial Crops and Products*, 58, 265-270.

Garcia, A and Pérez, E. (2009). Metabolismo secundario de plantas *Reduca*, 2(3), 119-145.

Gubitz, G. Mittelbach, M. Trabi, M. (1999). Exploitation of the tropical oil seed plant *Jatropha curcas L. Bioresource Technol*, 67(1), 37-42.

Georges K, Jayaprakasam B, Dalavoy S, Nair M. (2008). Pest-managing activities of plant extracts and anthraquinones from Cassia nigricans from Burkina Faso. *Bioresour Technol*, 99(6), 2037-45.

Gu K, Tian D, Mao H, Wu L, Yin Z. (2015). Development of marker-free transgenic *Jatropha curcas* producing curcin-deficient seeds through endosperm-specific RNAi-mediated gene silencing. *BMC Plant Biol*, 15(1), 242. doi: 10.1186/s12870-015-0625-z.

Gubitz G, Mittelbach M, Trabi M. (1999). Exploitation of the tropical oil seed plant *Jatropha curcas* L. *Bioresource Technol*, 67(1), 37-42.

Gurib-Fakim A. (2006). Medicinal plants: traditions of yesterday and drugs of tomorrow. *Molecular Aspects of Medicine*, 27(1), 1-93.

Guedes, R. Ameida, F. Cypriano, M. Oliveira, L. Castro, R. Fernades M. (2014). Detoxification of *Jatropha curcas* seed. *Industrial Crops and Products*, 52, 537-543.

Haas, W. Sterk, H. Mittelbach, M. (2002). Novel 12-Deoxy-16-hydroxyphorbol diesters Isolated from the seed oil of *Jatropha curcas. J Nat Prod*, 65(10), 1434-1440.

Hamarneh, A. Heeres, H. Broekhuis, A. Picchion, F. (2010). Extraction of *Jatropha curcas* proteins and application in polyketone-based wood adhesives. *International J Adhesion Adhesive*, 30(7), 615-25.

Heng, W. Hung, C. Ming, C. Ho, Y. Hsiang, D. (2009). Supercritical carbon dioxide extraction of triglycerides from *Jatropha curcas* L. seeds. *J Supercritical Fluid*, 51(2), 174-180.

Hidayat, Ch. Hastuti, P. Kusuma, A., Santika, L. (2014). Method of phorbol ester. *Journal of Bioscience and Bioengineering*, 117(3), 372-374.

Igbinosa O, Igbinosa E and Aiyegoro OA. (2009). Antimicrobial activity and phytochemical screening of stem bark extracts from *Jatropha curcas* (Linn). *African Journal of Pharmacy and Pharmacology*, 3(2), 58-62.

Igbinosa O, Igbinosa IH, Chigor VN. (2011). Polyphenolic contents and antioxidant potential of stem bark extracts from *Jatropha curcas* (Linn). *International Journal of Molecular Sciences*, 12(5), 2958-2971.

Igbinosa, Efosa F. Oviasogie, Etinosa O. Igbinosa. (2013). Effects of Biochemical Alteration in Animal Model after Short-Term Exposure of *Jatropha curcas* (Linn) Leaf Extract Osamuyimen O. *The Scientific World Journal*, 2013, 1-5.

Janovka D, Kubikova K, Kokoska L. (2003). Screening for antimicrobial activity of some medicinal plant species of traditional Chinese medicine. *Czech J Food Sci*, 21(3), 107-10.

Joon, J. Damayani, K. Ta, W. Taufiq, H. (2011). Biodiesel production from *Jatropha* oil by catalytic and non-catalytic approaches: An overview. *Bioresource Technol*, 102(2), 452-460.

Joshi C, Mathur P, Khare S. (2011). Degradation of phorbol esters by Pseudomonas aeruginosa PseA during solid-state fermentation of deoiled *Jatropha curcas* seed cake. *Bioresour Technol*, 102(7), 4815-9.

Jucá, T. Oliveira, A. Azevedo, R. Viana C. Souza, J. Batista, B. Ramos, M. (2015). A new peptide from *Jatropha curcas* seeds: Unusual sequence and insights into its synthetic analogue that enhances proteolytic activity of papain. *Process Biochemistry*, 50(9), 1434-1440.

King A, He W, Cuevas J, Freudenberger M, Ramiaramanana D, Graham I. (2009). Potential of *Jatropha curcas* as a source of renewable oil and animal feed. *J Experimental Botany*, 60(10), 2897-905.

Kongmany, S. Furuta, M. Matsuura, H. Okuda, S. Imamura, K. Maeda Y. (2014). Degradation of phorbol 12,13-diacetate in aqueous solution by gamma irradiation. *Radiation Physics and Chemistry*, 105, 98-103.

Kovendan K, Murugan K, Vincent S, Kamalakannan S. (2011). Larvicidal efficacy of *Jatropha curcas* and bacterial insecticide, Bacillus thuringiensis, against lymphatic filarial vector, Culex quinquefasciatus Say (Diptera: Culicidae). *Parasitol Res*, 109(5), 1251-7.

Kumar, A,. Sharma, S. (2008). An evaluation of multipurpose oil seed crop for industrial uses (*Jatropha curcas* L.): A review. *Industrial Crops Products*, 28(1), 1-10.

Li, Y. Chen, L. Lin, Y. Fang, Z. Che, L. Xu S. Wu, D. (2015). Effects of replacing soybean meal with detoxified *Jatropha curcas* kernel meal in the diet. *Animal Feed Science and Technology*, 204, 18-27.

Lin, J. Zhou, X. Wang, J. Jiang, P. Tang, K. (2010). Purification and characterization of curcin, a toxic lectin from the seed of *Jatropha curcas*. *Preparative Biochemistry Biotechnol*. 40(2), 107-118.

Liu S, Sporer M, Wink J. Jourdane R, Henning Y, Ruppel A. (1997). Anthraquinones in Rheum palmantumand Rumex dentatus (Polygonaceae), and phorbol esters in *Jatropha curcas* (Euphorbiaceae) with molluscicidal activity against the schistosome vector snails Oncomelania, Biornphalariaand Bulinus. *Trop Med Int Health*, 2(2), 179-88.

Liu, J-Q. Yang, Y-F. Wang, C-F. Li,Y. Qiu M-H. (2012). Three new diterpenes from *Jatropha curcas*. *Tetrahedron*, 68(4), 972-976.

Liu, J-Q. Yang, Y-F. Li, X-Y. Liu, E-Q. Li, Z-R. Zhou, L. Li, Y. Q-M-H. (2013). Cytotoxicity of naturally occurring rhamnofolane diterpenes from *Jatropha curcas*. *Phytochemistry*, 96, 265-272.

Liu, J. Yang, Y. Xia, J. Li, X. Li, Z. Zhou, L. Qiu, M. (2015). Cytotoxic diterpenoids. *Phytochemistry*. 117, 462-468.

Lyamaha PC and Idub M. (2015). Ethnomedicinal survey of plants used in the treatment of malaria in Southern Nigeria. *Journal of Ethnopharmacology*, 173, 287-302.

Makkar JH, Becker K, Sporer F and Wink M. (1997). Studies on nutritive potential and toxic constituents of different provenances of *Jatropha curcas*. *Journal of Agricultural and Food Chemistry*, 45(8), 3152-3157.

Mariz SR, Arauojo WC, Cerqueira GC. (2008). Avaliacao histopatologica em ratros apos tratamento agudo com o extrato etanolico de partes aereas de Jatropha gossypifolia L. *Brazilian Journal of Pharmacognsosy*, 18, 213-216.

Márquez N, Calzado MA, Sánchez-Duffhues G, Pérez M, Minassi A, Pagani A. (2008). Differential effects of phorbol-13-monoesters on human immunodeficiency virus reactivation. *Biochemical Pharmacol*, 75(6), 1370-80.

Martinez, J. Siddhuraju, P. Francis, G. Davila, G. Becker, K. (2006). Chemical composition, toxic/antimetabolic constituents, and effects of different treatments on their levels, in four provenances of *Jatropha curcas* L. from Mexico. *Food Chemistry*, 96(1), 80-89.

Martínez-Díaz, Y. González-Rodríguez, A. Delgado-Lamas, G. Espinosa-García, F. (2015). Geographic structure. *Industrial Crops and Products*, 74, 63-68.

Matsuse IT, Lim Y, Hattori M, Correa M and Gupta MP. (1998). A search for anti-viral properties in Panamanian medicinal plants. The effects on HIV and its essential enzymes. *Journal of Ethnopharmacology*, 64(1), 15-22.

Matsuse T, Lim YA, Hattori M, Correa M, Gupta MP (1999). A search for anti-viral properties in Panamanian Medicinal Plants - The effect on HIV and essential enzymes. *J. Ethnopharmacol*, 64, 15-22.

Méndez, L. Rojas, J. Izaguirre C. Contreras, B. Gómez, R. (2014). *Jatropha curcas* leaves analysis, reveals it as mineral source for low sodium. *Food Chemistry*, 165, 575-577.

Moshi M, Mbwambo Z, Nondo R, Masimba P, Kamuhabwa A, Kapingu M. (2006). Evaluation of ethnomedical claims and Brine shrimp toxicity of some plants used in Tanzania as traditionalmedicines. *Afri J Trad CAM*, 3(3), 48-58.

Mungman, S., Thippornwong, M., Tohtong, R. (2005). Anti-metastatic effect of curcusone B, a deterpene from *Jatropha curcas*. *In Vivo*, 19, 265-268.

Mujumdar AM, Misar AV, Salaskar M and Upadhye AS. (2001). Antidiarrhoeal effect of an isolated fraction (JC) of *Jatropha curcas*roots in mice. *Journal of Natural Remedies*, 1, 89-93.

Mujumdar A, Misar A. (2004). Anti-inflammatory activity of *Jatropha curcas* roots in mice and rats. *J Ethnopharmacol*, 90(1), 11-5.

Mishra, S.B., Vijayakumar, M., Ojha, S.K., Verma, A. (2010). Antidiabetic effect of *Jatropha curcas* L. levels extract in normal and alloxan-induced diabetic rats. *International Journal of Pharmaceutical Sciences*, 12, 482-487.

Nakao, M. Hasegawa, G. Yasuhara T. Ishihara, Y. (2015). Degradation of *Jatropha curcas* phorbol esters derived from Jatropha oil. *Ecotoxicology and Environmental Safety*, 114, 357-364.

Oliveira, J. Leite, P. Souza, L. Mello, V. Silva, E. Rubim, J. et al. (2009) Characteristics and composition of *Jatropha gossypiifolia* and *Jatropha curcas* L. oils and application for biodiesel production. *Biomass Bioenergy*, 33(3), 449-453.

Oliver-Bever B. (1986). Medicinal Plants in Tropical West Africa, Cambridge University Press, London, UK, 1986.

Openshaw K. (2000). A review of *Jatropha curcas*: an oil plant of unfulfilled promise. *Biomass Bioenergy*, 19(1), 1-15.

Osoyini, O. Onajobi, F. (2003). Coagulant and anticoagulant activities in *Jatropha curcas* latex. *J Ethnopharmacol*, 89(1), 101-105.

Oskoueian, E, Abullah, N., Saad, W.Z., Omar, A, Ahmed, S., Kuan, W.B., Zolkifli, N.A., Hendra, R, Ho, Y.W. (2011). Antioxidant, anti-inflammatory and anticancer activities of methanolic extracts from *Jatropha curcas* Linn. *Journal of Medicinal Plant Research*, 5, 49-57.

Ovando-Medina AS-Ge I, Adriano-Anaya L, Espinosa-Garcia F, Núñez-Farf[an J, Salvador-Figueroa M. (2011). Genetic Diversity of Population of *Jatropha curcas* in the State of Chiapas, Mexico. *Diversity*, 3(4), 641-659.

Pabón LC and Hernández-Rodríguez P. (2012). Importancia química de *Jatropha curcas* y sus aplicaciones biológicas, farmacológicas e industrials (*Jatropha curcas* importance of chemical and biological, pharmacological and industrial applications). *Revista Cubana de Plantas Medicinales* (Cuban Journal of Medicinal Plants), 17(2), 1-9.

Parawira, W. (2010) Biodiesel production from *Jatropha curcas*: A review. *Scientific Research Essays*, 5(1), 1796-808.

Pecina-Quinteroa V, Anaya-Lópeza JL, Zamarripa Colmenerob A, Montes Garcíac N, Núñez Colína LA, Solis Bonillab JL, Aguilar-Rangela MR, Gill Langaricac HR, Méjia Bustamantea DJ. (2011). Molecular characterisation of *Jatropha curcas* L. genetic resources from Chiapas, México through AFLP markers. *Biomass and Bioenergy*, 35(5), 1897-1905.

Phengnuam, T. Suntornsuk, W. (2013). Detoxification and anti-nutrients. *Journal of Bioscience and Bioengineering*, 115(2), 168-172.

Rahuman A, Gopalakrishnan G, Venkatesan P, Geetha K. (2008). Larvicidal activity of some Euphorbiaceae plant extracts against Aedes aegypti and Culex quinquefasciatus (Diptera: Culicidae). *Parasitol Res*, 102(5), 867-73.

Rakshit K, Harinder P, Klaus B. (2001). Jatropha diterpenes: a Review. *J Am Oil Chem Soc*, 88(3), 301-22.

Rakshit, K. Darukeshwara, J. Rathina, K. Narasimhamurthy, K. Saibaba, P. Bhagya, S. (2008). Toxicity studies of detoxified *Jatropha* meal (*Jatropha curcas*) in rats. *Food Chemical Toxicol*, 46(12), 3621-3625.

Ravindranath, N. Ravinder, M. Mahende, G. Ramu, R. Ravi, K. (2004) Deoxypreussomerins from *Jatropha curcas*: are they also plant metabolites. *Phytochemistry*. 65(16), 2387-2390.

Rakshit, K. Harinder, P. Klaus, B. (2011). *Jatropha* diterpenes: a Review. *J Am Oil Chem Soc*. 88(3), 301-322.

Roach, J. Devappa, R. Makkar, H. Becker, K. (2012). Isolation, stability. *Fitoterapia*, 83(3), 586-592.

Rousseaux CG and Schachter H. (2003). Regulatory issues concerning the safety, efficacy and quality of herbal remedies. *Birth Defects Research B*, 68(6), 505-510.

Rodrigues, J. Miranda, I. Gominho, J. Vasconcelos, M. Barradas, G. Pereira, H. Bianchi, F. Ferreira S. (2013a). Variability in oil. *Industrial Crops and Products*, 50, 828-837.

Rodrigues, J. Albino, S. Pereira, D. Dias, M. Soares, J. Cuquetto, H. Mengumi, M. (2013b). Production of edible mushroom. *LWT - Food Science and Technology*, 50(2), 575-580.

Rug M, Ruppel A. (2000). Toxic activities of the plant *Jatropha curcas* against intermediate snail hosts and larvae of schistosomes. *Trop Med Int Health*, 5(6), 423-30.

Runyoro D, Matee M, Ngassapa O, Joseph C, Mbwambo Z. (2006). Screening of Tanzanian medicinalplants for anti-Candida activity. *BMC Complement Altern Med*, 6, 11.

Sabandar C. Review. Secondary Metabolite Compounds from *Jatropha* species. Carlasabandar's Blog. 2010 Available in: http://carlasabandar.wordpress.com/2010/06/29/secondary-metabolite-compounds-from-jatropha-species/.

Sabandarm C. Ahmat N. Jaafar M. Sahidin I. (2013). Medicinal property, phytochemistry and pharmacology. *Phytochemistry*, 85, 7-29.

Saetae D, Worapot S. (2010). Antifungal activities of ethanolic extract from *Jatropha curcas* seed cake. *J Microbiol Biotechnol*, 20(2), 319-24.

Saetae, D. Worapot, S. (2011) Toxic compound, anti-nutritional factors and functional properties of protein isolated from detoxified *Jatropha curcas* seed cake. *International. J Molecular Sciences*, 12(1), 66-77.

Salinas N. (2010). Estudio de la variabilidad genética de tres poblaciones de piñón (*Jatropha curcas* L.) en la provincia de Manabí mediante la utilización de marcadores moleculares issr's [Tesis] (Study of the genetic variability of three pinion (*Jatropha curcas* L.) in the province of Manabi issr's using molecular markers [Thesis]). Escuela Politécnico del Ejército (Polytechnic School of the Army).

Sarmento RA, Rodrigues DM, Faraji F, Erasmo EA, Lemos F, Teodoro AV, Kikuchi WT, Dos Santos GR, Pallini A. (2011). Suitability of the predatory mites Iphiseiodes zuluagai and Euseius concordis in controlling Polyphagotarsonemus latus and Tetranychus bastosi on *Jatropha curcas* plants in Brazil. *Exp Appl Acarol*, 53(3), 203-14.

Seeff LB. (2007). Herbal hepatotoxicity. *Clinics in Liver Disease*, 11(3), 577-596.

Soares AM, Carvalho LP, Melo EJ, Costa HP, Vasconcelos IM, Oliveira JT. (2015). A protein extract and a cysteine protease inhibitor enriched fraction from *Jatropha curcas* seed cake have *in vitro* anti-Toxoplasma gondii activity. *Exp Parasitol*, 153, 111-7.

Sousa, T. Castro, N. Firmino, M. Danta,s P. Mendonça, S. Aquino, J. Medeiros, E. (2015). Fatty acid *Industrial Crops and Products*, (73), 106-108.

Staubmann, R. Schubert, M. Hiermann, A. Kartnig, T. (1999a). A complex of 5-hydroxypyrrolidin-2-one and pyrimidine-2,4-dione isolated from *Jatropha curcas. Phytochemistry*, 50(2), 337-338.

Staubmann, R. Ncube, I. Gubitz, G. Steine, W. Read, J. (1999b) Esterase and lipase activity in *Jatropha curcas* L. seeds. *J Biotecnol.* 75(2-3), 117-126.

Tang JL, Liu BY and Ma KW. (2008). Traditional Chinese medicine. *The Lancet*, 372(9654), 1938-1940.

Tang L, Zhang ZY, Guo XL, Liu BY, Chen F. (2011). Genetic diversity and genetic relationship of *Jatropha curcas* between China and Southeast Asian revealed by amplified fragment length polymorphisms. *Afr J Biotechnol*, 10(15), 2825-2832.

Theoduloz C, Rodríguez J.A, Pertino, M, Schmeda-Hirschmann, G. (2009). Antitumor activity of jatrophone and jatropholone derivatives. *Planta Medica*, 75, 1520-1522.

Thomas R, Sah NK, Sharma PB. (2008). Therapeutic biology of *Jatropha curcas*: a mini review. *Curr Pharm Biotechnol*, 9(4), 315-24.

Tongpoothorn, W. Chanthai, S. Sriuttha, M. Saosang, K. Ruangviriyachai, Ch. (2012). Bioactive properties and chemical constituents of methanolic extract and its fractions from *Jatropha curcas* oil. *Industrial Crops and Products* 36, 437-444.

Valdés-Rodríguez OA, Sánchez-Sánchez O, Pérez-Vázquez A, Caplan J and Danjon F. (2013). *Jatropha curcas* L. Root Structure and Growth in Diverse Soils. *The Scientific World Journal*, 2013, 1-9.

Villegas L, Fernández I, Maldonado H, Torres R, Zabaleta A, Vaisberg A, Hammond G. (1997). Evaluation of the wound-healing activity of selected traditional medicinal plants from Peru. *J Ethnopharmacol*, 55(3), 193-200.

WHO. Traditional Medicine: Growing Needs and Potential. WHO Policy Perspectives on Medicines. Geneva: World Health Organization; 2002. p. 1-6.

Widiyastuti, T. Rahardjo, T. Hendroko R. (2015). Evaluation. *Energy Procedia*, 65, 362-367.

Yao, L. Han, Ch. Chen, G. Song, X. Chang, Y. Zang W. (2012). A new asymmetric diamide from the seed *Fitoterapia*, 83(8), 1318-1321.

Yan, C. Rakshit, L. Devappa, K. Xin, J. Min, J. Makkar, H. et al. (2010). Toxicity of *Jatropha curcas* phorbol esters in mice. *Food Chemical Toxicol*. 48(2), 620-625.

Yang, Y-F. Liu, J-Q. Li, Z-R. Qiu, M-H. (2013). New eudesmenoic acid. *Fitoterapia* (89), 278-284.

Zhang, X-Q. Li, F. Zhao Z-G. Liu, X-L. Tang Y-X. Wang, M-K (2012). Diterpenoids from the root. *Phytochemistry Letters*, 5(4), 721-724.

In: The Promising Future of *Jatropha curcas*
Editors: Maira R. Segura-Campos et al.
ISBN: 978-1-63484-989-0
© 2016 Nova Science Publishers, Inc.

Chapter 5

PHYSIC NUT (*JATROPHA CURCAS* L.) EXTRACTS AND ACTIVE COMPOUNDS WITH BIOLOGICAL ACTIVITY AGAINST PEST INSECTS AND PATHOGENIC FUNGI

Liliana C. Córdova-Albores[1], Silvia Bautista-Baños[1*], Laura L. Barrera-Necha[1], Miguel Salvador-Figueroa[2], Ma. de Lourdes Adriano-Anaya[2] and Isidro Ovando-Medina[2]
[1]Centro de Desarrollo de Productos Bióticos-IPN, Yautepec, Morelos C.P., México
[2]Centro de Biociencias, Universidad Autónoma de Chiapas, Tapachula, Chiapas, México

ABSTRACT

Jatropha curcas belongs to the family Euphorbiaceae. This plant is native to tropical America, and it can be found abundantly in many tropical and subtropical regions. Because of its tremendous industrial potential, it is cultivated in various countries and its use has been the subject of numerous studies. The genus *Jatropha* is composed of approximately 175 species. Among the species with pesticidal and fungicidal properties is *J. curcas*. Studies have shown that leaf and stem extracts, seed oil and the chemical constituents of this plant species control numerous pest insects of agricultural importance such as *Helicoverpa armigera, Callosobruchus maculatus, Sitophilus zeamais*, *Fusarium oxysporum, Botrytis cinerea* and *Rhizoctonia solani*, among others. Of the isolated active compounds of *J. curcas*, curcin and diterpenes, including phorbol esters, have shown notable biological activity against pathogenic pests and diseases. In this chapter, the possible mechanism of action associated with disruptions to the cell wall structure and organelles is also shown, although more information on this matter should be generated. As shown in the literature review, extracts of *J. curcas* plant organs, including seeds, leaves, stem and their derivatives, have active potential for possible pesticide and

* Corresponding author: S. Bautista-Baños. Centro de Desarrollo de Productos Bióticos-IPN. Carretera Yautepec-Jojutla km. 6, calle CEPROBI No. 8, Col. San Isidro, Yautepec, Morelos C.P. 62731. México. Tel.: + 55 735 3942020 ext. 82571, fax: + 55 735 3941896. E-mail address: sbautis@ipn.mx.

fungicide use in various horticultural commodities; however, more information about the possible toxicological effects should be obtained and analyzed.

Keywords: action mechanisms, Euphorbiaceae, phorbol esters, seed oil

INTRODUCTION

The family Euphorbiaceae is constituted of 8,700 species and 320 genera mostly located in the tropical region of Africa and America (Webster, 1994). In Mexico, 50 genera and 8470 species have been reported (Martínez et al., 2007; Martinez-Gordillo et al., 2006; Martínez-Gordillo et al., 2002). Members of this family can be trees, shrubs or herbs being characterized, everyone, for producing latex, milky or colorless, irritating to skin and eyes. The latex, and different parts of plants, contains various metabolites of medical interest and industrial (Betancur et al., 2003; Özcan and Safa, 2004; Shi et al., 2005). Chemical data are available for various genera, particularly Euphorbia, where more than 120 species have been investigated. A survey of the data showed that the triterpenoids, followed by flavonoids and alkaloids, are the main classes of substances of interest in phytochemistry.

However, the presence of other substances such as coumarins, cyanogenic glycosides and tannins has also been reported. The Euphorbiaceae are important sources of toxins and compounds used in medicine. Some diterpenoids possess therapeutic potential as antihypertensive (Tirapelli et al., 2010), antiretroviral (Gyuris et al., 2009), anti-inflammatory (Khiev et al., 2011), analgesic (Khan et al., 2009), insect growth inhibition (Aiyelaagbe and Gloer, 2008), and antibacterial agents (Aiyelaagbe et al., 2007; Igbinosa et al., 2009; Kalimuthu et al., 2010; Neogi et al., 2008). The genus *Jatropha* belongs to the tribe Joannesieae within the family Euphorbiaceae and contains over 170 known species. The word *Jatropha* is derived from the Greek "*latros*" meaning doctor and "*trophé*" meaning food, implying medicinal and food uses (Kumar and Sharma, 2009; Makkar and Becker, 1999). In other languages it is known as: "pourghère," "pignon d'Inde" (French); "purgeernoot" (Dutch); "Purgiernuß," "Brechnuß" (German); "purgueira" (Portuguese); "fagiola d'India" (Italian); "dand barrî," "habel meluk" (Arabic); "yu-lu-tzu" (Chinese); "sabudam" (Thai); "mupuluka" (Angola); "butuje" (Nigeria); "makaen" (Tanzania); "purgeerboontjie" (Africa); "pinol," "pinon," "piñón," "piñón botija," "piñón criollo," "piñón lechero," "piñón purgante," "piñón vóci" (Spanish) (Makkar and Becker, 1999; Martínez-Herrera, 2006; Devappa et al., 2010). Dehgan and Webster (1979) reviewed the subdivision made by Pax (1910) in which they distinguished two subgenera of the genus *Jatropha*, with ten sections and ten subsections, in order to group together the Old and New World species, with *J. curcas* being the most primitive form of the genus.

BOTANICAL CHARACTERISTICS OF *J. CURCAS*

Linnaeus (1753) was the first to name the physic nut *J. curcas* according to the binomial nomenclature of "*Species Plantarum*" and it is still used today. *Jatropha* species are drought-resistant and able to grow in poor soil conditions (Brittaine and Lutaladio, 2010; Heller, 1996). It is commonly planted along property boundaries as living fences, in gardens and in mixtures with other crops such as vanilla and coffee (Makkar et al., 1997; Martínez et al., 2007).

In countries like Mexico there are 48 reported species of the genus *Jatropha*, of which 39 are endemic (Martínez-Gordillo et al., 2006; Steinmann, 2002), highlighted by *J. multifida*, used as an ornamental (Steinmann, 2002), *J. cuneata*, reported to have antimicrobial properties (Ruiz-Bustos et al., 2009) and *J. curcas*, used as a multipurpose plant (food, antimicrobial, biodiesel and living fence) (Makkar and Becker, 1999; Makkar et al., 1998; Martínez-Herrera et al., 2006; Ovando-Medina et al., 2011; Ovando-Medina et al., 2007; Prasad et al., 2012; Salvador-Figueroa et al., 2007).

Jatropha curcas is native to tropical America, and it can be found abundantly in many tropical and subtropical regions, both in dry and wet equatorial areas up to an elevation of 1,500 meters above sea level. *Jatropha curcas* is a deciduous shrub with milky sap that measures 1-8 meters in height; stem diameter is 14-18 centimeters, the scaly bark is light brown, and the branches contain bitter-tasting whitish latex. These shrubs or trees can have up to two flowering and fruiting seasons per annual cycle. Propagation of the species occurs in two ways: one sexual, through seeds which take 30 days to germinate, and the other asexual, through cuttings measuring 25-30 cm high and 3-5 cm in diameter, with budbreak beginning at 20 days. The leaves are approximately 6 cm long and 15 cm wide (Figure 1A), and arranged alternately with three short, petiolated lobes (Dehgan and Webster, 1979 cited in Heller, 1996; Makkar et al., 1998). Inflorescences are formed terminally on branches. The plant is monoecious and flowers are unisexual, although occasionally hermaphrodite flowers occur (Figure 1B). Pollination is by insects. After this is done, trilocular ellipsoidal fruit is formed (Figure 1C) (Dehgan and Webster, 1979 cited in Heller, 1996). The exocarp remains fleshy until the seeds are mature. The fruit is a capsule with two, three or four seeds, and measures approximately 4-5 cm long and 3-4 cm wide; they are green when still immature and yellow when mature, in the mature stage they become black-brownish (Figure 2A). The seed coat is exotegmic, and contains both a dark, ellipsoidal-shaped testa and a tegmen, both of which originate from outer and inner integuments of the maternal ovary respectively. The nut is white, and it has a thick endosperm and an embryo with two creamy-white foliaceous cotyledons (Figure 2B). *Jatropha curcas* is a diploid species with 2n = 22 chromosomes (Aker, 1997; Brittaine and Lutaladio, 2010; He et al., 2011; Heller, 1996; Kumar and Sharma, 2009; Makkar et al., 1998).

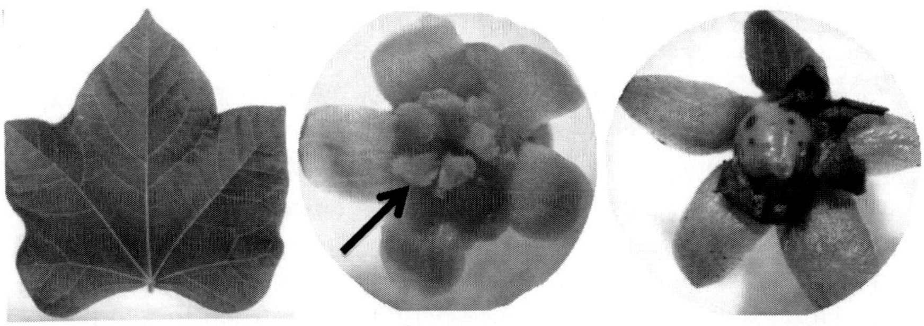

Figure 1. *Jatropha curcas* leaf (A), flower (B) and fruit (C).

a

b

Figure 2. *Jatropha curcas* fruits at different stage of maturity (A) and seeds (B).

USES OF *J. CURCAS*

Jatropha curcas has shown tremendous industrial potential, which is why it is cultivated in various countries and why its use has been the subject of numerous studies. Popluechai (2010) refers to the uses that have been given to *J. curcas* (Table 1). Seed, leaf and bark preparations have been used in traditional medicine (Omale et al., 2011; Mbakwem-Aniebo et al., 2012) and for veterinary purposes (Abdel-Shafy et al., 2011; Sanis et al., 2012). The oil has a strong purgative action, and it is also widely used to treat skin diseases and to relieve pain such as that caused by rheumatism. A leaf decoction is used against cough (Isawumi, 1978). The sap is used to stop bleeding. Nath and Dutta (1992) demonstrated the healing properties of curcin, which is isolated from latex. *Jatropha curcas* also possesses insecticidal and antimicrobial properties contained in the seed oil, concentrated fractions, as well as in extracts from all parts of the plant.

Table 1. Uses of *J. curcas*

Part plant	Uses	References
Seeds	Wastewater copper biosorption by seed coat	Jain et al. (2008)
	Seed diterpenes and toxins as antimicrobial	Goel et al. (2007)
	Antitumor, molluscidal and insecticidal	Luo et al. (2007); Liu et al. (1997)
	Jatropholone used to prevent gastric lesions	Pertino et al. (2007)
Seed oil	Biodiesel	Kaushik et al. (2007); Prasad et al. (2012)
	Resins and varnishes	Patel et al. (2008)
	Soap, lubricants and illuminants	Roy, (1998)
	Antimicrobial	Eshilokun et al. (2007)
	Biopesticidal Mosquito repellent through combustion Fungicidal	Goel et al. (2007); Asmanizar et al. (2012) Cantrell et al. (2011) Córdova-Albores et al. (2014); Devappa et al. (2012)
Seed cake	Protein rich feed	Devappa and Swamylingappa (2008); Martinez-Herrera et al. (2006)
	Organic fertilizer	Mendoza et al. (2007); Srinophakun et al. (2012)
	Lipase and protease production	Mahanta et al. (2008)
	Activated carbon Antifungal Termiticidal	Sricharoenchaikul et al. (2008) Saetae and Suntornsuk (2010) Sharma et al. (2011)
Fruit husk	Fuel through combustion	Singh et al. (2008)
	Pyrolyzed for biogas	Vyas and Singh (2007)
Fruit pericarp/Seed coat	Activated charcoal	Okeola et al. (2011)
Leaf	Feed for silkworms in silviculture	Grimm et al. (1997)
	Larvicidal	Rahuman et al. (2008)
	Bactericidal	Akinpelu et al. (2009); Kalimuthu et al. (2010)
	Fungicidal	Onuh et al. (2008); Siva et al. (2008)
Stem bark	Dark blue dye	Srivastava et al. (2008)
	Waxes and tannins	Burkill (1994)
	Antineoplastic	Biehl and Hecker (1986)
	Antimicrobial	Igbinosa et al. (2009)
	Jatrophone against snake bite	Brum et al. (2006)
Root	Antimicrobial	Aiyelaagbe et al. (2007a)

Adapted from Popluechai, 2010.

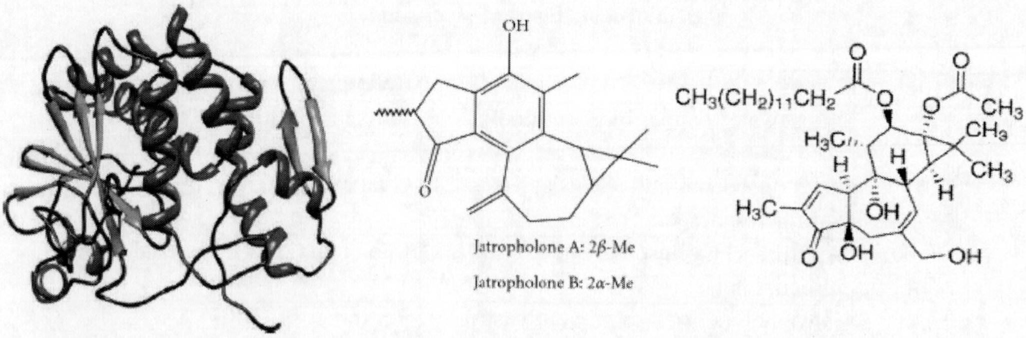

a b c

Source: Alves dos Santos et al. (2012); Srivastava et al. (2012); Sigma-Aldrich (2013).

Figure 3. Chemical structure of (A) Curcin, (B) Diterpene: Jatropholone A and B, (C) Phorbol ester.

CHEMICAL COMPOUNDS ASSOCIATED WITH THE TOXICITY OF *J. CURCAS* AND BIOLOGICAL ACTIVITY

The seeds of *J. curcas* have toxic qualities and are mainly due to the following components: a toxic protein named curcin (Figure 3A), diterpene esters (Figure 3B) and tannins (Figure 3C). Curcin is similar to ricin; a toxic protein of the castor oil plant (*Ricinus communis*) and it was Felke (1913) the first to isolate curcin (Heller, 1996). Wei et al. (2004) extracted curcin from *Jatropha* seeds and evaluated its antifungal activity against various fungi including *Pyricularia oryzae*, *Pestalotiopsis funerea*, *Rhizoctonia solani* and *Sclerotinia sclerotiorum.* Results showed a 100% inhibition of spore formation on *P. oryzae* and 83.8% inhibition on *P. funerea* at 50 μg mL^{-1} concentration. These authors suggest that the inhibitory effect may be due to the inhibition of protein synthesis of the hyphae and spores.

Moreover, It is reported that most of the compounds synthesized by *Jatropha* species are diterpenes, of which there are at least 20 reported. Diterpenes have been isolated from the seeds and roots of this plant species (Adolf et al., 1984; Chen et al., 1988; Naengchomnong et al., 1994; Naengchomnong et al., 1986). Table 2 shows a synthesis of the name and biological activity of diterpenes isolated from *J. curcas* (Devappa et al., 2011). Among the diterpenes, a highly toxic group of compounds known as phorbol esters stands out. These are defined as polycyclic compounds in which two hydroxyl groups on neighboring carbon atoms are esterified to fatty acids (Goel et al., 2007). With respect to its pesticide activity, Ratnadass et al. (2009) assessed contact toxicity on 3rd and 4th instar larvae of the cotton bollworm (*H. armigera*) by applying 1 μL of phorbol ester solutions at concentrations of 0.003 to 1.75 μg per insect. These authors reported that the contact toxicity of phorbol esters was very low. However, they also conducted ingestion toxicity tests on the 1st to 4th larval instars. The results showed insecticidal activity on all instars, being more pronounced in juveniles. Saetae and Suntornsuk (2010) assessed the antifungal activity of ethanolic extracts of *J. curcas* seed cake and the importance of the phorbol esters present in the extracts on *Fusarium oxysporum*, *F. semitectum*, *Colletotrichum capsici*, *C. gloeosporioides*, *Pythium aphanidermatum*, *Lasiodiplidia theobromae* and *Curvularia lunata*. They used concentrations from 0 to 10,000 mg L^{-1}, with the concentrations starting at 6,000 mg L^{-1} showing 100% mycelial growth inhibition for all pathogens tested. These authors are the first to report that phorbol esters are responsible for the fungicidal activity of extracts since they note that the removal of these compounds from the extracts resulted in no significant differences ($p > 0.05$) compared to the control treatment. In further studies, Devappa et al. (2012) evaluated the fungicidal activity of phorbol esters contained in the oil content of *J. curcas* on the mycelial growth of 7 fungi (*Aspergillus niger, Botrytis cinerea, F. oxysporum, F. moniliforme, A. flavus, Curvularia lunata* and *Penicillium notatum*) at concentrations from 28.6 to 143.3 $\mu g\ mL^{-1}$). The results showed an inhibitory effect on all species tested; however, *F. oxysporum, F. moniliforme* and *B. cinerea* were the most susceptible with 100% inhibition. These authors conclude that the *Fusarium* species used for testing could serve as models to assess the presence of phorbol esters in *J. curcas* oil. Other diterpene identified as Curcusone B isolated from *J. curcas* bark also showed a notable diameter inhibition of fungi such as *Aspergillus niger, Penicillium* sp and *Rhizopus* sp (Sahidin et al., 2011).

Table 2. Diterpenes isolates from *J. curcas* and biological activity

Diterpene	Biological activity	Reference
15-O-acetyl-15-epi-(4E)-jatrogrossidentadion	ND	Ravindranath et al. (2004)
(14E)-14-O-acetyl-5.6-epoxyjatrogrossidentadion	ND	Ravindranath et al. (2004)
3β-acetoxy-12-methoxy-13- methyl-podocarpa- 8,11,13-trien-7-one ND		Ravindranath et al. (2004)
3β,12-dihydroxy-13-methylpodocarpane-8,10,13-triene ND		Ravindranath et al. (2004)
Curcasone A	Anti invasive effects in tumor cells	Naengchomnong et al. (1986)
Curcasone B	Anti invasive effects in tumor cells	Muangman et al. (2005)
Curcasone C	Anti invasive effects in tumor cells	Naengchomnong et al. (1986)
Curcasone D	Anti invasive effects in tumor cells	Naengchomnong et al. (1986)
Jatropherol	Insecticide, Rodenticide	Jing et al. (2005)
Curculathyrane A	ND	Naengchomnong et al. (1986)
Curculathyrane B	ND	Naengchomnong et al. (1986)
Jatrophol	ND	Naengchomnong et al. (1994)
Jatropha factor C1	Cytotoxic, molluscidal, rodenticidal	Haas et al. (2002); Li et al. (2010)
Jatropha factor C2	Cytotoxic, molluscidal, rodenticidal	Haas et al. (2002); Li et al. (2010)
Jatropha factor C3	Cytotoxic, molluscidal, rodenticidal	Haas et al. (2002); Li et al. (2010)
Jatropha factor C4	Cytotoxic, molluscidal, rodenticidal	Haas et al. (2002); Li et al. (2010)
Jatropha factor C5	Cytotoxic, molluscidal, rodenticidal	Haas et al. (2002); Li et al. (2010)
Jatropha factor C6	Cytotoxic, molluscidal, rodenticidal	Haas et al. (2002); Li et al. (2010)
Heudolotinone	ND	Ravindranath et al. (2004)
Jatrophalactam	Cytotoxic	Wang et al. (2009)

Adapted from Devappa et al., 2011.
ND = Not detected.

With respect to tannins, they are phenolic substances associated with toxic and antinutritional effects such as reduced food intake, growth retardation and impaired nutrient absorption (Butler et al., 1986 in Devappa et al., 2010). In this sense studies have focused only on feeding trials that have demonstrated the toxicity of the seeds in mice, rats, goats, calves and chickens (Adam and Magzoub, 1975; Ahmed and Adam, 1979; El-Badwi et al., 1995; Liberalino et al., 1988). To our knowledge, no information about their insecticidal and fungicidal potential of this chemical constituent obtained from this plant species has been published, however; the phytochemical analysis of the plant latex of *J. curcas* obtained with ethanolic and methanolic extracts, showed the presence of various bioactive compounds such as alkaloids, glycosides, phenols, saponines, steroids, tannins and terpenoids (Sibi et al. 2013).

Activity of *J. curcas* Extracts and Seed Oil against Pest Insects

Various studies show that this plant can be successfully used in many cases to control different pests (Table 3). On this matter, pesticidal activity has been recorded for various insects, including among others cotton bollworm (*Helicoverpa armigera*), cotton aphid (*Aphis gossypii*), pink bollworm (*Pectinophora gossypiella*), leafhopper (*Empoasca biguttula*), Chinese bean weevil (*Callosobruchus chinensis*), maize weevil (*Sitophilus zeamays*), potato tuber moth (*Phthorimaea opercullela*), pink stalk borer (*Sesamia calamistis*), maize stalk borer (*Busseola fusca*), American cockroach (*Periplaneta americana*), German cockroach (*Blattella germanica*), milkweed bug (*Oncopeltus fasciatus*), tobacco hornworm (*Manduca sexta*), rice weevil (*Sitophilus zeamais*) and seed beetle (*Callosobruchus maculatus*) (Wink et al., 1997; Adebowale and Adedire 2006; Asmanizar et al. 2012).

Of all the types of extracts used to control various genera of insects, *J. curcas* seed oil stands out. According to the published literature, overall the efficacy of *J. curcas* as a pesticide depends on various factors, including the type of solvent used for oil extraction, concentration applied, target insect and insect life stage. In this regard, in study about the efficacy of petroleum ether and aqueous extracts, it was concluded that control of flea beetle (*Podagrica* spp) on okra was more effective when 5% petroleum ether was applied (Emosairue and Uguru, 1999). In other studies on the effect of *J. curcas* extracts for controlling two insects namely rice leaf folder and cotton boll worm of muskmelon, by means of using different solvents, it was reported that those obtained from methanol, petroleum ether and ethanol showed the highest antifeedant and pesticidal activity, of up to 80%, in comparison with ethyl acetate, benzene, hexane and water solvents where control was from 50 to 65% (Dowlathabad et al., 2010). Overall, the insecticidal activity of *J. curcas* is closely associated with the highest concentrations (Musa et al., 2011; Khani et al., 2011; Asmanizar et al., 2012; Azzaz and Khalifa, 2012). For example, control of the weevil pest of rice, maize and wheat increased with increasing concentration. Khani et al. (2011) reported that the percentage of adult mortality was almost 70% at the highest concentration of 10 $\mu l\ g^{-1}$ compared with 6% control at the lowest concentration of 2 $\mu l\ g^{-1}$, while Musa et al. (2011) reported a dramatic increased as the adult mortality as concentration increases from 0.2 to 1.0

ml. Similarly, Katoune et al. (2011) in a study on reported heteropteran bugs of legume cowpea a significantly higher inhibition at 10% compared with 2.5% concentration; however, in this study, when extracts where tested on the aphid (*Aphis craccivora*), control was not significant among treatments and it was similar to that of the pesticide Deltamethrin. Field studies carried out by Habou et al. (2011) also demonstrated the efficacy of *J. curcas* oil extracts for controlling numerous pests of cowpeas including aphids, bugs and thrips. In a study by Agboka et al. (2009), the magnitude of the pesticide effect of *J. curcas* on the life cycle of the maize cob borer (*Mussidia nigrivenella*) was well documented In that study, *J. curcas* oil application at 5% was sufficient to deter oviposition of the insect and egg hatch, but larvae survival was not affected.

Table 3. *Jatropha curcas* application on various crops to control different insect pests

Crop	Pest	Type of extract	Effective dose	Reference
Bean	*Callosobruchus chinensis*	Seed oil	ND	Wink et al. (1997)
Cowpea	*Megalurothrips* *Clavigralla tomentosicollis* *Aphis craccivora*	Seed oil	7.5-10% 2.5-10% 5-10%	Katoune et al. (2011)
	Aphis fabae	Seed oil	15%	Habou et al. (2011)
	Callosobruchus maculatus	Seed oil	6.66 μL g^{-1}	Boateng and Kusi (2008)
	Callosobruchus maculatus	Seed oil	0.5-2%	Adebowale and Adedire (2006)
Cumin	*Meloidogyne incognita*	Leaf	0.5 g mL^{-1}	Sharma and Trivedi (2002)
Maize	*Mussidia nigrivenella*	Seed oil	1.6%	Agboka et al. (2009)
	Sitophilus zeamais	Seed oil	1%	Musa et al. (2011)
	Busseola fusca	Seed oil	ND	Wink et al. (1997)
Musk Melon	*Cnaphalocrocis medinalis* *Helicoverpa armigera*	Seed oil Seed oil	500 ppm 500 ppm	Dowlathabad et al. (2010) Dowlathabad et al. (2010)
Okra	*Podagrica spp*	Seed oil	5%	Emosairue and Uguru (1999)
	Meloidogyne incognita	Leaf	100%	Umeh and Ndana (2010)
Potato	*Phthorimaea opercullela*	Seed oil	ND	Wink et al. (1997)
Rice	*Corcyra cephalonica*	Seed oil	20 μl g^{-1}	Khani et al. (2011)
	Sitophilus zeamais	Seed oil	0.4%	Asmanizar et al. (2012)
Tomato	*Meloidogyne incognita*	Water soluble fraction of deoiled seed cake	20%	Kalaraiasan et al. (2007)
Wheat	*Sitophilus granaries*	Seed oil	1.5%	Nabil and Yasser (2012)
Various crops	*Plutella xylostela* *Spodoptera frugiperda*	Seed oil Seed oil	1% 1%	Wink et al. (1997) Wink et al. (1997)
	Mycus persicae	Seed oil	1%	Wink et al. (1997)

ACTIVITY OF *J. CURCAS* EXTRACTS AGAINST PLANT PATHOGENIC MICROORGANISMS

Respect to the fungicidal activity of *J. curcas*, various studies have confirmed the potential of *J. curcas* extracts to control fungal pathogens (Table 4). Likewise, the inhibitory effect of *J. curcas* extracts may be dependent on concentration, solvent used for extraction, plant organ and target microorganism (Bassey et al., 2013). Overall, the fungicidal activity has been reported in leaves and seeds of this plant species and their antifungal properties affects include numerous pathogenic fungi that belong to, among others, the genera *Aspergillus*, *Fusarium*, *Colletotrichum* and *Alternaria* (Gaikwad et al., 2012; Bassey et al., 2013; Sibi et al., 2013) On this matter, studies have focused on fungal development in *in vitro* tests demonstrated that mycelial growth of several fungi, including *Alternaria alternata*, *Aspergillus flavus*, *A. fumigans*, *A. niger* and *Fusarium chladmydosporum*, is inhibited when grown on seed oil and crude seed extracts at concentrations of 100 and 500 μl (Srivastava et al., 2012). Córdova-Albores et al. (2014a) also evaluated the antifungal potential of *J. curcas* oil at concentrations of 2.5, 5 and 10 mg mL^{-1} on the mycelial growth of various species of the genus *Fusarium* isolated from gladiolus corm and demonstrated that all *Fusarium* species tested were sensitive to the 5 mg mL^{-1} oil concentration. However, in another study, a higher concentration of 8 mg mL^{-1} was needed to obtain the maximum growth inhibition of *F. oxysporum* (Sachdeva et al., 2011). Working on this same fungus but isolated from eggplants (*Solanum melongena*), Siva et al. (2008) reported that *J. curcas* leaf extracts obtained either with water, ethanol or acetone at 50% concentration gave a complete control of this fungus. Similar success was also reported on *F. oxysporum* f sp. *cumini* when grown on nutrient agar mixed with 1% aqueous leaf extracts of this plant species (Sharma and Trivedi, 2002). In another study, fungal species such as *Rhizoctonia solani* and *Trichoderma viride* were also affected when grown on media amended with seed oil extracts in a concentration range from 5 to 30% (Gaikward et al., 2012). These same authors demonstrated that sporulation of these and other fungal species such as *F. oxysporum*, *A. alternata*, *A. flavus* and *A. niger* was also affected by extracts obtained from *J. curcas* leaves, roots, stems and seeds. Conidial germination of the fungus *Drechslera heveae* was significantly reduced by 70% after the application of aqueous *J. curcas* extracts at at 100% concentration (Ogbebor and Adekunle, 2008). In an attempt to evaluate the fungicidal effect of different plant organs of *J. curcas*, it was found that seeds followed by the fruit pulp and the whole fruit had significant inhibition of the mycelia of the fungus *Colletrotrichum gloeosporioides* isolated from papaya fruits (Rahman et al., 2011). In another study, the five fractions obtained from *J. curcas* leaves and stems, labeled as A, B, C, D and E, showed complete inhibition of the fungus *Penicillium digitatum* at the concentration of 1.0 mg mL^{-1} (Ayambipe et al., 2009). In this study, as the concentration of the fractions increased, fungal growth also increased inhibition. The chemical characterization of the oil revealed that triacylglycerols were the major compounds since the synergy of triacylglycerol inhibited the *in vitro* development of *F. oxysporum* f. sp. *gladioli* (Córdova-Albores et al., 2014b).

Table 4. Activity of *J. curcas* extracts against phytopathogenic fungi

Fungi	Type of extract/Active compound	Active dose	Effect	Reference
Aspergillus niger *A. flavus* *A. fumigatus* *Fusarium oxysporum*	Ethanolic of leaves	10%	Mycelial growth inhibition	Bassey et al. (2013)
A. fumigatus *A. niger* *A. terreus* *F. solani* *Penicillium digitatum* *Rhizopus arrhizus*	Petroleum ether latex extract	20 µL per plate	Mycelial growth inhibition	Sibi et al. (2013)
Alternaria alternata *A. flavus* *A. niger* *F. oxysporum* *Rhizoctonia solani* *Trichoderma viride*	Leaves, steam, root, latex, oil	20%	Mycelial growth inhibition and decrease in sporulation	Gaikwad et al. (2012)
A. alternata *A. flavus* *A. fumigates* *A. niger* *F. chlamydosporum* *P. glabrum*	Seed oil	500 µL per plate	Mycelial growth inhibition	Srivastava et al. (2012)
F. verticillioides *A. flavus*	Deolied seed extract/Crude seed extract	-	Mycelial growth inhibition *in vitro* and *in vivo* on Yam	Makun et al. (2011)
Colletotrichum gloeosporioides	Methanolic fruit/seed extract	10 mg mL^{-1}	Mycelial growth inhibition	Rahman et al. (2011)
A. alternata *F. solani*	Stem bark	8 mg mL^{-1}	Mycelial growth inhibition	Sachdeva et al. (2011)
A.niger *Rhizopus* sp. *Penicillium* sp.	Curcusone, jatrophone and stigmasterol	10000 µg mL^{-1}	Mycelial growth inhibition	Sahidin et al. (2011)
F. oxysporum *F. solani* *F. proliferatum*	Seed oil	5 mg mL^{-1}	Mycelial growth inhibition, decrease in sporulation	Córdova-Albores et al. (2014a)
Cladosporium sp. *A. niger* *A. terreus* *P. expansum*	Methanolic of leaves	1 mg mL^{-1}	Mycelial growth inhibition	Ayanbimpe et al. (2009)
Drechslera heveae	Aqueous of leaves	50%	Conidial germination inhibition	Ogbebor and Adenkule (2008)
F. oxysporum f. sp. *melongenae*	Aqueous, ethanolic, acetonic of leaves	50%	Mycelial growth inhibition	Siva et al. (2008)
Colletotrichum musae	Leaves	50%	Mycelial growth inhibition *in vitro* suppressed antracnose disease *in vivo* on bananas	Thangavelu et al. (2004)
F. oxysporum f. sp. *cumini*	Aqueous of leaves	60 µL mL^{-1}	Mycelial growth inhibition	Sharma and Trivedi (2002)

The few studies focused on demonstrating the fungicidal activity of *J. curcas* in *in situ* studies have also shown the potential of this plant to control various plant diseases. In this respect, Thangavelu et al. (2004) reported a notable reduction of anthracnose disease caused by the fungus *C. musae* on banana fruits at the yellow ripe stage of various cultivars after dipping fruit in aqueous leaf extract at 50% concentration. In another study, Ogbebor and Adenkule (2008) found a 45% disease index in rubber (*Hevea brasiliesins*) trees treated with aqueous extracts of Jatropha leaves, whereas the untreated plants showed a disease index of 50%. Siba et al. (2008) also reported a significant control of *F. oxysporum* in eggplants when treated with aqueous extracts of *J. curcas* leaves at 40% concentration. In another study, Makun et al. (2011) also reported a significant 80% reduction of *F. verticillioides* growth on fresh-cut yam tubers treated with Jatropha seed extracts and oil.

POSSIBLE ACTION MECHANISMS OF VARIOUS COMPOUNDS ISOLATED FROM *J. CURCAS*

There are many reports on the biological activity of crude extracts of *J. curcas* for the control of pests and diseases, but few describe the action mechanisms of a particular compound of *J. curcas*. In this regard, Wei et al. (2005) evaluated the antifungal activity of the enzyme β-1,3-glucanase isolated from this plant species on the fungus *Rhizoctonia solani* and *Giberella zeae* and found in observations made through transmission electron microscopy, that the cell wall of the treated hyphae with the enzyme was thin and weak. The authors also revealed that these changes might be because PR-2 proteins hydrolyze the structural β-1,3-glucan present in the fungal cell wall, or it may inhibit the synthesis of the fungal cell wall structure or disrupt cell wall structure. For their part, Jing et al. (2005) isolated from *J. curcas* the diterpene Jatropherol-I and evaluated the toxic effect, the influence on the enzymatic activity on the digestive tract and the ultrastructure of the epithelial cells in the silkworm (*Bombyx mori*). These authors reported that compared to control cells, exposure to the compound changes were observed in columnar cells at 12 h, the endoplasmic reticulum showed destruction and started to vesiculate at 48 h, and most organelles were destroyed after 72 h. These ultrastructural changes indicated the disruption of cells. According to these authors, the compound Jatropherol-I might be the cause of the disruption of the epithelial cells, eventually associated with the production of digestive enzymes and other enzymes.

CONCLUSION

Interest in the *Jatropha* crop has been increasing due to its possible use as a biodiesel fuel. On this matter, the growing demand for *Jatropha* cultivation for oil production and bioenergy transformation leads to the use of the plant (leaves, stems, roots and seeds) and possibly its residues from the transformation process for the management of pests and diseases in agriculture. In addition, in the present review, we have documented some work related to the use of *J. curcas* and its active compounds, contained in both the oil and the plant, to control some pests and microorganisms of agricultural interest; however, most reports are from *in vitro* work; therefore, much research still to be developed regarding the use of this plant species and metabolites, in applied related research. We believe that its potential in the agriculture industry, specifically as an alternative pest and disease control method, has yet to be assessed in a wide range of horticultural products.

It has been reported that the toxicity of *J. curcas* is due to the presence of phorbol esters and controversy exists regarding the use of this product for the control of pests and diseases in horticultural crops. The use of *J. curcas* derivatives in the field, as well as the toxicological risks of long-term use for both humans and the environment, has not yet been well documented, and there are no reports on the use of non-toxic varieties of *J. curcas*. Nonetheless, there is enough evidence showing the potential of this plant in agriculture, to warrant further research on this matter.

REFERENCES

Abdel-Shafy, S., Nasr, S.M., Abdel-Rahman, H.H. and Habeeb, S.M. (2011). Effect of various levels of dietary *Jatropha curcas* seed meal on rabbits infested by the adult ticks of *Hyalomma marginatum marginatum* I. Animal performance, anti-tick feeding and haemogram. *Tropical Animal Health Production*, 43, 347-357.

Adam, S.E. and Magzoub, M. (1975). Toxicity of *Jatropha curcas* for goats. *Toxicology*, 4, 347-354.

Adebowale, K.O. and Adedire, C.O. (2006). Chemical composition and insecticidal properties of the underutilized *Jatropha curcas* seed oil. *African Journal of Biotechnology*, 5, 901-906.

Adolf, W., Opferkuch, H.J. and Hecker, E. (1984). Irritant phorbol derivatives from four *Jatropha* species. *Phytochemistry*, 23, 129-132.

Agboka, K., Mawufe, A.K., Tamò, M. and Vidal, S. (2009). Effects of plant extracts and oil emulsions on the maize cob borer *Mussidia nigrivenella* (Lepidoptera: Pyralidae) in laboratory and field experiments. *International Journal Tropical Insect Science*, 29, 185-194.

Ahmed, O.M. and Adam S.E. (1979). Effects of *Jatropha curcas* on calves. *Veterinary Pathology*, 16, 476-482.

Aiyelaagbe, O.O., Adesogan, K., Ekundayo, O. and Gloer, J.B. (2007). Antibacterial diterpenoids from *Jatropha podagrica* (Hook). *Phytochemistry*, 68, 2420-2425.

Aiyelaagbe, O.O., Adeyini, B.A., Fatunsin, O.F. and Arimah, B.D. (2007a). *In vitro* antimicrobial activity and phytochemical analysis of *Jatropha curcas* roots. *International Journal of Pharmacology*, 3, 106-110.

Aiyelaagbe, O.O. and Gloer, J.B. (2008). Japodic acid, A novel aliphatic acid from *Jatropha podagrica* Hook. *Records of Natural Products*, 2, 100-106.

Aker, C. L. (1997). Growth and reproduction of *J. curcas*. In: Gübitz G.M., Mittelbach M. and Trabi, M. (Eds.), *Biofuels and Industrial Products from Jatropha curcas*. pp. 2-18. DBV Graz, Austria.

Akinpelu, D.A., Aiyegoro, O.A. and Okoh, A.I. (2009). The bioactive potentials of two medicinal plants commonly used as folklore remedies among some tribes in West Africa. *African Journal of Biotechnology*, 8, 1660-1664.

Alves dos Santos, E., Carvalho, C.M., Costa, A.L.S., Conceição, A.S., Prado Moura, F.B. and Goulart Santana, A.E. (2012). Bioactivity evaluation of plant extracts used in indigenous medicine against the snail, *Biomphalaria glabrata*, and the larvae of *Aedes aegypti*. *Evidence-Based Complementary and Alternative Medicine*, 1, 1-10.

Asmanizar Djamin A. and Idris A.B. (2012). Evaluation of *Jatropha curcas* and *Annona muricata* seed crude extracts against *Sitophilus zeamais* infesting stored rice. *Journal of Entomology*, 9, 13-22.

Ayanbimpe, G.M., Ojo, T.K., Afolabi, E., Opara, F., Orsaah, S. and Ojerinde, O.S. (2009). Evaluation of extracts of *Jatropha curcas* and *Moringa oleifera* in culture media for selective inhibition of saprophytic fungal contaminants. *Journal of Clinical Laboratory Analysis*, 23, 161-164.

Azzaz, N.A.E. and Khalifa, Y.A.M. (2012). *Jatropha Curcas* Oil as insecticide and germination promoter. *Journal of Applied Sciences Research*, 8, 668-675.

Bassey, I.N., Ogbemudia, F.O., Harold, K.O. and Idung, K.E. (2013). Combined antifungal effects of extracts of *Jatropha curcas* and *Chromolaena odorata* on seed borne fungi of *Solanum gilo* Raddi. *Bulletin of Environment, Pharmacology and Life Sciences*, 2, 13-17.

Betancur, L., Checa, J. and Estornell, E. (2003). Jatrophane diterpenes from the latex of *Euphorbia obtusifolia* with inhibitory activity on the mammalian mitochondrial respiratory chain. *Planta Medica*, 69, 177-178.

Biehl, J. and Hecker, E. (1986). Irritant and antineoplastic principles of some species of the genus *Jatropha* (*Euphorbiaceae*). *Planta Medica*, 52, 430-430.

Boateng, B.A. and Kusi, F. (2008). Toxicity of *jatropha* seed oil to *Callosobruchus maculatus* (Coleoptera: Bruchidae) and its parasitoid, *Dinarmus basalis* (Hymenoptera: Pteromalidae). *Journal of Applied Sciences Research*, 4, 945-951.

Brittaine, R. and Lutaladio, N. (2010). *Jatropha*: A Smallholder Bioenergy Crop. The Potential for Pro-Poor Development. *Integrated Crop Management* Vol. 8. FAO. Rome, Italy.

Brum, R.L., Cardoso, C.A.L., Honda, N.K., Andreu, M.P. and Viana, L.L.D.S. (2006). Quantitative determination of jatrophone in "Cachaqa" prepared with *Jatropha elliptica*. *Chemical and Pharmaceutical Bulletin*, 54, 754-757.

Burkill, H.M. (1994). *The useful plants of west tropical Africa*, Vol. 2. *Royal Botanical Gardens*. Vol. 2 pp. 90-94 Kew, Richmond, United Kingdom.

Butler, L., Rogler, J., Mehansho, H. and Carlson, D. (1986). Dietary effects of tannins. In: Cody V., Middleton E., Harborne J. B., (Eds), *Plant Flavonoids In Biology and Medicine; Biochemical, Pharmacological, and Structure Activity Relationships*. pp. 141-157. A. R. Liss. New York, US.

Cantrell, C., Ali, A., Duke, S. and Khan, I. (2011). Identification of mosquito biting deterrent constituents from the Indian folk remedy plant *Jatropha curcas*. *Journal of Medical Entomology*, 48, 836-845.

Chen, M. J., Hou L. L. and Zhang G. W. (1988). The diterpenoids from *Jatropha curcas* L. *Acta Botanica Sinica* 30, 308-311.

Córdova-Albores, L., Bautista-Baños, S., Martínez-Herrera, J., Barrera-Necha L., Hernández-López, M. and Cruz-Hernández, A. (2014a). Morphological and molecular characterization of pathogenic isolates of *Fusarium* spp. obtained from gladiolus corms and their sensitivity to *Jatropha curcas* L oil. *African Journal of Microbiology Research*, 8, 724-733.

Cordova-Albores, C.L., Rios, M.Y., Barrer-Necha, L. and Bautista-Baños, S. (2014b). Chemical compounds of a native *Jatropha curcas* seed oil from Mexico and their antifungal effect on *Fusarium oxysporum* f. sp. *gladioli. Industrial Crops and Products*, 62, 166-172.

Dehgan, B. and Webster, G.L. (1979). Morphology and infrageneric relationships of the genus *Jatropha* (Euphorbiaceae). *Publications in Botany*, Vol. 74. University of California, US.

Devappa, R.K., Makkar, H.P.S. and Becker, K. (2010). Jatropha toxicity-A review. *Journal of Toxicology and Environmental Health*, 13, 476-507.

Devappa, R.K., Makkar, H.P.S. and Becker, K. (2011). Jatropha diterpenes: A review. *Journal American. Oil Chemistry. Society*, 88, 301-322.

Devappa, R.K., Rajesh, S.K., Kumar, V., Makkar, H.P.S. and Becker, K. 2012. Activities of *Jatropha curcas* phorbol esters in various bioassays. *Ecotoxicology and Environmental Safety*, 78, 57-62.

Devappa, R.K. and Swamylingappa, B. (2008). Biochemical and nutritional evaluation of *Jatropha* protein isolate prepared by steam injection heating for reduction of toxic and antinutritional factors. *Journal of the Science of Food and Agriculture*, 88, 911-919.

Dowlathabad, M.R., Sreeyapureddy, A., Adhikari, A., Bezawada, K. and Nayakanti, D. (2010). Pharmaceutical investigation and biopesticidal activity *of Jatropha curcas* L. seed oil on digestive enzymic profiles of *Cnaphalocrocis medinalis* (rice leaf folder) and *Helicoverpa armigera* (cotton boll worm). *International Research Journal Pharmaceutical*, 1, 194-200.

El-Badawi, S.M., Adam, S.E. and Hapke, H.J. (1995). Comparative toxicity of *Ricinus communis* and *Jatropha curcas* in Brown Hisex chicks. *Dtsch Tierarztl Wochenschr* 102, 75-77.

Emosairue, S.O. and Uguru, E.I. (1999). Field trial of aqueous and petroleum ether extracts of *Monodora myristica* (Gaertn.) dunal and *Jatropha curcas* L. for the control of okra flea beetles, *Podagrica* spp. *Journal Applied Chemistry Agriculture Research*, 6, 100-104.

Eshilokun, A.O., Kasali, A.A., Ogunwande, I.A., Walker, T.M. and Setzer, W.N. (2007). Chemical composition and antimicrobial studies of the essential oils of *Jatropha integerrima* Jacq (leaf and seeds). *Natural Products Communications*, 2, 853-855.

Felke J. (1913). Über die Giftstoffe der Samen von *Jatropha curcas* (About the toxins the seeds of *Jatropha curcas*). *Landw. Vers. Stat.* 82, 427-467.

Gaikwad, R.S., Kakde, R.B., Kulkarni, A.U., Gaikwad, D.R. and Panchal, V.H. (2012). *In vitro* antimicrobial activity of crude extracts of *Jatropha* species. *Current Botany*, 3, 9-15.

Goel, G., Makkar, H.P.S., Francis, G. and Becker, K. (2007). Phorbol esters: Structure, biological activity and toxicity in animals. *International Journal of Toxicology*, 26, 279-288.

Grimm, C., Somarriba, A. and Foidl, N. (1997). Development of eri silkworm *Samiacynthia ricini* (Boisd.) (Lepidoptera: *Saturniidae*) on different provenances of *Jatropha curcas* leaves. In: Gübitz, G.M., Mittelbach, M., Trabi, M. (Eds.), *Biofuels and Industrial Products from Jatropha curcas*. pp. 233-240. DBV Graz, Austria.

Gyuris, A., Szlávik, L., Minárovits, J., Vasas, A., Molnar, J. and Hohmann, J. (2009). Antiviral activities of extracts of *Euphorbia hirta* L. against HIV-1, HIV-2, and SIV_{mac251}. *In vivo*, 23, 429-432.

Haas, W., Strerk, H. and Mittelbach, M. (2002). Novel 12-Deoxy-16-hydroxyphorbol diesters isolates from the seed oil of *Jatropha curcas*. *Journal Natural Products*, 65, 1434-1440.

Habou, Z.A., Haougui, A., Mergeai, G., Haubruge, E., Toudou, A. and Verheggen, F.J. (2011). Insecticidal effect of *Jatropha curcas* oil on the aphid *Aphis fabae* (Hemiptera: Aphididae) and on the main insect pests associated with cowpeas in Niger. *Tropicultura*, 29, 225-229.

He, W., King, A.J., Khan, M.A., Cuevas, J.A. and Ramiaramanana, D. (2011). Analysis of seed phorbol-ester and curcin content together with genetic diversity in multiple provenances of *Jatropha curcas* L. from Madagascar and Mexico. *Plant Physiology, and Biochemistry* 49, 1183-1190.

Heller, J. (1996). Physic nut. *Jatropha curcas* L. Promoting the conservation and use of underutilized and neglected crops. Institute of Plant Genetics and Crop Plant Research, Gatersleben/International Plant Genetic Resources Institute, Rome.

Igbinosa, O.O., Igbinosa, E.O. and Aiyegoro, O.A. (2009). Antimicrobial activity and phytochemical screening of stem bark extracts from *Jatropha curcas* (Linn). *African Journal of Pharmacy and Pharmacology*, 3, 58-62.

Isawumi, M. A. (1978). Nigerian chewing sticks. *The Nigerian Field* 43, 111-121.

Jain, N., Joshi, H. C., Dutta, S.C., Kumar, S. and Pathak, H. (2008). Biosorption of copper from wastewater using *Jatropha* seed coat. *Journal of Scientific and Industrial Research*, 67, 154-160.

Jing, L., Fang, Y., Ying, X., Wenxing, H., Meng, X., Syed, M.N. and Fang, C. (2005). Toxic impact of ingested Jatropherol-I on selected enzymatic activities and the ultrastructure of midgut cells in silkworm, *Bomboxy mori* L. *Journal Applied Entomology*, 129, 98-104.

Kalaiarasan, P., Senthamarai, M., Ramesh, D. and Sudheer, M.J. (2007). *Jatropha*: An efficient organic amendment for the management of root-knot nematode, *Meloidogyne incognita* in tomato. *Indian Journal Nematology*, 37, 115-118.

Kalimuthu, K., Vijayakumar, S. and Senthilkumar, R. (2010). Antimicrobial activity of the biodiesel plant, *Jatropha curcas* L. *International Journal of Pharma and Bio Sciences*, 1, 1-5.

Katoune, H.I., Malam Lafia, D., Salha, H., Doumma, A., Yaye Drame, A., Pasternak, D. and Ratnadass, A. (2011). Physic nut (*Jatropha curcas*) oil as a protectant against field insect pests of cowpea in Sudano-Sahelian cropping systems. *Journal SAT Agriculture Research*, 9, 1-6.

Kaushik, N., Kumar, K., Kumar, S. and Roy, S. (2007). Genetic variability and divergence studies in seed traits and oil content of *Jatropha* (*Jatropha curcas* L.) accessions. *Biomass and Bioenergy*, 31, 497-502.

Khan, M. T. H., Ather, A., Pinto, A.C. and Maciel, M.A.M. (2009). Potential benefits of the 19-*nor*-clerodane *trans*-dehydrocrotonin on the central nervous system. *Brazilian Journal of Pharmacognosy*, 19, 7-13.

Khani, M., Awang, R.M., Dzolkhifli, O., Rahmani, M. and Rezazadeh, S. (2011). Tropical medicinal plant extracts against rice weevil, *Sitophilus oryzae* L. *Journal of Medicinal Plants Research*, 5, 259-265.

Khiev, P., Oh, S.R., Chae, H.S., Kwon, O.K., Ahn, K.S., Chin, Y.W. and Lee, H.K. (2011). Anti-inflammatory diterpene from *Thyrsanthera suborbicularis*. *Chemistry Pharmacy Bulletin*, 59, 382-384.

Kumar, A. and Sharma, S. (2009). An evaluation of multipurpose oil seed crop for industrial uses (*Jatropha curcas* L.) *A review. Journal of Industrial Crops and Products*, 28, 1-10.

Li, C.Y., Devappa, R.K., Liu, J.X., Makkar, H.P.S. and Becker, K. (2010). Toxicity of *Jatropha curcas* phorbol esters in mice. *Food Chemistry Toxicology*, 48, 620-625.

Liberalino, A.A.A., Bambirra, E.A., Moraes –Santos, T. and Viera, C.E. (1988). *Jatropha curcas* L. seeds. Chemical analysis and toxicity. *Archives of Biology and Technology*, 31, 539-550.

Linnaeus, C. (1753). Species plantarum. In: *Jatropha*. Impensis Laurentii Salvii (Printed Kiesewetteri), Stockholm, pp. 1006-1007.

Liu, S.Y., Sporer, F., Wink, M., Jourdane, J., Henning, R., Li, Y.L. and Ruppel, A. (1997). Anthraquinones in *Rheum palmatum* and *Rumex dentatus* (Polygonaceae) and phorbol esters from *Jatropha curcas* (Euphorbiaceae) with molluscicidal activity against the schistosomias vector snails *Oncomelania*, *Biomphalaria* and *Bulinus*. *Tropical Medicine and International Health*, 2, 179-188.

Luo, M.J., Liu, W.X., Yang, X.Y., Xu, Y., Yan, F., Huang, P. and Chen, F. (2007). Cloning, expression, and antitumor activity of recombinant protein of curcin. *Russian Journal of Plant Physiology*, 54, 202-206.

Mahanta, N., Gupta, A. and Khare, S.K. (2008). Production of protease and lipase by solvent tolerant *Pseudomonas aeruginosa* PseA in solid state fermentation using *Jatropha curcas* seed cake as substrate. *Bioresource Technology*, 99, 1729-1735.

Makkar, H.P.S., Becker, K., Sporer, F. and Wink, M. (1997). Studies on nutritive potential and toxic constituents of different provenances of *Jatropha curcas*. *Journal Agricultural Food Chemistry*, 45, 3152-3157.

Makkar, H.P.S. and Becker, K. (1999). Nutritional studies on rats and fish (carp *Cyprinus carpio*) fed diets containing unheated and heated *Jatropha curcas* meal of a non-toxic provenance. *Plant Foods for Human Nutrition,* 53, 183-192.

Makkar, H.P.S., Becker, K. and Schmook, B. (1998). Edible provenances of *Jatropha curcas* from Quintana Roo state of Mexico and effect of roasting on antinutrient and toxic factors in seeds. *Plant Foods for Human Nutrition*, 52, 31-36.

Makun, H. A., Anjorin, S.T., Adeniran, L.A., Onakpa M.M. and Muhammad, H.L. (2011). Antifungal activities of *Jatropha curcas* and *Ricinus cumunis* seeds on *Fusarium verticilliodes* and *Aspergillus flavus* in yam. ARPN *Journal of Agricultural and Biological Science*, 6, 22-27.

Martínez, M. A., Evangelista, V., Basurto, F., Mendoza, M. and Cruz-Rivas, A. (2007). Flora útil de los cafetales en la sierra norte de Puebla, México (useful flora of the coffee plantations in the mountains north of Puebla, Mexico). *Revista Mexicana de Biodiversidad*, 78, 15-40.

Martínez-Gordillo, M., Jiménez-Ramírez, J., Cruz-Durán, R., Juárez-Arriaga, E., García, R., Cervantes, A. and Mejía-Hernández, R. (2002). Los géneros de la familia Euphorbiaceae en México (The genera of Euphorbiaceae family in Mexico). *Anales del Instituto de Biología, Universidad Nacional Autónoma de México, Serie Botánica* (Annals of the Institute of Biology, National Autonomous University of Mexico, Botany Series), 73, 155-281.

Martínez-Gordillo, M., Steinmann, V., Jiménez, J., Cervantes, A., Ramírez, Y. and Ramírez, A. (2006). Catálogo de autoridades taxonómicas de las euforbiáceas (Euphorbiaceae: Magnoliopsida) de México (Catalog taxonomic authorities Euphorbiaceae

(Euphorbiaceae: Magnoliopsida) of Mexico). Facultad de Ciencias, UNAM. Base de datos SNIB-CONABIO, proyecto CS006 (Faculty of Sciences, UNAM. Database SNIB - CONABIO, CS006 project). México.

Martínez-Herrera, J. (2006). Diversidad genética y nutricional del piñón (*Jatropha curcas* L.) en México (genetic and nutritional diversity pinion (*Jatropha curcas* L.) in Mexico). PhD Thesis. Instituto Politécnico Nacional (National Polytechnic Institute). México.

Martínez-Herrera, J., Siddhuraju, P., Francis, H., Dávila-Ortiz, G. and Becker, K. (2006). Chemical composition, toxic/antimetabolic constituents, and effect of different treatments on their levels, in four provenances of *Jatropha curcas* L. from Mexico. *Food Chemistry*, 96, 80-89.

Mbakwem-Aniebo, C., Okoyomo, E.P., Ogugbue, C.J. and Okonko, I.O. (2012). Effects of *Jatropha curcas* leaves on common dermatophytes and causative agent of *Pityriasis versicolor* in Rivers State, Nigeria. *Nature and Science*, 10, 151-157.

Mendoza, T.C., Castillo, E.T. and Aquino, A.L. (2007). Towards making *Jatropha curcas* (tubing bakod) a viable source of biodiesel oil in the Philippines. *Philippine Journal of Crop Science*, 32, 29-43.

Muangman, S., Thippornwong, M. and Tohtong, R. (2005). Antimetastatic effects of curcusone B, a diterpene from *Jatropha curcas*. *In Vivo*, 19, 265-268.

Musa, A.K., Belewu, M.A., Lawal, A.M., Adekola, F.O., Olarewaju, B O. and Ibraheem, S.O. (2011). Costs analysis and toxicity of *Jatropha curcas* L. on maize weevil, *Sitophilus zeamais* Motsch. *African Journal of Plant Science*, 5, 233-236.

Nabil, A.E.A. and Yasser, A.M.K. (2012). *Jatropha curcas* oil as insecticide and germination promoter. *Journal Applied Science Research*, 8, 668-675.

Naengchomnong, W., Tarnchompoo, B. and Thebtaranonth, Y. (1994). (+)-Jatrophol, (+)-marmesin, propacin and jatrophin from the roots of *Jatropha curcas* (Euphorbiaceae). *Journal Science Society Thailand*, 20, 73-83.

Naengchomnong, W., Thebtaranonth, Y., Wiriyachitra, P., Okamoto, K.T. and Clardy, J. (1986). Isolation and structure determination of four novel diterpenes. *Tetrahedron Letters*, 27, 2439-2442.

Nath, L.K. and Dutta S. K. (1992). Wound healing response of the proteolytic enzyme curcain. *Indian Journal Pharmacology*. 24, 114-115.

Neogi, U., Saumaya, R., Mishra, R.K. and Raju, K.C. (2008). Lipid content and *in vitro* antimicrobial activity of oil seeds of some Indian medicinal plants. *Current Research in Bacteriology*, 1, 1-6.

Ogbebor, O.N. and Adekunle, A.T. (2008). Inhibition of *Drechslera heveae* (Petch) M. B. Ellis, causal organism of Bird's eye spot disease of rubber (*Hevea brasiliensis* Muell Arg.) using plant extracts. *African Journal of General Agriculture*, 4, 19-26.

Okeola, F.O., Odebunmi, E.O. and Ameen, O.M. (2011). The conversion of the fruit pericarp and seed coat of *Jatropha curcas* to activate charcoal. *Advances in Biotechnology and Chemical Processes*, 1, 17-21.

Omale, J., Unekwojo, E.G. and Ojochenemi, A.A. (2011). Assessment of biological activities: A comparison of *Pergularia daemia* and *Jatropha curcas* leaf extracts. *British Biotechnology Journal*, 1, 85-100.

Onuh, M.O., Ohazurike, N.C. and Emeribe, E.O. (2008). Efficacy of *Jatropha curcas* leaf extract in the control of brown blotch disease of cowpea (*Vigna unguiculata*). *Biological Agriculture and Horticulture*, 25, 201-207.

Ovando-Medina, I., Albores-Flores, V., Toledo-García, M., Ruiz-González, S., Vázquez-Ovando, A., Jiménez-Escobedo, J.C., Adriano-Anaya, M.L. and Salvador-Figueroa, M. 2007. Producción de Biodiesel en El Soconusco, Chiapas a partir de Higuerilla (*Ricinus sp.*) y Piñón (*Jatropha sp.*) (Biodiesel production in Soconusco, Chiapas from Castor (Ricinus sp.) And pinion (Jatropha sp.)) Universidad Autónoma de Chiapas. Tuxtla Gutiérrez, Chiapas, Mexico.

Ovando-Medina, I., Espinosa-García, F.J., Nuñez-Farfan, J. and Salvador-Figueroa, M. (2011). Genetic variation in Mexican *Jatropha curcas* L. estimated with seed oil fatty acids. *Journal of Oleo Science*, 60, 301-311.

Özcan, A. and Safa, A. (2004). Comparison of supercritical fluid and Soxhlet extraction for the quantification of hidrocarbon from *Euphorbia macroclada*. *Talanta*, 64, 491-495.

Patel, V.C., Varughese, J., Krishnamoorthy, P.A., Jain, R.C., Singh, A.K. and Ramamoorty, M. (2008). Synthesis of alkyd resin from *Jatropha* and Rapeseed oils and their applications in electrical insulation. *Journal of Applied Polymer Science*, 107, 1724-1729.

Pax, F. (1910). Euphorbiaceae-Jatropheae. In: Das Pflanzenreich IV (The plant kingdm IV). 147(42) Engler, A. (ed.). pp. 1-148. Verlag von Wilhelm Engelmann, Leipzig.

Pertino, M., Schmeda-Hirschmann, G., Rodriguez, J.A. and Theoduloz, C. (2007). Gastroprotective effect and cytotoxicity of semisynthetic jatrpholone derivatives. *Planta Medica*, 73, 1095-1100.

Popluechai, S. (2010). Molecular Characterization of *Jatropha curcas*; Towards an understanding of its potential as a non-edible oilseed-based source of biodiesel. Thesis. Newcastle University. United Kingdom.

Prasad, L., Pradhan, S., Das, L.M. and Naik, S.N. (2012). Experimental assessment of toxic phorbol ester in oil, biodiesel and seed cake of *Jatropha curcas* and use of biodiesel in diesel engine. *Applied Energy*, 93, 245-250.

Rahman, M., Ahmad, S.H., Mohamed, M.T.M. and Rahman, M.Z.A. (2011). Extraction of *Jatropha curcas* fruits for antifungal activity against anthracnose (*Colletotrichum gloeosporioides*) of papaya. *African Journal of Biotechnology*, 10, 9796-9799.

Rahuman, A.A., Gopalakrishnan, G., Venkatesan, P. and Geetha, K. (2008). Larvicidal activity of some Euphorbiaceae plant extracts against *Aedes aegypti* and *Culex quinquefasciatus* (Diptera: Culicidae). *Parasitology Research*, 102, 867-873.

Ratnadass, A., Togola, M., Cissé, B. and Vassal, J.M. (2009). Potential of sorghum and physic nut (*Jatropha curcas*) for management of plant bugs (Hemiptera: Miridae) and cotton bollworm (*Helicoverpa armigera*) on cotton in an assisted trap-cropping strategy. *Journal of SAT Agricultural Research*, 7, 1-7.

Ravindranath, M., Ravinder Reddy, C., Ramesh, R., Ramu, A., Prabhakar, B., Jagadeesh, B. and Das, B. (2004). New lathyrane and podocarpane diterpenoids from *Jatropha curcas*. *Chemistry Pharmacy Bulletin*, 52, 608-611.

Roy, S. (1998). Biofuel production from *Jatropha curcas*. 10th European Conference and Technology Exhibition on Biomass for Energy and Industry. Wurzburg, Germany. June 8-11.

Ruiz-Bustos, E., Velázquez, C., Garibay-Escobar, A., García, Z., Plascencia-Jatomea, M., Cortez-Rocha, M.O., Hernández-Martínez, J. and Robles-Zepeda, R.E. (2009). Antibacterial and antifungal activities of some Mexican medicinal plants. *Journal of Medicinal Food*, 12, 1398-1402.

Sachdeva, K., Kumar, S.N., Manmohan, S., Birendra, S. and Charanjeet, S. (2011). Evaluation of analgesic and antimicrobial activity of stem bark extract of *Jatropha curcas*. *International Journal of Research in Ayurveda and Pharmacy* 2, 572-576.

Saetae, D. and Suntornsuk, W. (2010). Antifungal activities of ethanolic extract from *Jatropha curcas* seed cake. *Journal of Microbiology and Biotechnology*, 20, 319-324.

Sahidin Ardiansyah Muhammad, T. and Marianti, M. (2011). Terpenoids from the stem bark of *Jatropha* plants and their biological activities. *Makara Sains*, 15, 106-110.

Salvador-Figueroa, M., Ovando-Medina, I., Albores-Flores, V., López, M., Ruiz, S., Velázquez, G., Jarquín-Gálvez, R. and Adriano-Anaya, M.L. (2007). Desarrollo de tecnología para la producción sustentable de piñón (*Jatropha* spp.) como alternativa para obtener bioenérgeticos e incrementar los índices de desarrollo de los agricultores (Development of technology for the sustainable production of pinion (Jatropha spp.) As an alternative for bioenergy and increase the rates of development of farmers). Universidad Autónoma de Chiapas. Tuxtla Gutiérrez, Chiapas, México.

Sanis, J., Ravindran, R., Ramankutty, S.A., Gopalan, A.K.K., Nair, S.N., Kavillimakkil, A.K., Bandyopadhyay, A., Rawat, A.K.S. and Ghosh, S. (2012). *Jatropha curcas* (Linn) leaf extract a possible alternative for population control of *Rhipicephalus* (*Boophilus*) *annulatus*. *Asian Pacific Journal of Tropical Disease*, 1, 225-229.

Sharma, N. and Trivedi, P.C. (2002). Screening of leaf extracts of some plants for their nematicidal and fungicidal properties against *Meloidogyne incognita* and *Fusarium oxysporum*. *Asian Journal Experimental Science*, 16, 21-28.

Sharma, S., Verma, M., Prasad, R. and Yadav, D. (2011). Efficacy of non-edible oil seedcakes against termite (*Odontotermes obsesus*). *Journal of Scientific and Industrial Research*, 70, 1037-1041.

Shi H., Williams I., Sung H., Zhu H. and Min Z. (2005). Cytotoxic diterpenoids from the roots of *Euphorbia ebracteolata*. *Planta Medica*, 71, 349-354.

Sibi, G., Wadhavan, R., Singh, S., Shukla, A., Dhananjaya, K., Ravikumar, R.K. and Mallesha, H. (2013). Plant latex: A promising antifungal agent for post harvest disease control. *Pakistan Journal of Biological Sciences*, 16, 1737-1743.

Sigma-Aldrich (2013). Product information of phorbol 12-myristate, 13-acetate. http://www.sigmaaldrich.com/etc/medialib/docs/Sigma/Product_Information_Sheet/1/p8139pis.Par.0001.File.tmp/p8139pis.pdf (last accessed on 19.09.13).

Singh, R.N., Vyas, D. K., Srivastava, N.S.L. and Narra, M. (2008). SPRERI experience on holistic approach to utilize all parts of *Jatropha curcas* fruit for energy. *Renewable Energy*, 33, 1868-1873.

Siva, N., Ganesan, S., Banumathy, N. and Muthuchelian. (2008). Antifungal effect of leaf extract of some medicinal plants against *Fusarium oxysporum* causing wilt disease of *Solanum melogena* L. *Ethnobotanical Leaflets*, 12, 156-163.

Sricharoenchaikul, V., Pechyen, C., Ahtong, D. and Atong, D. (2008). Preparation and characterization of activated carbon from the pyrolysis of physic nut (*Jatropha curcas* L.) waste. *Energy and Fuels*, 22, 31-37.

Srinophakun, P., Tiapiwatanakun, B., Sooksathan, I. and Punsuvon, V. (2012). Prospect of deoiled *Jatropha curcas* seed cake as a fertilizer for vegetable crops. A case study. *Journal of Agricultural Science*, 4, 211-226.

Srivastava, S.K., Tewari, J.P. and Shukla, D.S. (2008). A folk dye from leaves and stem of *Jatropha curcas* L. used by Tharu tribes of Devipatan division. *Indian Journal of Traditional Knowledge*, 7, 77-78.

Steinmann, V.W. (2002). Diversidad y endemismo de la familia Euphorbeaceae en México (Diversity and endemism of the family in Mexico Euphorbeaceae). *Acta Botánica Mexicana*, 61, 61-93.

Thangavelu, R., Sundararaju, P. and Sathiamoorthy, S. (2004). Management of anthracnose disease of banana caused by *Colletotrichum musae* using plant extracts. *Journal of the Horticultural Science and Biotechnology*, 79, 664-668.

Tirapelli, C.R., Ambrosio, S.R., De Oliveira, A.M. and Tostes, R.C. (2010). Hypotensive action of naturally ocurrin diterpenes: A therapeutic promise for the treatment of hypertension. *Fitoterapia*, 81, 690-702.

Umeh, A. and Ndana, R.W. (2010). Effectiveness of *Jatropha curcas* and *Jatropha gossypiifolia* plant extracts in the control of *Meloidogyne incognita* on okra. *International Journal of Nematology*, 20, 226-229.

Vyas, D.K. and Singh, R.N. (2007). Feasibility study of *Jatropha* seed husk as an open core gasifier feedstock. *Renewable Energy*, 32, 512-517.

Wang, X.C., Zheng, Z.P., Gan, X.W. and Hu, L.H. (2009). Jatrophalactam, a novel diterpenoid lactam isolated from *Jatropha curcas*. *Organic Letters*, 11, 5522-5524.

Webster, G.L. (1994). Synopsis of the Genera and Suprageneric Taxa of Euphorbiaceae. *Annals of the Missouri Botanical Garden*, 81, 33-144.

Wei, Q., Liao, Y., Chen, Y., Wang, S.H., Xu, Y., Tang, L. and Chen, F. (2005). Isolation, characterisation and antifungal activity of β-1,3-glucanase from seeds of *Jatropha curcas*. *South African Journal of Botany*, 71, 95-99.

Wei, Q., Liao, Y., LiJun, Z., JinXia, Z., ShengHua, W. and Chen, F. (2004). Antifungal activity of curcin from *Jatropha curcas* seeds. *Chinese Journal of Oil Crop Sciences*, 26, 71-75.

Wink, M., Koschmieder, C., Sauerweien, M. and Sporer, F. (1997). Phorbol esters of *J. curcas*, Biological activities and potential applications. In: Gübitz, G.M., Mittelbach, M., Trabi, M. (Eds.), *Biofuels and Industrial Products from Jatropha curcas*. pp. 160-166. DBV Graz, Austria.

In: The Promising Future of *Jatropha curcas*
Editors: Maira R. Segura-Campos et al.

ISBN: 978-1-63484-989-0
© 2016 Nova Science Publishers, Inc.

Chapter 6

FUNCTIONAL POTENTIAL OF *JATROPHA CURCAS* L. IN PREVENTION AND TREATMENT OF NONCOMMUNICABLE DISEASES

Edwin Enrique Martínez Leo and Maira Rubi Segura-Campos*
Facultad de Ingeniería Química, Universidad Autónoma de Yucatán, Mérida, Yucatán, México

ABSTRACT

The Noncommunicable Diseases (NCD) are the principal cause of death worldwide (WHO, 2012). This all gathers a group of chronic-inflammatory diseases characterized by the cell redox status imbalance which takes to an oxidative stress potential and to an accumulation of cytokines and pro-inflammatory markers which aggravates these diseases and their complications (Johnson et al., 2012). The most common NCD are obesity, Diabetes Mellitus 2, cardiovascular diseases, dyslipidemia and cancer (WHO, 2012). The potential of developing new kind of food, whose nutritive contribution to diet takes to an additional therapeutic effect, is a new strategy for the prevention and treatment of NCD. Those active biological compounds with antioxidant and anti-inflammatory potential are an ideal alternative for the treatment of NCD by affecting two principal compounds of this group of diseases. Recently, several studies *in vitro* and *in vivo* have described the antioxidant and anti-inflammatory potential of several type of food. One of them would be *Jatropha curcas* L. (Oskoueian, et al., 2011). This activity is attributed to a high content of antioxidant compounds and its action by inhibiting pro-inflammatory prostaglandins biosynthesis (Othman et al., 2015). This is the reason why the obtaining of functional ingredients is a new and trendy alternative for consumers with a NCD and for food design and specific dietetic portfolios for the prevention and treatment of NCD and inflammatory kind.

Keywords: antioxidant, oxidative stress, inflammation

---

* Maira Rubi Segura Campos: Facultad de Ingeniería Química, Universidad Autónoma de Yucatán, Periférico Norte Km. 33.5, Tablaje Catastral 13615, Col. Chuburná de Hidalgo Inn, 97203, Mérida, Yucatán, México. Telephone: 52 999 946-09-56, fax: 52 999 946-09-94, email: maira.segura@correo.uady.mx.

NCD, A GREAT GLOBAL MULTIFACTORIAL PROBLEMS

During the last decades the lifestyle has suffered great modifications, changes in dietetic patterns, reduction of the physical work, sedentary lifestyle, increase of alcohol and tobacco consumption, stress and an increase of the consumption of high energetic density, simple sugars and saturated fats have been registered (WHO, 2011).

The NCD are the principal cause of death, out of the 57 million of deaths that occurred around the world in 2008, 36 million – almost 2/3 were caused by NCD, mainly caused by cardiovascular disease, cancer, diabetes and pulmonary chronic disease (WHO, 2012).

In 2011, the cardiovascular disease killed more than 17 million people around the world, and nowadays WHO estimates that during the following 10 years more than 20.7 million of deaths will occur due to the same reason, and 2.4 of these would be attributed to high blood pressure (WHO, 2012).

The NCD are catalogued as complex due to different factors which are involved in their etiology. The development of these diseases is associated to the imbalance of the oxidative systems and the antioxidant mechanisms in the body which takes to the production of big quantities of free radicals (FR) and the reduction of the neutralization speed of these radicals as it can be observed in Figure 1.

The imbalance of the oxidative-reductive reactions in our system leads to an increase of Reactive Oxygen Species (ROS), Reactive Oxygen and Nitrogen Species (RNS) and FR which lead to the restoration of oxidative stress when the antioxidant system is beaten. This can be seen by the metabolic and cellular disorders which lead to the development of a NCDS.

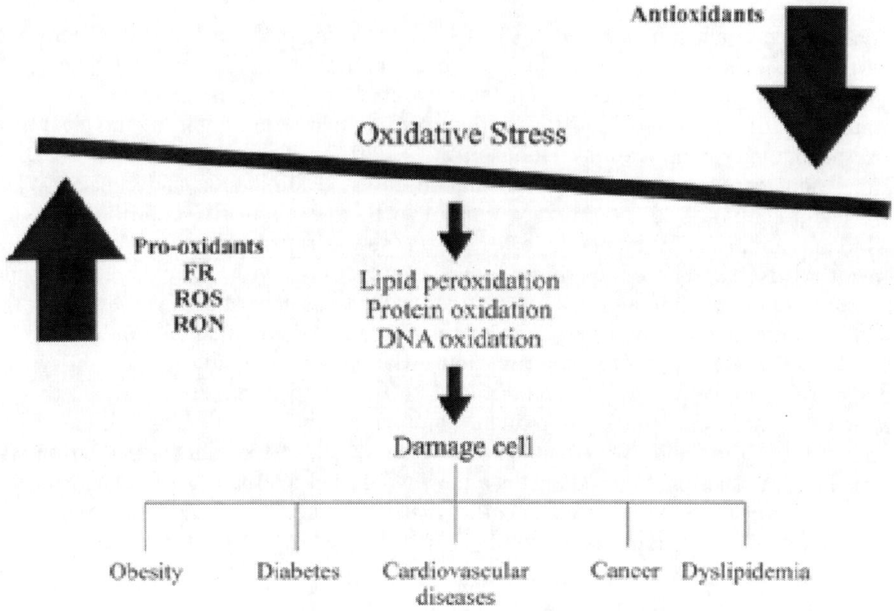

Figure 1. The oxidative stress as a producer of a Non Communicable Disease.

The elevation of the pro-oxidant systems before the antioxidant ones, leads to the instauration of oxidative stress characterized by the exacerbated production of FR and ROS, this leads to the development of oxidative damage caused by the oxidation of biomolecules and the reduction of biological functions (Barbosa et al., 2008).

The oxidative stress is a cell status where the intracellular redox homeostasis is altered. This imbalance is produced due to a great production of reactive species and the deficiency of antioxidant mechanisms (Lozada and García 2009).

The reactive species, where FR, ROS and RNS are included, are produced as a consequence of physiologic processes and play a very important role in the cell signal pathways, the gens transcription and the immune response. The ROS include superoxide radical (O_2^-), hydroxyl radical (OH^-), hydrogen peroxide (H_2O_2) and the hypochlorous acid ($HClO^-$). On the other hand, The RON refers to nitric oxide (NO), and the molecules derivated from it as nitrogen dioxide (NO_2) and peroxynitrite ($ONOO^-$). The RON have a life time longer than the ROS which makes them more harmful (Andresen et al., 2006).

The FR are atoms or group of atoms which have an unpaired or free electron, so they are very reactive and they tend to gain an electron from stable molecules so that they can reach their electrochemical stability. Once the FR has obtained the electron it needs, the stable molecule which gives the electron turns into a FR itself. This way a new chain reaction has begun and this will take in some time to cell damage (Andresen et al., 2006).

The abnormal increase of FR produces alterations in the mechanisms and cellular functions, which leads to different pathological responses. These responses depend on the type and quantity of FR involved, the biochemical cellular composition and specially, the cellular aptitude to fight against these aggressive elements. This can occur through several mechanisms, which are better kwon as the antioxidant defense system (Da Costa et al., 2012).

The mechanisms of oxidative damage induced by FR are the mainly: The activation of monocyte-macrophage system and neutrophils, the activation of intracellular mitochondria and the ischemia and reperfusion mechanism (Filomeni et al., 2015).

The O_2^- is produced by all kind of cells, however, in a patient with a NCD, the O_2^- is generated mainly by the activated polymorphonuclear neutrophils, due to the inflammatory processes which are an important characteristic of NCD (Filomeni et al., 2015).

O_2^- acts as a pro-inflammatory agent, by recruiting neutrophils, inducing the liberation of chemotactic factors and other pro-inflammatory mediators such as TNF-α and IL-1, and also generating damage on DNA and starting lipid peroxidation. On the other hand, H_2O does not have unpaired electrons; however, it is considered as a RNS due to its capability of inactivating enzymes and neutrophils, going through cellular membranes and using the Fenton reaction to produce OH^- by reacting with Fe and Cu atoms (Andresen et al., 2006).

The OH^- Radical is a strong FR, much more reactive than O_2. It produces damage mainly on DNA, proteins and lipids, inducing the lipidic peroxidation, and more damage on cellular membranes (Filomeni et al., 2015; Da Costa et al., 2012).

The lipid peroxidation is the most frequent oxidative damage on the polyunsaturated fatty acids from cell membranes. First, a fatty acid oxidizes due to the exit of a hydrogen atom from a metilenic group to OH- which acts as an oxidative agent producing a peroxyl radical. This radical repeats this process as a chain until the tissue damage is irreversible (Galicia and Gutiérrez, 2014).

INFLAMMATION AS RESPONSE TO OXIDATIVE AND TISSUE DAMAGE IN NCD

Inflammation is the response from a vascularized tissue to damage. It can be caused by microbial infections, physical or chemical agents, necrotic tissue o immune reactions. The irreversible cellular damage is one of the principal causes of inflammation and tissue damage in patients with a NCD (Sprung et al., 2015).

ATP consuming is present in this type of damage as a product of the FR aggression to mitochondrial membrane which takes to ischemia and cellular necrosis. This generates inflammatory processes on response to tissue damage (Richar et al., 2007).

Before a lesion, the inflammatory response generates a great accumulation of immune cells. Initially, the antigen presenting cells activate lymphocytes which proliferate and secrete molecules such as IL-2, IL-4, IFN-γ and several chemotactic factors. These induce the neutrophils, macrophages and lymphocytes' recruitment and activation. These, liberate to the environment highly reactive oxygen derivatives, lysozymes, prostaglandins, leukotrienes, and cytokines such as TNF-α, IL-1 and IL12 which amplify the inflammatory response with a special harmful potential for the surrounding tissues (Barros et al., 2011).

The TNF-α and IL-1induce the local expression and secretion of high quantities of IL-6, this increases the concentration of this cytokine in the systemic circulation. It has been described that TNF-α and IL-6 interfere directly with diverse mechanisms of cellular signal, insulin for instance. This leads to an abnormality on the glucose metabolism on the adipose tissue and skeleton muscle (Rotter et al., 2003; Solini et al., 2010).

Obesity is one of the principal diseases with an inflammatory origin and it unleashes other metabolic disorders such as insulin resistance, diabetes, and cardiovascular disease. Chronic systemic inflammation is associated to an accumulation of pro-inflammatory macrophages and cytokines on adipose tissue. This chronic systemic inflammation is related to the disorder of several mechanisms and signals which alter the correct cell function (Wensveen et al., 2015).

Several studies have demonstrated that inflammation and oxidative stress are presented in obese people (Olusi, 2002; Myara et al., 2003). Skrha et al., (2009) reported that the MDA quantity is 35% higher in obese people than non-obese ones. Davi et al., (2002) reported that PCR concentrations are 63% higher in obese people than in non-obese ones, also TNF-α and IL-6 de 1.67 and 1.78 vs 0.40 and 0.32 mg/L in comparison to non-obese ones.

The high levels of adipokines in obesity and the leukocyte infiltration increase ROS formation. This leads to a major oxidative stress status.

The TNF-α, IL-6 and PCR plasmatic concentrations are associated to adiposity level so, the more quantity of adipose cells, the more quantity of pro-inflammatory cytokines (Cao, 2014) (Figure 2).

Hyperglycemia, the metabolic increase of ROS production, the reduction of the antioxidant defenses, and the lipid oxidation are mechanisms which are presented in the physiopathology of people who live with obesity and they produce an increase of the inflammatory status producing more and worse complications and abnormalities on the metabolic pathways (Vincent and Taylor, 2006).

Diabetes is characterized by TNF-α, IL6, PCR high plasmatic concentrations and PAII (Plasminogen-1 activator inhibitor) (Nakanishi et al., 2008).

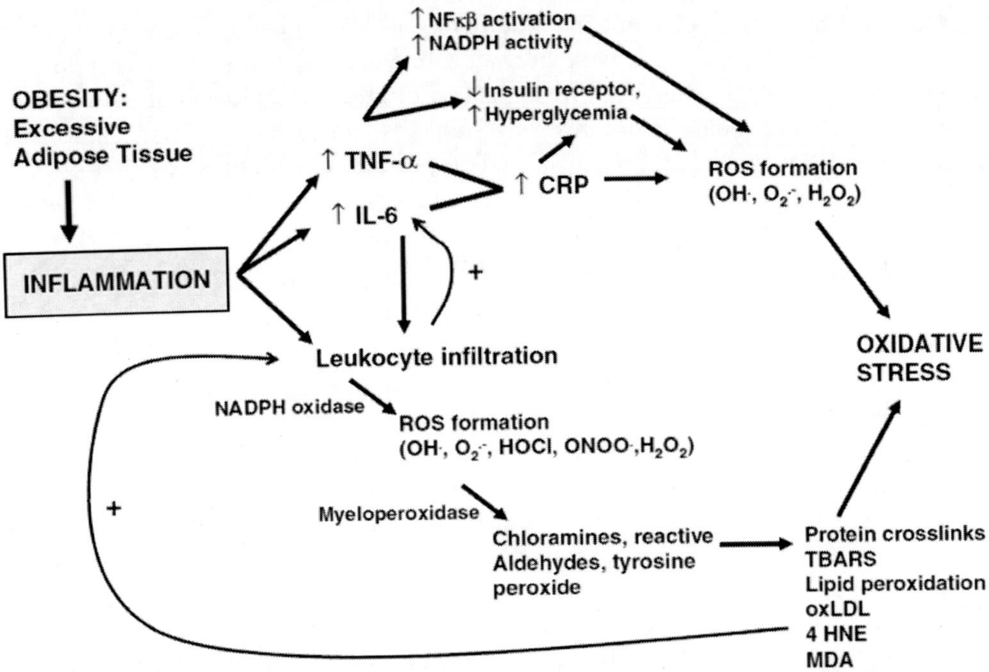

Adipokine levels increase, indirectly causing ROS formation via several intracellular signaling pathways and insulin receptor impairment. Leukocyte infiltration causes enzymatic formation of ROS. Both pathways generate ROS and oxidative damage (Vincent and Taylor, 2006).

Figure 2. Inflammatory pathways that induce oxidative stress in obesity.

Several studies confirm the theory that an inflammatory status is present in Diabetes Mellitus Type II (the type which is associated to obesity). Also there is clear evidence which points out that inflammation is an important process in these disease through the use of TNF-α, IL1, and IL6 (Miranda and Reza, 2008).

TNF-α elevated levels are associated directly to terminal kidney disease in patients with DM2, without any relation to clinical markers or protein presence (Niewczas et al., 2012).

In cardiovascular disease, diverse studies point out that the TNF-α, IL-6 and IFN-γ elevation produces a structural and functional alteration in endothelial cells. These alterations produce an increase of the oxidative stress and factors which induce cell necrosis and endothelial dysfunction.

The presence of chemotactic cytokines as VEGF (vascular endothelial growth factor), TGF-beta1 (Transforming Growing Factor Beta 1) and IL-8are important angiotensin regulators induced by inflammation and directly modulated by Nitric Oxide (Kofler et al., 2005).

Antioxidant Defense Mechanism

Inflammation and oxidative stress contribute the development of chronic disease and the additional production of reactive species, so, human system has antioxidant defense mechanisms to counter the attack of FR, RON and ROS.

The antioxidant gives an electron and oxidizes itself when it collides head on with a FR, transforming itself into a non-toxic weak free radical. In some cases (such as vitamin E) this weak free radical can regenerates itself by the action of other antioxidants. It is important to mention that not all antioxidants work this way. Some of them, called enzymatic antioxidants, can catalyze or accelerate chemical reactions and they also react with FR (Day, 2014).

A New Paradigm in the Nutritional Therapy: Functional Foods

With the beginning of the new century, the technological development and the genetic code deciphering, the food science researches took another point of view. Now, it is believed that the daily diet compounds do have an implication on the human genome.

Nowadays, the nutritional genomics have accepted the importance of the interaction between food and the physiological function of the system, and have also set the functional foods researches as a preventive and therapeutic action for the disease (Samaranayaka and Li-Chan, 2011).

The American National Science Academy defines as a functional food to any food or modified ingredient which independently of its nutritional contribution presents an additional physiologic benefit to the consumer. This way, a great variety of foods and their antioxidant, anticarcinogenic, hypoglycemic, antihypertensive, and hypolipidemic effects have been studied and documented as potential biological effects and also as a new therapeutic alternative for a disease (Herrera et al., 2014).

Nowadays functional foods can be found in diverse markets and supermarkets under a wide range of compounds. These compounds can be attributed to delaying or preventing disease effects. Some companies and industries such as Nestle®, Kellogs®, Nutrigem®, USANA®, offer to the consumer several products for different diseases such as diabetes, obesity, hypertension and cancer. Also, diverse foods have demonstrated in a big number of clinical, *in vitro* and *in vivo* studies, important biological functions in human system due to their chemical compounds.

The functional power of foods on health has an ancient origin. Along oriental culture history, food has been considered as important as medicine for the prevention and treatment of diseases. The relation food-medicine has been known since 1000 B.C. for Chinese culture (Drago et al., 2006). The therapeutic effect of food is due to the chemical constitution of them. These substances have a biological effect which is denominated as bioactive compounds. Several of these chemical substances with a biological activity are produced by the secondary metabolisms of plants, and are considered as non-nutritive compounds due to it does not contribute with any energetic contribution (Herrera et al., 2014).

All chemical substance contained in a food which gives a healthy benefit furthermore the basic nutritional considerations is known as a bioactive compound. These compounds are generally found in vegetal products and foods with a high content of polyunsaturated fatty acids and are contained in small quantities and this situation gives them a functional value. Nowadays one of the new global trends in nutrition is focused on the consumption of foods that more than having a nutritional content, have also extra benefits for health (Díaz et al., 2008).

The research of new agents and natural bioactive compounds with an antioxidant and anti-inflammatory potential have led to the development of new researches for the discovery of new therapeutic ways for chronic and inflammatory diseases.

The NCD have an important oxidative and inflammatory imbalance as etiology, so the dietetic use of antioxidant and anti-inflammatory foods is an important approach for the prevention and treatment of this high prevalence group of diseases.

Nowadays there are some internationally recognized alimentary schemes such as Mediterranean diet, for being the first on their kind to include food with antioxidant and anti-inflammatory effect and even more interesting there is clinical and epidemiological evidence of their beneficial effects on human health by reducing the risk for developing cardiovascular disease and some types of cancer (Sosa, 2010).

Jatropha curcas: A Potential Food for the Prevention and Treatment of NCD

There are plenty of recent studies which are related to *Jatropha curcas'* bioactive compounds where its therapeutic effects on the prevention and control of NCD have been described.

The *Jatropha L.* gender (Gr. *Iatros*: medicine; *trophe*: food) belongs to Plantae kingdom, Subreino Tracheobionta subkingdom, Magno liophyta division, Magnoliopsida class, Rosidae subclass, geraniales order, Euphor biaceae family, Crotonoideae subfamily and it was determined by Linneo (1753-1754) also including seven species, two of them included in *Cnidoscolus,* a segregated specie such as Manihot gender, and other specie referred to Aleurites Gender. The other 3 species are part of the *J. gossypifolia, J. multifida and J. curcas* gender (Font, 2003).

The *Jatropha* gender members are mostly little shrubs that present a great morphological variation. Another gender close to *Jatropha* is *Cnidoscolus,* and its species were part of *Jatropha* for a long period of time, however, they are currently considered as a different gender. This gender can be differentiated from *Jatropha* by the white-green flowers it has, and also by the itching stem, leaves and flowers (Toral et al., 2008) (Figure 3).

This gender is used as a plague control because of its insecticide and fungicide effects. There have been several studies for the medical effects in different species of this gender, specially for its effects on the treatment of skin infections, sexually transmitted infections, jaundice and fever (Pabón and Hernández, 2012).

In *J. curcas* leaves there have been identified metabolites of apigenin, vitexin, and isovitexin which can be used in the treatment of malaria, rheumatism and muscular pain. Latex is used as a disinfectant for buccal infections and it has been also established that it has compounds with anticarcinogenic effects such as jatrophina, jatropha, and curcaina (Pabón and Hernández, 2012).

An antimicrobial activity of *Jatropha* against *Staphylococcus aureus* and *Escherichia coli* has been observed (Pabón and Hernández, 2012).

Figure 3. *Jatropha curcas* seed and leaf.

Several studies have declared that *in vitro Jatropha curcas* have an antioxidant effect (Subroto et al., 2013), among these studies the most referred is the *in vitro* study in *J. curcas* extracts (Linn) where an antioxidant activity on the inhibition of DPPH and ABTS radicals is shown and this is associated to an important content of phenolic, flavonoids and anthocyanins compounds.

According to Igbinosa et al., (2011) and Lei et al., (2015) a higher DPPH radical inhibition percentage is present in methanolic extracts (91.5%), with a 80.5% watery extracts are found, and the ethanolic extracts with 78.2% in comparison with the activity presented by the BHT control with 98.5% in a 1.0 g/ml concentration.

Similar information is reported about the inhibition of ABTS radical of methanolic extracts (89%), watery extracts (86.6%) and ethanolic extracts (87.7%) in comparison to the BHT reference control (96.6%). This information clearly shows the *J. curcas*' natural antioxidant potential (Lei et al., 2015).

J. curcas also reports inhibitory activity on H_2O_2 radicals and superoxide anion. A 95.82%, 89.93% and 83.92% radical inhibitory percentage is reported in methanolic, watery and ethanolic extracts respectively in 1 mg/mL of *J. curcas* concentrations. The superoxide radical inhibition reports 85.69%, 80.29% and 79.23% respectively in 1 mg/mL concentrations (Igbinosa et al., 2011).

Gallegos et al. (2015) studied the antioxidant activity on the cellular line Caco-2 of *J. curcas* L. of hydrolyzed protein, which shows an antioxidant activity non-dependent of protein concentrations (100, 200, and 500 μg); however after the hydrolyzed protein was ultrafiltrated the fractions lower than 1 kDa presented a greater antioxidant power even with quantities of 50 μg of protein, so the oxidative stress was even 4 times lower in contrast with the ABAP positive control (2, 2'-Azobis (2-amidinopropane) dihydrochloride) (Gallegos et al., 2015).

According to Oskoueian et al. (2011), the antioxidant activity of *J. curcas* is also related to the anti-inflammatory potential. iNOS (Nitric Oxide Synthase) is inhibited in 92.3% in concentrations of 250 μM, which is an important enzyme in the inflammatory process generated by the immune system. It has a dose-dependent response in 264.7 RAW cellular line macrophages with a previous inflammation induction with LPS and IFN-γ.

The *J. curcas* watery extract in a dose of 150 mg/Kg shows an anti-inflammatory activity in an edema rat model. *J. curcas* extracts show a 60.7% inhibition of inflammation in comparison to the anti-inflammatory effect produced by aspirin (64.3%) and indomethacin (83.9%) in this edema rat model. Also the same concentration of *J. curcas* has an analgesic activity with a writhing inhibition effect of 39% in comparison to the positive control, acetaminophen (57.9%) (Olukunle et al., 2011).

These results show clearly the therapeutic activity of *J. curcas* extracts on the inflammatory origin disease. Another research of Baroroh et al., (2014) shows that *J. curcas* ethanolic extracts have an anti-arthritic effect evaluated in Wistar male rats with induced rheumatoid arthritis. The results showed that with a dose of 150 mg/kg, 300 mg/kg and 600 mg/kg inflammation can be reduced as much as the effect of diclofenac at the same doze (Figure 4).

The extract reduces inflammation, edema, and cartilage destruction on joints associated to a reduction of IL 1-B and TNFα levels, which are two cytokines implied in the development of rheumatoid arthritis.

On the other hand the inflammatory response is favored by the part of the plant which is used to produce the extract and also the time to produce it. According to Nayak (2010), the ethanolic extracts of *J. curcas* roots in comparison to the extracts produced from the stem and leaves present a higher inhibition percentage of cyclooxygenase (enzyme responsible of the prostaglandins, thromboxane and leukotrienes synthesis) without any significant difference with the diclofenac (Figure 5) with a dose-response curve that increases with time.

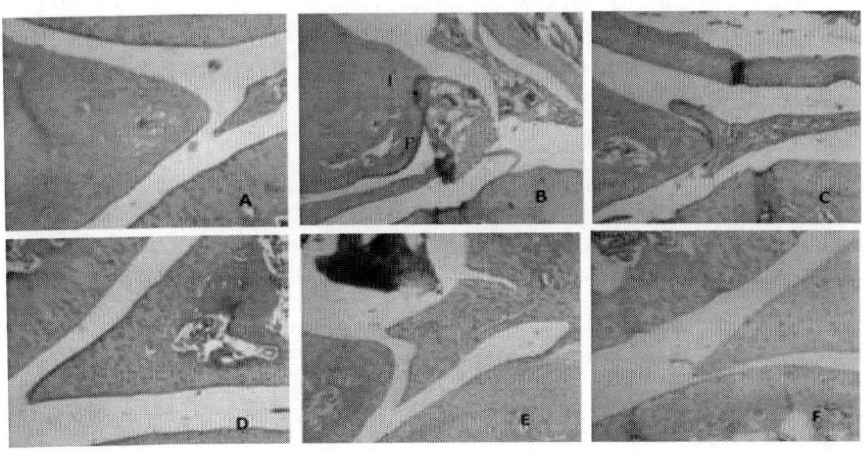

(A) normal rat (B) control rat (C) diclofenac (6.75 mg/kg) treated rat (D) J. curcas 150 mg/kg treated rat, (E) *J. curcas* 300 mg/kg treated rat, (F) *J. curcas* 600 mg/kg treated rat. Images (X 40 magnification) are typical and representative of each study group, (I) Section of joint cavity of arthritic rats showing inflammation and (P) proliferation with granulation tissue adjacent to the damaged articular cartilage. Section of joint cavity of rats on standard drug, *J. curcas* 150 mg/kg, 300 mg/kg and 600 mg/kg showing decreasing inflammation, edema in arthritic joints and cartilage destruction (Baroroh et al., 2014).

Figure 4. Histopathological representation of joints stained with H and E.

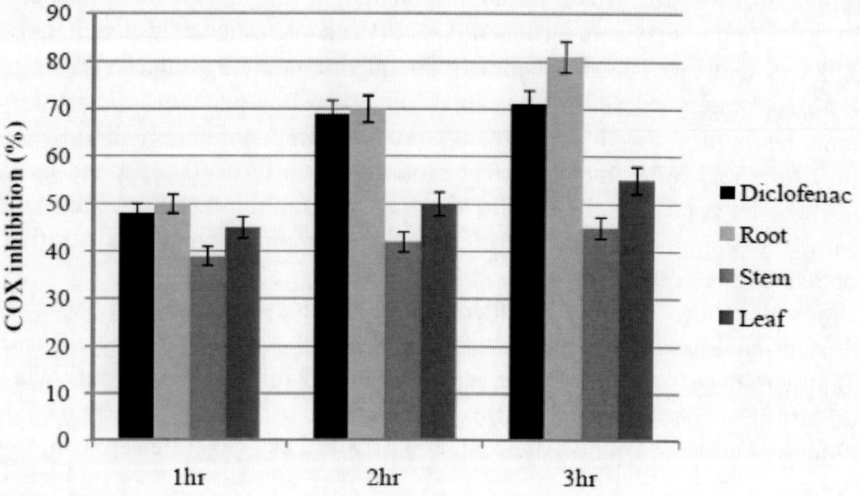

Nayak and Patel 2010.

Figure 5. Percentage COX inhibition in root, stem and leaf *Jatropha curcas*.

Conclusion

The development of researches for the discovery of new therapeutic trends for the prevention and control of NCD have allowed the exploration of the biological effect of diverse foods. *J. curcas* presents an important antioxidant and anti-inflammatory activity and it can be incorporated to nutritional schemes for people who live with a NCD for an integral treatment and a greater positive effects for their health.

Acknowledgments

The author would like to thank José Ceballos for the English revision of the paper.

References

Andresen, M., Regueir, T., Leighton, F. (2006). Oxidative stress in critically ill patients. *Revista Médica de Chile*, 134, 649-656.

Barbosa, K., Bressan, J., Martínez, J.A. (2008). Influencia de la dieta sobre marcadores plasmáticos de estrés oxidativo en humanos (Influence of diet on plasma markers of oxidative stress in humans). *Annales del Sistema Sanitario de Navarra*, 31, 259-280.

Baroroh, H., Sobri, I., Nur, E., Hertiani, T., Ikawati, Z. (2014). *Jatropha curcas* leaves exert anti-arthritic activity on adjuvant-induced arthritis in rats. *Universa Medicina*, 33, 3-10.

Barros, C., Kimiko, R., Machado, A., Gerola, L., Salomao, R. (2011). Citocinas y dolor (Cytokines and pain). *Revista Brasileña de Anestesiología*, 6, 137-142.

Cao, H. (2014). Adipocytokines in obesity and metabolic disease. *Journal of Endocrinology*, 220, T47-T59.

Da Costa, L., Badawi, A., El, A. (2012). Nutrigenética y modulación del estrés oxidativo (Nutrigenetics and modulating oxidative stress.) *Annales de Nutrición y Metabolismo*, 60, 27-36.

Davi, G., Guagnano, MT., Ciabattoni, G., Basili, S., Falco, A., Marinopiccoli, M. et al. (2002). Platelet activation in obese women: role of inflammation and oxidant stress. *Journal of the American Medical Association*, 288, 2008-2014.

Day, B. (2014). Antioxidant therapeutics: Pandora's box. *Free Radical Biologic Medicine*, 8, 58-64.

Díaz, LE., Duarte de Prato, A., Marcos, A. (2008). Alimentos nutracéuticos, suplementos dietéticos y plantas medicinales (nutraceuticals, dietary supplements and herbal). En: Alimentos Funcionales. *Aproximación a una nueva alimentación* (Functional Foods. A new approach to food), Ed. Dirección General de Salud Pública (Directorate General of Public Health), 46-75.

Drago, M., López, M., Saínz T. (2006). Componentes bioactivos de alimentos funcionales de origen vegetal (bioactive components of functional foods of plant origin). *Revistas científicas de América Latina*, 37, 58-68.

Filomeni, G., De Zio, D., Cecconil, F. (2015). Oxidative stress and autophagy: the clash between damage and metabolic needs. *Cell Death and Differentiation*, 22, 377-388.

Font, F. (2003). Las especies del género Jatropha L (Euphorbiacea, Crotonoideae) en Argentina (Species of the genus Jatropha L (Euphorbiacea, crotonoideae) in Argentina). *Revista del Círculo de Coleccionistas de Cactus y Crasas de la República Argentina*, 2, 4-20.

Galicia, M., Gutiérrez, G. (2014). Papel del estrés oxidativo en el desarrollo de la enfermedad hepática alcohólica (Role of oxidative stress in the development of alcoholic liver disease). *Revista de gastroenterología de México*, 79, 135-144.

Gallegos, S., Torres, C., Solorza, J., Alaiz, M., Girón, J., Martínez, A., Chel, L., Vioque, J. (2015). Antioxidant and Chelating Activity of Nontoxic *Jatropha curcas* L. Protein Hydrolysates Produced by *In Vitro* Digestion Using Pepsin and Pancreatin. *Journal of Chemistry*, 1-9.

Herrera, F., Betancur, D., Segura, M. (2014). Compuestos bioactivos de la dieta con potencial en la prevención de patologías relacionadas con sobrepeso y obesidad: péptidos biológicamente activos (Biologically active peptides: bioactive dietary compounds with potential in the prevention of pathologies related to overweight and obesity). *Nutrición Hospitalaria*, 29, 10-20.

Igbinosa, O., Igbinosa, I., Chigor, V., Uzunuigbe, O., Oyedemi, S., Odjadjare, E., Okoh, A., Igbinosa, E. (2011). Polyphenolic Contents and Antioxidant Potential of Stem Bark Extracts from *Jatropha curcas* (Linn). *International Journal of Molecular Sciences*, 12, 2958-2971.

Johnson, A.R., Milner, J.J., and Makowski, L. (2012). The inflammation highway: metabolism accelerates inflammatory traffic in obesity. *Immunological Reviews*, 249, 218-238.

Kofler, S., Nickel, T., and Weis, M. (2005). Role of cytokines in cardiovascular diseases: a focus on endothelial responses to inflammation. *Clinical Science*, 108, 205-213.

Lei, W., Zhang, W., Yin, L., Yan, F., Xu, Y., Chen, F. (2015). Extraction optimization of total triterpenoids from *Jatropha curcas* leaves using response surface methodology and evaluations of their antimicrobial and antioxidant capacities. *Electronic Journal of Biotechnology*, 18, 88-95.

Lozada, S., García L. Oxidative stress and antioxidants: how to keep the balance (Oxidative stress and antioxidants: how to keep the balance). *Revista de la Asociación Colombiana de Dermatología*, 17, 172-179.

Miranda, L., and Reza, A. (2008). Obesidad, inflamación y diabetes (Obesity, inflammation and diabetes). *Gaceta Médica de México*, 144, 39-45.

Myara, I., Alamowitch, C., Michel, O., Heudes, D., Bariety, J., Guy-Grand, B. et al. (2003). Lipoprotein oxidation and plasma vitamin E in nondiabetic normotensive obese patients. *Obesity Research*, 11, 112-120.

Nakanishi, S., Yamane, K., Ohishi, W., Nakashima, R., Yoneda, M., Nojima, H. et al. (2008). Manganese superoxide dismutase Ala16Val polymorphism is associated with the development of type 2 diabetes in Japanese-Americans. *Diabetes Research and Clinical Practice*, 81, 381-385.

Nayak, B., Patel, K. Anti-inflammatory screening of *Jatropha curcas* Root, stem and leaf in albino rats. *Journal Biology Plant*, 55, 9-13.

Niewczas, M., Gohda, T., and Skupien, J. (2012). Los receptores circulantes 1 y 2 de TNF ¿pueden predecir el desarrollo de enfermedad renal terminal en la diabetes mellitus tipo 2? (Circulating receptors 1 and 2 of TNF can they predict the development of end-stage renal disease in diabetes mellitus type 2?). *Nefrología clínica*, 23, 507-515.

Olukunle, J., Adenubi, O., Oladele, G., Sogebi, E., Oguntoke, P. (2011). Studies on the anti-inflammatory and analgesic properties of *Jatropha curcas* leaf extract. *Acta Vet. Brno*, 80, 259-262.

Olusi, SO. (2002). Obesity is an independent risk factor for plasma lipid peroxidation and depletion of erythrocyte cytoprotectic enzymes in humans. *International Journal of Obesity and Related Metabolic Disorders*, 26, 1159-1164.

Oskoueian, E., Abdullah, N., Saad, W., Ahmad, S. (2011). Antioxidant, anti-inflammatory and anticancer activities of methanolic extracts from *Jatropha curcas* Linn. *Journal of medicinal plant research*, 5, 49-57.

Othman, A., Abdullah, N., Ahmad, S., Ismail, I., Zakaria, M. (2015). Elucidation of in-vitro anti-inflammatory bioactive compounds isolated from *Jatropha curcas* L. plant root. *BMC Complementary and Alternative Medicine*, 15, 1-10.

Pabón, L., Hernández, P. (2012). Chemical importance of *Jatropha curcas* and its biological. *Revista Cubana de Plantas Medicinales*, 17, 194-209.

Richar, M. Vinay, K. Abul, A. Nelson F. (2007). Compendio de Robins y Cotran Patología estructural y funcional (Compendium of Robins and Cotran structural and functional pathology). 7ª ed. España. Ed. Elseiver. 3-57.

Rotter, V., Nagaev, I., and Smith, U. (2003). Interleukin-6 (IL-6) induces insulin resistance in 3T3–L1 adipocytes and is, like IL-8 and tumor necrosis factor-alpha, overexpressed in human fat cells from insulin-resistant subjects. *Journal of Biological Chemistry*, 278, 45777-45784.

Samaranayaka, A., and Li-Chan, E. (2011) Food derived peptidic antioxidants: A review of their production, assessment, and potential aplications. *Journal of Functional Foods*, 3, 229-254.

Skrha, J., Sindelka, G., Kvasnicka, J., Hilgertova, J. (2009). Insulin action and fibrinolysis influenced by vitamin E in obese type 2 diabetes mellitus. *Diabetes Research and Clinical Practice*, 44, 27-33.

Solini, A. Muscelli, E. Stignani, M. Melchiorri, L. Santini, E. Rossi, C. et al. (2010). Soluble human leukocyte antigen-g expression and glucose tolerance in subjects with different degrees of adiposity. *Journal of Clinical Endocrinology a Metabolism*, 95, 3342-3346.

Sosa, J. (2010). Efecto de la dieta DASH sola, en combinación con ejercicio físico y la pérdida de peso sobre la presión arterial y biomarcadores en hombres y mujeres con hipertensión (Effect of the DASH diet alone, in combination with exercise and weight loss on blood pressure and biomarkers in men and women with hypertension). *Anales de la Facultad de Medicina*, 71, 241-248.

Sprung, C., Ivashkevich, A., Forrester, H., Redon, C., Georgakilas, A., Martin, O. (2015). Oxidative DNA damage caused by inflammation may link to stress-induced non-targeted effects. *Cancer Letters*. 356. 72-81.

Subroto, E., Manurung, R., Jan Heeres, H., Broekhuis, A. (2013). Screening of antioxidants as stabilisers for *Jatropha curcas* L. *European Journal of Lipid Science and Technology*, 115, 909-920.

Toral, O., Iglesias, JM., Montes de Oca, S., Sotolongo, JA., Torsti, M. (2008). *Jatropha curcas* L., a tree species with energetic potential in Cuba. *Pastos y Forrajes*, 31, 191-207.

Vincent, HK., and Taylor, AG. (2006). Biomarkers and potential mechanisms of obesity-induced oxidant stress in humans. *International Journal of Obesity*, 30, 400-418.

Wensveen, F., Jelenčić, V., Valentić, S., Šestan, M., Wensveen, T., Theurich, S., Glasner, A., Mendrila, D., Štimac, D., Wunderlich, F., Brüning, T., Mandelboim, O., andPolić, B. (2015). NK cells link obesity-induced adipose stress to inflammation and insulin resistance. *Nature Immunology*.

WHO. (2011). Obesity, Diabethes and High Blood Pressure. Report of a WHO Consultation. *WHO Technical Report Series 894*. Ginebra: Organización Mundial de la Salud (Geneva: World Health Organization).

WHO. (2012). Obesity: preventing and managing the global epidemic. Report of a WHO Consultation. *WHO Technical Report Series 894*. Ginebra: Organización Mundial de la Salud (Geneva: World Health Organization).

In: The Promising Future of *Jatropha curcas*
Editors: Maira R. Segura-Campos et al.
ISBN: 978-1-63484-989-0
© 2016 Nova Science Publishers, Inc.

Chapter 7

BIOTECHNOLOGICAL APPLICATIONS OF PROTEIN HYDROLYSATES AND CURCIN FROM *JATROPHA CURCAS* L SEEDS

L. P. Jaramillo-Quintero[1,*] ***and A. L. Martínez-Ayala***[2,†]
[1]Universidad Autónoma de Tlaxcala, Facultad de Ciencias Básicas, Ingeniería y Tecnología, Calzada Apizaquito S/N, Apizaco, Tlaxcala, México
[2]Instituto Politécnico Nacional. Centro de Desarrollo de Productos Bióticos, Carretera Yautepec-Jojutla, Yautepec, Morelos, México

ABSTRACT

The extraction of oil from *Jatropha curcas* seeds for the production of biodiesel generates a protein-rich seed cake like by-product. The main protein fractions are globulin, glutelin and curcin, an albumin that is characteristic of this species. This composition facilitates the use of this plant for biotechnological applications of its proteins. In fact, the protein isolates from *J. curcas* are a good substrate for the production of hydrolysates with improved functional and nutritional properties. The peptides released upon enzymatic hydrolysis of *J. curcas* seeds have higher antioxidant and iron-chelating activity than the original hydrolysates. Several functional proteins and cyclic peptides with important biological activities, since plant protection until pharmaceutical properties, have been isolated from *Jatropha curcas*. Among them, curcin and curcacycline have shown potential toxicity against tumor cells, antimalarial activity, and inhibition of the classical activation pathway of human complement and the proliferation of human T-cells. In fact, cell lines exposed to curcin exhibited significant metabolic arrest by several mechanisms in a dose-dependent manner. The protein has been conjugated to nanoparticles or antibodies in order to enhance the targeting and the anti-tumor activity of curcin. Conjugated curcin has shown significant inhibitory effects on the proliferation of cancer cell lines that over-express the target receptor and had less of an inhibitory effect on cells with lower levels of target receptor expression. In this way, the conjugates open up the possibility of using this ribotoxin as a therapeutic against

* E-mail: pjaramillo4@hotmail.com.
† Corresponding author: A. L. Martínez-Ayala. E-mail address: alayala@ipn.mx.

cancers of diverse origins. It is expected that the information collated will open avenues for new applications of the proteins in the *J. curcas* plant, thereby enhancing the financial viability and sustainability of a *Jatropha*-based biodiesel industry.

Keywords: curcin, bioactive peptides, cancer, immunotoxin

INTRODUCTION

Since the energy-producing potential of *Jatropha curcas* L. was recognized because of its high oil production, its resistance to adverse environmental conditions and its condition of non-food culture, large plantations have been cultured in India, Indonesia, Thailand, China and some countries in America with the prime aim of producing biodiesel from the seed oil (King et al. 2009). Mechanical pressing or solvent extraction of the oil results in kernel meal (defatted shell-free kernel) or seed cake (screw-pressed whole seeds) by-products, which represent around 65% of the seed. This by-product could be used as livestock feed because of its high protein content (58–60%) and its good amino acid balance. However, the presence of antinutritional compounds such as trypsin inhibitors, phytate, saponins, curcin lectin and phorbol esters prevent its use in animal food. Trypsin inhibitors and lectins can be inactivated by heat treatments, whereas phytate is heat stable (Makkar et al. 1997; Martínez-Herrera et al. 2006). In addition to this potential use, the proteins present in the seed cake have several potential biotechnological applications (Makkar et al. 1997; Makkar et al. 2009).

Selje-Assmann et al. (2007) and Martinez Ayala et al. (2002) studied the storage proteins in extracted protein fractions from defatted kernel meals of different genotypes of *J. curcas.* Both reported that the glutelins, globulins, and albumins made up the majority of the total recovered protein, but with different contents of glutelins (56.9 vs. 39.8%) and globulins (27.4 vs. 44.4%), whereas prolamins and non-extracted residues were present in minor quantities. Selje-Assmann et al. (2007) found the same polyacrylamide gel electrophoresis band patterns for each of the protein fractions of the different genotypes of *Jatropha* studied (both nontoxic and toxic genotypes). Martinez Ayala et al. (2002) reported that the albumin fraction was composed of four main components with molecular masses less than 30 kDa, and the globulin fraction had six major bands: four between 30 and 70 kDa and two less than 20 kDa. The prolamins were fractionated into three bands, whereas glutelins consisted of a band of 33 and one of 27 kDa.

In 2012, Peralta-Flores et al. reported that the protein digestibility was highest for glutelins and globulins, with values of 81 and 80%, respectively. Electrophoretic analysis showed that globulins and glutelins exhibited similar polypeptide profiles. Electrophoresis patterns suggested a structural relationship among the 2S, 7S and 11S storage proteins from plant sources. According to the FAO/WHO reference, the protein fractions had acceptable levels of most of the essential amino acids for pre-school children and adults, but its globulins and glutelins were low in lysine and tryptophan (Martínez-Herrera et al. 2006). With this information, the authors concluded that the development of possible applications of protein fractions from *J. curcas* will require further interdisciplinary research, such as the functional and toxicological evaluation of these proteins, but so far, a description of its different fractions has been achieved.

Although the *Jatropha* cake by-product obtained after oil extraction is rich in proteins, it is not used as animal feed because of its toxicity, which is attributed to the presence of phorbol esters. However, nontoxic *J. curcas* genotypes that do not contain phorbol esters have been reported in Mexico (Makkar et al. 2008), where traditional dishes are prepared using the roasted fruits of these nontoxic genotypes.

With this background, Martínez et al. reported in 2012 the cholesterol content and protein quality of nontoxic kernel flour from *J. curcas* L. in rats. In this work, the experimental diets contained flour, flour-lysine (1%), flour-phytase (500 FTU) and two control diets: nitrogen-free and casein. The rats were fed for 28 days and the protein quality, including the protein efficiency ratio (PER), net protein ratio (NPR) and true digestibility (TD) of nontoxic genotype defatted flour was assessed using Wistar rats. The PER and NPR obtained values for diets with *J. curcas* flour were lower than those obtained for casein, and no statistical differences were found in TD. Regarding the cholesterol effect, there was a significant decrease in LDL cholesterol concentration with a diet of *J. curcas* meal with 1% lysine and a slight, but definite, increase in HDL-cholesterol concentration. This change could be considered beneficial in relation to its effect on cardiovascular disease.

It is evident that *J. curcas* flour is a valuable protein-rich material, which may be considered for incorporation in food and feed, as their PER, NPR and TD values were higher than those of other seed meals. Diets containing *J. curcas* flour did not cause any adverse effects, suggesting it to be safe as a food or feed component. It also should be noted that *J. curcas* seeds from nontoxic genotypes have been consumed by humans for centuries in some regions of Mexico. Heat treatment improves the quality of the *J. curcas* proteins, as is the case for many seeds, increasing their digestibility and absorption in addition to completely inactivating trypsin inhibitors and lectins that may have adverse health effects. The authors recommend heat treatment of *J. curcas* and the addition of lysine and phytase in order to achieve a better quality product for use of the flour for food. Supplementation with lysine provided a higher PE value. Finally, they suggest it would be appropriate to conduct further studies on the biological assessment of lysine and phytase supplemented *J. curcas* meal/flour, which is expected to increase the PER.

The production of protein isolates has been proposed as a method to reduce antinutritional and toxic factors (Devappa et al. 2008). Protein isolates can be further processed in order to produce protein hydrolysates that have better functional and nutritional properties. In addition, protein hydrolysates are also a source of bioactive peptides that may have beneficial biological effects (Vioque et al. 2000). Thus, bioactive peptides with antihypertensive, immunomodulatory, opioid, antioxidant, hypocholesterolemic, and metal chelating activities have been described (Gobbetti et al. 2002).

As a whole, *J. curcas* proteins can be hydrolyzed to get bioactive peptides with chemical and biological activities, or they can be used to obtain bioplastics as substitutes for petrochemical-based plastics. On the other hand, consumption of the toxic albumin curcin, present in seeds and leaves of this plant, can cause severe damage when it is consumed. The clinical symptoms include vomiting, diarrhea, abdominal pain and circulatory collapse, among others. Besides these harmful effects on individuals, purified curcin can inhibit translation in the reticulocyte lysate system and in *in vitro* cells, with the result of eliminate cells. This effect of curcin has been used to eliminate microorganism and mammalian cancer cells. Recently, interest has grown in the use of curcin in the development of antitumor drugs that selectively target tumor cells (Jaramillo-Quintero et al. 2015).

This chapter aims to review the potential applications of protein hydrolysates and curcin from *J. curcas* L seeds in order to increase functional or nutritional properties of the seed cake and moreover that may have pharmaceutical or biological effects.

Bioactive Peptides and Protein Hydrolysates from *Jatropha curcas* Seeds

Recently, cyclopolypeptides and related congeners have been shown to have potential pharmacological applications. Cyclic peptides isolated from latex, seeds, and roots of many plants possess various biological activities such as cytotoxic activity, immunosuppressive activity, antimalarial activity, vasorelaxant activity and cyclooxygenase, acetylcholine esterase, and tyrosinase inhibitory activities. *Jatropha* species have been shown to be a rich source of bioactive cyclic peptides, which contain 7–10 residues and a high proportion of hydrophobic amino acids (Devappa et al. 2010).

Mahafacyclin was purified from a CH_2Cl_2/MeOH extract of dried *Jatropha mahafalensis* latex as one major component (mahafacyclin A) and one minor component (mahafacyclin B). The amino acid composition of mahafacyclin A (MW 688.40) was determined to be cyclo-(Gly-Thr-Ile-Leu-Gly-Val-Phe), whereas the amino acid sequence of mahafacyclin B (MW 804.37) was determined to be cyclo-(Gly-Thr-Phe-Phe-Gly-Phe-Phe). Both compounds exhibited antimalarial activity against *Plasmodium falciparum* with IC_{50} values of 16 and 2.2 μM, respectively (Baraguey et al. 2000; Baraguey et al. 2001).

Labaditin and biobollein are cyclic peptides isolated from *Jatropha multifida*. Labaditin is a decapeptide with an amino acid sequence of cyclo-(Ala-Gly-Val-Trp-Thr-Val-Trp-Gly-Thr-Ile) and molecular weight of 1071.2 g/mol. Labditin selectively inhibited the classical pathway of human complement activation *in vitro*. Biobollein is a cyclic nonapeptide, with an amino acid sequence of cyclo-(Ala-Ser-Ile-Leu-Gly-Leu-Gly-Trp-Ala). The activity of both cyclic peptides was suggested to be based on interaction with C-1 of the human complement system. Both peptides bind to aggregated and antigen-bound IgG and, most likely, block the antibody Clq acceptor site. This property is restricted to IgG subclass IgG1. The immunomodulatory activity of these peptides could be of pharmaceutical importance (Altei et al. 2008; Pieters et al. 1999; Rudi 1993).

With respect to *Jatropha curcas,* several peptides with biological activities have been identified; for example, jatrophidin (Figure 1), an octapeptide was isolated from the latex of the trees. The latex was partitioned with ethyl acetate, fractionated on Sephadex G15, eluted in solid phase extraction, and purified by HPLC to obtain jatrophidin I. The amino acid analysis, mass spectroscopy, and 1D/2D nuclear magnetic resonance (NMR) studies demonstrated that jatrophidin I exists as two conformers of a cyclic structure (Gly-Trp-Leu-Asn-Leu-Leu-Gly-Pro) with the conformational equilibrium of proline residues between cis and trans forms, indicating that this peptide has more than one conformational state in solution. The isolates of jatrophidin I had weak antifungal activity against the strains of *Candida albicans, Candida krusei, Candida parapsilosis*, and *Cryptococcus neoformans* and moderate activity as an acetylcholinesterase inhibitor, when compared with the standard galantamine.

Curcacycline A (Figure 1) was isolated from the ethanolic extract of *J. curcas* latex. It is a cyclic octapeptide (MW 766.97). The amino acid sequence was determined to be cyclo-(Gly-Leu-Leu-Gly-Thr-Val-Leu-Leu). Curcacycline A displayed a moderate inhibition of (i) the classical activation pathway of human complement and (ii) the proliferation of human T-cells.

Curcacycline B (Figure 1) is a cyclic nonapeptide (MW 863) isolated from the latex of *J. curcas*. The amino acid sequence was found to be cyclo-(Leu-Gly-Ser-Pro-Ile-Leu-Leu-Gly-Ile). The absolute stereochemistry of the amino acids was shown to be in the "L" configuration. It contains mostly hydrophobic residues and one proline, thus differing from the previously isolated cyclic peptides from *Jatropha* latex species, which do not contain proline. The structure of curcacycline B was suggested to be a substrate for peptidylprolyl cis-trans isomerase (PPIase), as it has some structural features in common with cyclosporin A (inhibitor of cyclophilins A and B). Curcacycline B enhances PPlase activity by 60% at 30 μM based on a R-chymotrypsin rotamase-coupled enzymatic experiment using human cyclophilin B, whereas no modification of cyclophilin B activity was observed in the presence of curcacycline A. Curcacycline B from *J. curcas* possesses antimalarial activity (IC_{50} < 10 mM) against *P. falciparum*.

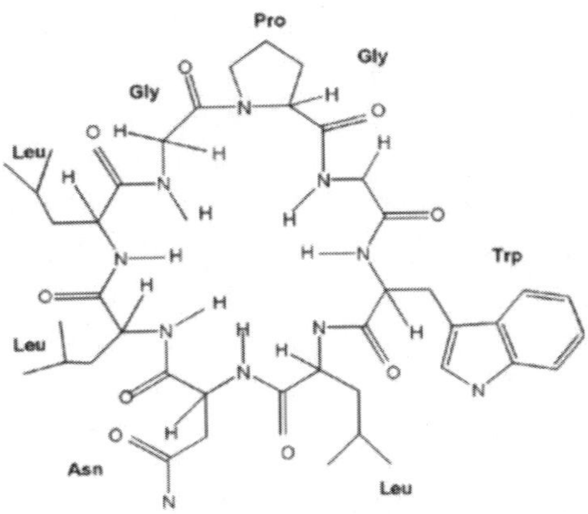

Jatrophidin

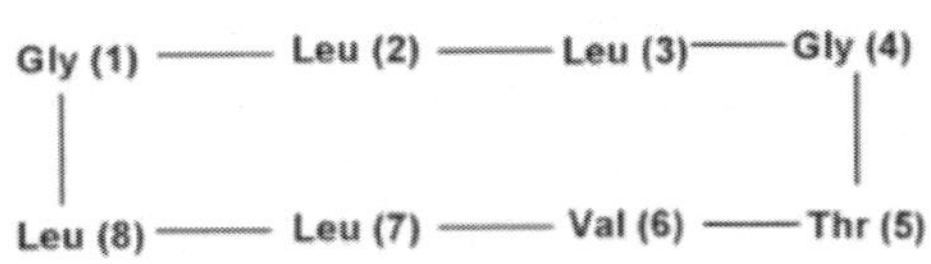

Curcacycline A

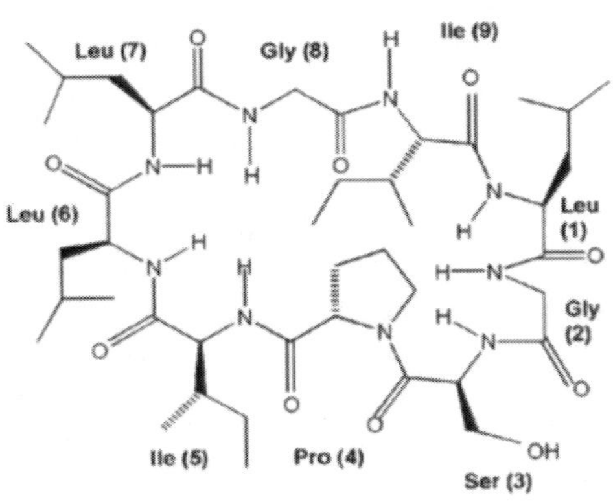

Curcacycline B

Figure 1. Structure of *Jatropha curcas* peptides.

With this background, Gallegos-Tintoré et al. (2011) evaluated the antioxidant and chelating activity of *Jatropha curcas* L. protein hydrolysates produced by treatment with the food-grade protease preparation alcalase. The goal was to revalorize this crop as a source of new, high-added-value bioactive peptides. They found that the 50 min protein hydrolysate, with a degree of hydrolysis of 31.7%, showed the highest antioxidant and chelating activity. These activities were also determined in six peptidic fractions that were separated by gel filtration chromatography of the 50 min hydrolysate. The lower molecular weight peptidic fractions had the highest antioxidant and chelating activities, which correlated with a higher content of antioxidant and chelating amino acids, such as tyrosine and histidine. Results showed that *J. curcas* represents a good source of bioactive peptides. This may be important for the revalorization of defatted *J. curcas* flour, a by-product resulting from oil extraction for biodiesel production. This is especially important in countries such as Mexico, because it would positively impact the cultivation of this plant in arid regions.

The chelating properties of bioactive peptides have been correlated to their antioxidant activity. Various metabolic pathways produce free radical species in living tissues. When these free radicals are generated in excess, or when cellular defenses are deficient, biomolecules may be damaged by oxidative stress, a phenomenon that seems to be involved in cellular aging, cardiovascular diseases, cancer, and neurodegenerative diseases. Oxidative reactions may also be responsible for alterations in foods. Thus, unsaturated fatty acids are prone to peroxidative degradation, resulting in adverse effects on nutritional and functional properties. Lipid peroxidation products may even have toxic effects on mammalian cells. Although copper and iron are essential cofactors for a variety of enzymes, their excess may also have deleterious prooxidant effects *in vivo*, by promoting the formation of reactive oxygen species. In addition, they also promote oxidative reactions in foods that have a negative impact on the flavor, texture, nutritive value, and shelf-life of food products. Chelating peptides may constitute a powerful tool for preventing these oxidative processes, both in foods and *in vivo*. Among chelating peptides, phosphopeptides derived from milk proteins are the best characterized. The positive effect of these peptides on the *in vivo* and *in vitro* absorption of minerals such as calcium, zinc, or iron has been reported.

In 2015, Gallegos et al. investigated the antioxidant and chelating properties of *J. curcas* protein hydrolysates obtained from a nontoxic genotype in order to determine their possible use for producing bioactive peptides that could be used for manufacturing functional foods. In addition, because pepsin and pancreatin are involved in gastrointestinal digestion in humans, the results might provide valuable information on the production of bioactive peptides with antioxidant and metal chelating properties, during physiological digestion of *J. curcas* proteins.

The hydrolysates were produced by treatment of a nontoxic genotype with the digestive enzymes pepsin and pancreatin *in vitro*, to mimic the physiological digestion of proteins, and then were characterized by fast protein liquid chromatography and reverse phase chromatography. Treatment with pepsin for one hour followed by treatment with pancreatin for two hours yielded the highest antioxidant and metal chelating properties. Peptidic fractions with higher radical scavenging activity were analyzed by matrix-assisted laser desorption/ionization mass spectrometry. The antioxidant activity was determined by measuring inhibition of the oxidative degradation of β-carotene and by measuring the reactive oxygen species (ROS) in Caco-2 cell cultures. Cu^{2+} and Fe^{2+} chelating activities were also determined. The hydrolysates inhibited the degradation of β-carotene and the formation of

ROS in Caco-2 cells. The lower molecular weight peptidic fractions from FPLC had stronger antioxidant activity in cell cultures compared with the hydrolysates, which correlated with a higher content of antioxidant and chelating amino acids. These fractions were characterized by a large presence of peptides with different molecular masses. The hydrolysates exhibited both Cu^{2+} and Fe^{2+} chelating activity. It was concluded that nontoxic *J. curcas* seed is a good source of naturally occurring antioxidant and metal chelating peptides of different molecular masses.

This information might increase the economic value of *J. curcas* defatted meal from nontoxic *J. curcas* genotypes, which could be used for producing functional food components and may have a positive impact on the economic value of this crop.

CURCIN AS A RIBOSOME-INACTIVATING PROTEIN AND ITS BIOLOGICAL EFFECTS

The toxic protein curcin that is found in *J. curcas* L. seeds was isolated, identified and named by Felke in 1914. Later, Schwarting and Stirpe associated this protein with ricin (from *Ricinus communis*) (Stirpe et al. 1976).

Subsequent comparison of amino acid sequences between curcin and ricin A-chain and trichosanthin (from *Trichosanthes kirilowii)* showed 54% and 57% identity respectively, which is a relatively high similarity among them. In fact, all amino acid residues forming the active sites of the A-chain of ricin and trichosanthin were found in curcin at corresponding positions, with only one exception; Gln-173 in ricin and thrichosanthin was replaced by Glu-163 in curcin (Lin et al. 2003b).

Since 1993, Barbieri classified the toxoalbumin curcin as a ribosome-inactivating protein (RIP) type I, because it consists of a single polypeptide chain of approximately 30 kDa. Although it was explored and analyzed from *J. curcas* seeds from China by Huang in 1991, it was not until 2003b that Lin et al. exposed the N-glycosidase activity of this protein when the endo-fragment was liberated in rat liver rRNAs treated with curcin and aniline (Lin et al. 2003b; Lin et al. 2010).

Curcin (Figure 2), a ribosome-inactivating protein (RIP), is an enzyme that belongs to the rRNA N-glycosidase family (EC 3.2.22) that is able to specifically and irreversibly inhibit protein synthesis. In fact, curcin exerts its toxic effect by depurinating a specific adenine residue (A_{4324} in the 28S rRNA of the 60S subunit of rat ribosomes) located in the universally conserved GAGA-tetraloop, also known as the α-sarcin/ricin loop, present in 23/26/28S rRNA. This results in the inability of the ribosome to bind elongation factor 2 (EF-2) in eukaryotes and elongation factor G (EF-G) in prokaryotes, and prevents polypeptide chain translocation from the aminoacyl to peptidyl region and therefore the elongation of the chain (Ferreras et al. 2011; Schrot et al. 2015). Although the N-glycosidase activity on A_{4324} in 28S rRNA is the main substrate of the RIPs, it is now known that they can release adenine from all adenine-containing polynucleotides, for which reason they are also known as polynucleotide:adenosine glycosidases (Barbieri et al. 1997). Besides the N-glycosidase activity, other activities associated with some RIPs are chitinase activity (Shih et al. 1997), topological activity on DNA (Huang et al. 1992), HIV integration inhibitory activity (Lee-

Huang et al. *1*995), superoxide dismutase activity (Li et al. 1997), DNase activity (Ruggiero et al. 2007) and lipase activity (Lombard et al. 2001: Morlon-Guyot et al. 2003).

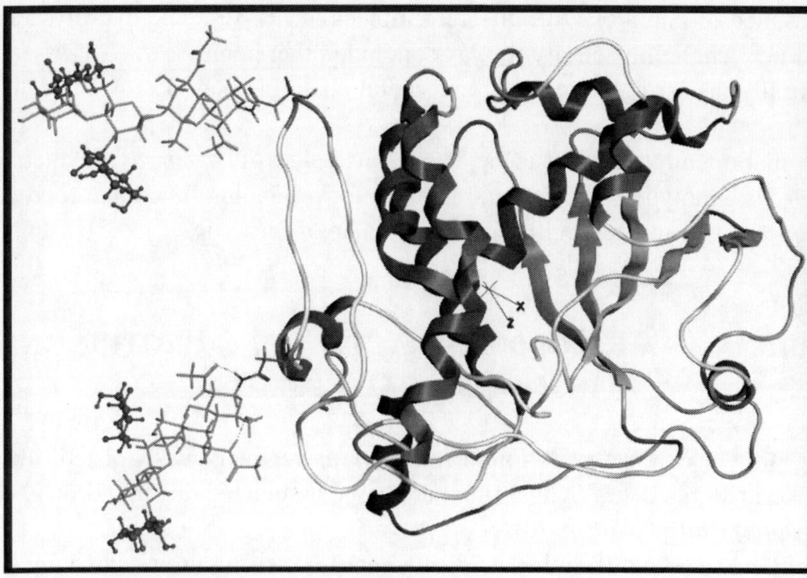

Figure 2. Glycosylated curcin (glycan GlcNAc2Man3Xyl at positions N-266 and N-274) (GlyProt).

Until now, no precise biological role has been assigned to RIPs, but the fact that they occur seasonally and can be induced by jasmonic acid or abscisic acid under stress conditions, such as unfavorable growing conditions, viral infections, microbial contamination or senescence, allowed researchers to postulate their involvement as a chemical defense mechanism against plant predators like fungi and viruses (Girbes et al. 2004; Stirpe 2004). Besides, some RIPs could play a role in plant senescence (Stirpe et al. 1996). The majority of the authors favor RIP investigation in agriculture because of its antiviral role and in anti-cancer therapy because of its possibility to eliminate malignant cells by apoptosis and necrosis (Puri et al. 2012).

It has been reported that curcin could be synthesized in two isoforms originating from two subfamilies: (a) in the endosperm of the seeds it is called curcin and has a mass of 28–29 kDa, or (b) in the leaves by stress conditions and microbes and called curcin 2. This second curcin was expressed in leaves of seedlings growing under simulated drought (polyethylene glycol treatment), temperature stress (4ºC or 50ºC) and fungal infection conditions (*Gibberelle zeae*, *Curvularia lunata* and *Pestalotia funerea*). The authors suggest that differential expression of curcin 2 might be due to the sensitivity or specificity of the leaves towards various treatments, and the leaves' different response to fungi species might be due to different resistances (Qin et al. 2005). With this information, Huang et al. (2008) investigated the biological significance of curcin 2 *in vivo*. Its sequence was introduced into the tobacco genome and the resulting resistance to viral and fungal infection of these transgenic plants was tested. They found that the T1 population of the transgenic lines showed an increased tolerance to tobacco mosaic virus (TMV) and the fungal pathogen *Rhizoctonia solani* by delaying the development of systemic symptoms of TMV and reducing the damage caused by the fungal disease. These tolerances corresponded to the curcin 2 level in the transgenic plants. Another isoform of curcin identified by Qin et al. (2010) and named curcin-L with a mass of 32 kDa, was expressed in the leaves and induced by certain conditions such as treatment with abscisic acid, salicylic acid and polyethylene glycol, exposure to 4° and 45°C, and fungal infection (*Gibberelle zeae, Pestalotia funerea).* Analysis of the 5' flanking regions of curcin and curcin-L revealed that the 5' flanking region of curcin-L has three major inserted fragments, which are not present in the corresponding region of the curcin. Comparison of characteristic cis-elements suggested the presence of several motifs that are involved in the endosperm-specific expression of the 5' flanking region of curcin, whereas in curcin-L some stress- and defense-responsive motifs are found to be mainly located in the three inserted fragments. Comparison of the antifungal activity of the two RIPs showed that the curcin-L antifungal activity is higher than that of curcin. Differences in the expression and activity of curcin and curcin-L suggest that these two RIPs have different functions (Qin et al. 2010).

Phylogenetic analysis of the predicted amino acid sequences of curcin and curcin-L in plants revealed that these belong to a major subfamily and are close to trichosanthin (TCS). Studies on the mRNA and protein levels showed that both curcin and curcin-L have an organ-specific expression pattern. Curcin is only expressed and accumulated in the endosperm; its expression begins in the globular embryo period and peaks during the mature embryo period. In contrast, curcin-L is only expressed in the leaves, but its expression is induced by certain conditions such as treatment with phytohormones or polyethylene glycol, exposure to high and low temperatures, and fungal infection. It is important to emphasize that the majority of the investigations concerning the biological activities of this RIP used curcin.

One of the first reports about the purification and biological activity of curcin was performed by Stirpe et al. (1976). They analyzed the effect of the crude extract and different fractions of *J. curcas* seeds on the protein synthesis in rabbit reticulocyte lysate and on toxicity in mice. They found that peak I was more toxic *in vivo*, whereas peak II was more active *in vitro*. They concluded that curcin enters cells with difficulty, because it lacks a carrier moiety or at least the galactose-binding groups by which ricin binds to cell membranes. This could be the reason that curcin is less toxic than ricin and abrin.

In 2003b, Lin et al. evaluated the effect of curcin and concluded that its activity is more potent than saporin (0.5 nmol/L), luffin A (1 nmol/L), luffin B (4 nmol/L) and trichosanthin (0.32 nmol/L), with an IC_{50} of 0.19 nmol/L for translation inhibition in the cell-free reticulocyte lysate system. Additionally, they investigated the sensitivity of a variety of cultured cancer cells and normal cells to curcin, to explore the application of curcin to the construction of immunotoxins. The IC_{50} values of curcin on SGC-7901, Sp2/0, and human hepatoma were 0.23 (0.15–0.32 mg/L), 0.66 (0.35–0.97 mg/L) and 3.16 (2.74–3.58 mg/L), respectively. Curcin also inhibited protein synthesis of HeLa cells and normal lung cells (MRC), but only at concentrations a thousand-fold higher than those in SGC-7901. Finally, they concluded that curcin had an obvious antitumor effect and its mechanisms were related to its N-glycosidase activity; this condition and the presence of one cysteine residue in their polypeptide chain make it a suitable molecule for the preparation of immunoconjugates with promising use as a chemotherapeutic agent for the treatment of various cancers.

This same group (2003a) cloned the coding region of curcin cDNA into the bacterial expression vector pQE-30, and the recombinant protein was expressed in *E. coli* strain M15. The purified recombinant curcin was able to inhibit protein synthesis in rabbit reticulocyte lysates with an IC_{50} of 0.32 nmol/L. Therefore, recombinant curcin was deemed a novel protein with potential application in biological, agricultural and medical fields, like chemotherapeutic agents as part of immunoconjugates for the treatment of various cancers or AIDS. The antitumor activity of this recombinant curcin was probed *in vitro* on tumor cell lines, such as cellular pulmonary cancer NCL-H446 cells and the gastric cancer cell lines SGC-7901 and S180. The refolded protein inhibited the growth of tumor cells at 5 μg/mL, and when the concentration was increased, the inhibitory effect was higher, with cells undergoing death by apoptosis; however, it had no effect on WISH cells or the vesicular stomatitis viruses. It was also evident that the maximal inhibition mediated by the recombinant protein was a higher than that of the native protein (Luo et al. 2006).

With this background, in 2010 Lin et al. described the purification of curcin to homogeneity, and studied its preliminary chemical, physical, and biological properties. They reported a pure curcin yield of 2.92 mg/100 g of seeds, with a molecular mass of 28.2 kDa, a pI of 8.54 as determined by isoelectric focusing. Schiff's reagent revealed that this was a glycoprotein with a total sugar content of 4.91%. Purified curcin had 251 amino acids with the following composition: Asx31, Thr15, Ser16, Glx22, Pro9, Gly15, Ala22, Val26, Met2, Ile14, Leu24, Tyr14, Phe12, Lys18, His2, Arg7, Cys1 and Trp1. The N-terminal thirty-two amino acid sequence of purified curcin was determined as A-G/Y-S/K-T/A-P/D-T-L-T-I-T-Y-D-A-T/A-A-D-K-K-N-Y-A-Q-F-I-K-D-LR-E-A-F/A-G. The N-terminal amino acid sequence of curcin compared with other RIPs showed significant homology; conserved amino acids were found in the N-terminal region with 19 of the RIPs reported to Genbank. Furthermore, the curcin amino acid residues Tyr-20(Y), Phe-23(F) and Arg-28(R) aligned perfectly with the consensus sequences found in the same positions for all of the examined RIPs. The secondary structure of curcin as analyzed by its Circular Dichroism (CD) spectrum showed that 22.3% is α-helix, 43.5% is β-sheet and 34.2% is random coil and corner. With respect to its biological activity, curcin was a hemagglutinin that can agglutinate rabbit erythrocytes at a concentration larger than 7.8 mg/L. The rRNA N-glycosidase activity of curcin was demonstrated on ribosomes with an endo-fragment of approximately 450 nt. The oral acute toxicity test showed a LD_{50} of 104.737 $\pm$ 29.447 mg/kg and the parenteral acute toxicity test showed a LD_{50} of 67.20 $\pm$ 10.445 mg/kg. Such results demonstrated that curcin was toxic according to the acute toxicity standards.

Although Luo et al. in 2006 showed that the morphology of the tumor cells treated with recombinant curcin resembled cells undergoing death by apoptosis, it was Zhao et al. (2012) who investigated the effect of curcin on apoptosis using mouse sarcoma-180 cells as a model. They found that after seven days of incubation, cell growth inhibition was over 40% at a curcin concentration of 100 μg/mL. Even on the third day, the inhibition rate was over 20% and curcin could significantly inhibit S-180 cell growth at concentrations as low as 10 μg/mL. In addition, they concluded that the incubation time was also an important factor in determining the extent of curcin-induced cell growth inhibition. They cultured cells treated with curcin and acridine orange and observed by fluorescence microscopy that the nucleus and cytoplasm were both yellow-green, which indicated that the chromatin had condensed, the nucleus had broken down, and apoptotic bodies had been released. Conversely, the cells of the control group had normal morphology and the nucleus exhibited uniform yellow-green fluorescence.

On the other hand, DanDan et al. (2012) probed the antitumor activity of a recombinant curcin 2 from the leaves of *J. curcas* and found that this protein had significant inhibition of proliferative effects on A549 (lung cancer) and HepG2 (liver cancer) tumor cells, with IC_{50}

values of 92.246 and 691.42 μg/mL, respectively, but had no significant inhibitory effects on HeLa cells (cervical cancer).

With all of this information about the toxic effects of curcin, Mohamed et al. (2014a) recently explored the toxicity of curcin against normal and cancer cell lines. They related morphological events to organelle degeneration and membrane protein disruptions during apoptosis. Cells exposed to curcin (100 μg/mL) for 72 h exhibited significant cellular metabolic arrest, with the cancer cell lines being more sensitive. In fact, fibroblast cells were the most resistant cells to curcin; the factor that could be responsible for this increased viability is their higher mitotic index as compared with the other cell types. The effect of curcin on cancer cells was profound, with glioma (brain cancer) registering below 20% viability at 100 μg/mL and about 65% at 1 μg/mL of the RIP. The mortality rate of MCF-7 (breast cancer) was nearly 70% at 100 μg/mL of curcin and around 45% at 10 μg/mL. The most resistant among the cancer cell lines, yet still susceptible, was the multi-drug resistant MDA-MB-453 (breast cancer) presenting around 45% viability at the highest and nearly 55% at the mid-concentration of curcin. The authors speculated that less sensitive MDA-MB-453, which is a triple negative multi-drug resistant variant, exhibited better drug efflux when compared to other cancer cell lines.

Besides the cytocidal effects, Mohamed reported that curcin-treated cell lines predominantly exhibited several morphological alterations similar to autophagy and apoptosis. The cells exhibited enhanced reactive oxygen species (ROS) production and mitochondrial damage, with ROS playing an additional role in the lipid peroxidation of erythrocytes. Loss of membrane integrity and consequentially the mitochondrial potential, elevated the NF-kB, disruption of the cytoskeletal scaffold leading to cellular collapse and decreased the anti-apoptotic survivin expression, which preserves the mitotic apparatus and allows normal mitotic progression, whereas its anti-apoptotic function is executed via its ability to prevent caspase activation. With curcin exposure, mitochondrial membrane potential is lost, leading to the possible release of caspases and cytochrome c. Evidently, mitochondria are crucial in the control of apoptosis, being involved not only in the intrinsic but also the extrinsic pathways.

The role of curcin as a cancer repressor was analyzed with the 3D colony assay. The 3D cancer mass regression and complete disruption by curcin provides concrete supplementary proof for its inevitable activity on malignant solid tumors, eliminating the core of the cancer mass and with it, the chance of future proliferation and recurrence. In fact, the cancer colonies failed to propagate in culture even after regular media supplementation (Mohamed et al. 2014a).

These authors project curcin as a highly cogent toxic RIP capable of multiple modes of action on shutting down cellular machinery, and propose it as a cytocidal phytomolecule in immunotoxins and in nanodrug delivery.

The cytotoxic effect and apoptosis of mammal cells are two aspects about curcin that open the possibility of its use as a pharmacological tool. Searching for carriers that can lead curcin to diseased cells instead of healthy cells is a field of study with high potential since this could enhance the targeting of the anti-tumor ability of curcin.

Zheng et al. published one of the first reports on this topic in 2013. They fused curcin and a transferrin receptor (TfR) binding peptide, codified as TfRBP9. The recombinant vector pQE-30-curcin-TfRBP9 was established and transferred into *E. coli* M15 and purified from the inclusion bodies. The specificity and cytotoxicity of the conjugate was probed on HepG2 (liver cancer), SKBR-3 (breast cancer) and LO-2 (normal liver) cells. Immuno-fluorescence, flow cytometry and the cell inhibition assay showed that TfRBP9 significantly enhanced the ability of curcin to bind the transferrin receptor (TfR) on HepG2 cells and consequently produce significant proliferation inhibition on this culture that over-expressed transferrin receptors. On the contrary, the proliferation of cells that expressed lower TfR (SKBR-3) or the lowest TfR (LO-2) was weaklier inhibited. Additionally, curcin-TfRBP9 induced higher apoptosis rates in the HepG2 cells than did curcin alone.

Mohamed et al. (2014b) developed a multifunctional conjugate that consisted of gold nanoparticles PEGylated and conjugated to folate, a transferrin antibody and curcin, to achieve a toxic dual targeted nanoformulation directed to gliomas (human brain glioblastoma). The objective was to combine chemo- and photo-thermal therapy against this type of cancer. The conjugate had pronounced specificity toward gliomas as compared with HCN-1A cells (human cortical neurons) and prominent toxic aspects of curcin were observed, such as ROS generation, mitochondrial and cytoskeletal destabilization, with 3% glioma viability after 48 h with 100 μg/mL of the conjugate and 90% HCN-1A cell viability under the same conditions. The photo-thermal ablation properties of gold nanoparticles completely disrupted the cancer colonies with significant precision.

With these results, Mohamed et al. (2014b) concluded that the dual mode of lethality achieved with the photo-thermal ablation combined with the effect of curcin proved extremely competent. They proposed this nano-conjugate system as a paragon for future cancer treatment strategies to restrict and destroy the cancer mass, negating the possibilities of future metastasis or recurrence.

In order to attack cancer cells that overexpress human epidermal growth factor 2 (Her2), a conjugate of curcin was constructed by chemically conjugating it with an antibody against Her2 (Jaramillo-Quintero et al. 2015). To this immunotoxin, reductive amination between curcin and a monoclonal antibody against Her2 involved the sugar residues on curcin, which were previously oxidized with $NaIO_4$, and the ε-amine group of the fragment crystallizable region (Fc) of the antibody. The *in vitro* cytotoxicity of curcin and the curcin:anti-Her2 conjugate was assessed on breast cancer cell lines SK-BR-3 ($Her2^+$) and MDA-MB-231 ($Her2^-$); the results clearly showed that curcin toxicity was not specific to these breast cancer cell lines with IC_{50} values of 15.5 ± 8.3 and 18.6 ± 2.4 μg/mL, respectively (Figure 3). However, the immunotoxin was more toxic to the SK-BR-3 cell line than to the MDA-MB-231 cell line, with IC_{50} values of 2.2 ± 0.08 and 147.6 ± 2.5 μg/mL respectively. These values showed that the immunotoxin was seven times more toxic to the SK-BR-3 cell line than curcin and eight times less toxic to the MDA-MB-231 cell line. According to these authors, the specificity of the immunotoxin curcin:anti-Her2 made it a therapeutic candidate against $Her2^+$ cancer, and the method of conjugation by reductive amination could be a rapid and "site-specific" method for constructing immunotoxins without compromising the active sites of the involved proteins, and in this way preserving the action of each protein in this hybrid molecule.

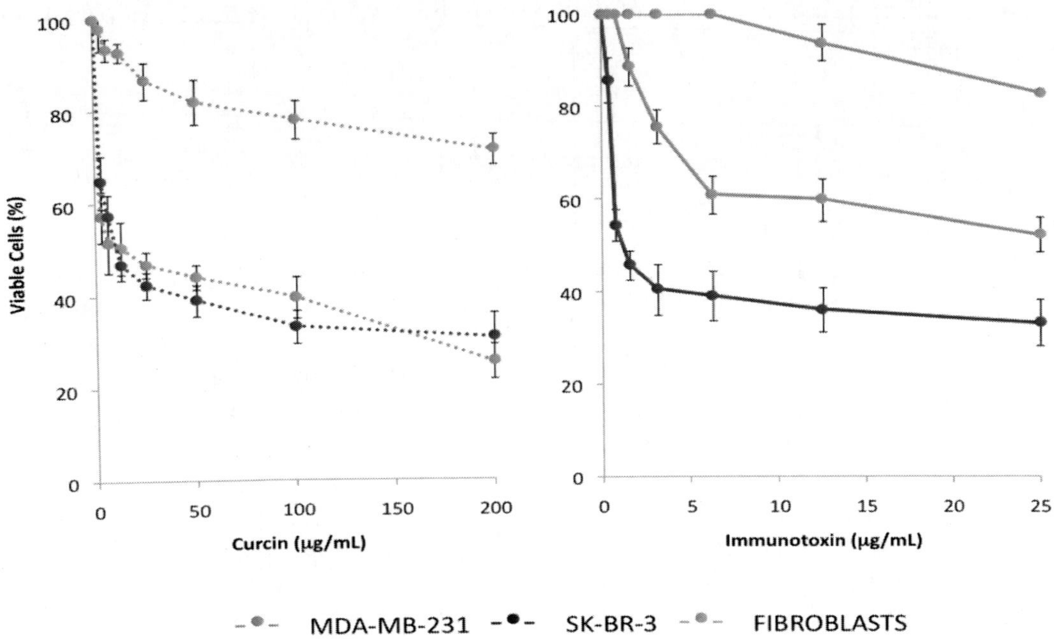

Figure 3. *In vitro* cytotoxic activity of curcin and the immunotoxin curcin:anti-Her2 on the cell lines SK-BR-3, MDA-MB-231 and FIBROBLASTS.

The antitumor properties of curcin on cancer cells by apoptosis make it a potential protein for pharmacological applications.

Another biological activity of curcin that has been analyzed by Kittikajhon et al. (2009) is its antimicrobial effect. The minimal inhibitory concentration values were determined against thirty-one strains of human-pathogenic microorganisms by the agar dilution technique. It was found that curcin could inhibit growth of 11 strains of bacteria, with the most potent inhibitory activity on *Staphylococcus epidermidis* ATCC12228 and *Bacillus subtilis* ATCC6633 with minimum inhibitory concentration at 78.1 mg/mL. Finally, the authors concluded that the use of the seed kernel as a natural source of anti-infective to both bacterial infections might be of interest.

Conclusion

The biofuel production of *Jatropha curcas* seed oil may potentially generate productive value-added products, such as proteins with nutritional, biochemical or pharmaceutical properties. From a nutritional perspective, *Jatropha curcas* proteins are comparable to soybean proteins. Its protein content, digestibility and amino acid composition make it a promising source of animal nutrition; however, a detoxification process, particularly of phorbol esters, would be required. The presence of bioactive proteins, particularly curcin, and the potential production of bioactive peptides with clinical efficacy from the *J. curcas* seed cake, opens the opportunity to research these proteins/peptides in the area of nutrition and pharmacy, beyond basic chemistry. It is important to look to genetic engineering, synthetic chemistry and chemical modification of these proteins to develop strategies for practical and effective applications.

References

Altei, W.F., Picchi, D.G., Barbosa, S.C., Cilli, E.M., Giannini, M.J., Cardoso-Lopes, E.M., Young, M.C.M., Torres, L.B., Giesel, G.M., Verli, H. and Bolzani, V.S. (2008) NMR studies, solid phase synthesis and MD/SA simulation as a tool for structural elucidation of new bioactive peptides from the latex of *Jatropha curcas* L. *Planta. Med., 74,* 1-338 (SL65) 7th Joint Meeting of the Association Francophone pour l'Enseignement et la Recherche en Pharmacognosie (AFERP), American Society of Pharmacognosy (ASP), Society for Medicinal Plant Research (GA), Phytochemical Society of Europe (PSE), and Societ Italiana di Fitochimica (SIF), Athens, Greece.

Barbieri, L., Valbonesi, P., Bonora, E., Gorini, P., Bolognesi, A. and Stirpe, F. (1997) Polynucleotide:adenosine glycosidase activity of ribosome-inactivating proteins: effect on DNA, RNA and poly(A). *Nucleic Acids Research, 25(3),* 518-522.

Baraguey, C., Blond, A., Cavelier, F., Pousset, J.L., Bodo, B. and Auvin-Guette, C. (2001) Isolation, structure and synthesis of mahafacyclin B, a cyclic heptapeptide from the latex of *Jatropha mahafalensis. J. Chem. Soc., Perkin Trans., 1,* 2098–2103.

Baraguey, C., Blond, A., Correia, I., Pousset, J.L., Bodo, B. and Auvin-Guette, C. (2000) Mahafacyclin A, a cyclic heptapeptide from *Jatropha mahafalensis* exhibiting β-bulge conformation. *Tetrahedron Lett., 41,* 325–329.

DanDan, X., Bo, Z., YaDong, H., QiHao, Z., ZhiJian, S., Hua, X. and Qing, Z. (2012) Expression of recombinant curcin in *Jatropha curcas* L and its anti-tumor activity *in vitro*. *Med. Plant., 3(8),* 68-71.

Devappa, R.K. and Swamylingappa, B. (2008) Biochemical and nutritional evaluation of Jatropha protein isolate prepared by steam injection heating for reduction of toxic and antinutritional factors. *Journal of the Science of Food and Agriculture, 88(5),* 911–919.

Felke J. (1914) The poisonous principles of the seeds of *Jatropha curcas* Linn. *Landw Versuchsw, 82,* 427-30.

Ferreras, J., Citores, L., Iglesias, R., Jiménez, P. and Girbés, T. (2011) Use of ribosome-inactivating proteins from Sambucus for the construction of immunotoxins and conjugates for cancer therapy. *Toxins, 3*, 420-441.

Gallegos-Tintoré, S., Torres-Fuentes, C., Martínez-Ayala, A.L. Solorza-Feria, J., Alaiz, M., Girón-Calle, J. and Vioque, J. (2011) Antioxidant and chelating activity of *Jatropha curcas* L. protein hydrolysates. *J Sci Food Agric, 91,* 1618-1624.

Gallegos, T.S., Torres, F.C., Solorza, F.J., Alaiz, M., Girón, C., J., Martínez, A.A.L, Chel, G.L. and Vioque, J. (2015) Antioxidant and chelating activity of nontoxic *Jatropha curcas* L. protein hydrolysates produced by *in vitro* digestion using pepsin and pancreatin. *J of Chemistry* Volume 2015, Article ID 190129, 9 pages http://dx.doi.org/10.1155/ 2015/190129.

Girbes, T., Ferreras, J.M., Arias, F.J. and Stirpe, F. (2004) Description, distribution, activity and phylogenetic relationship of ribosome-inactivating proteins in plants, fungi and bacteria. *Mini Rev. Med. Chem., 4*, 461–476.

Gobbetti, M., Stepaniak, L., DeAngelis, M., Corsetti, A. and Di Cagno, R. (2002) Latent bioactive peptides in milk proteins: proteolytic activation and significance in dairy processing. *Critical Reviews in Food Science and Nutrition, 42(3),* 223–239.

Huang, P.L., Chen, H.C., Kung, H.F., Huang, P.L., Huang, P., Huang, H.I. and Lee-Huang, S. (1992) Anti-HIV plant proteins catalyze topological changes of DNA into inactive forms. *Biofactors, 4,* 37–41.

Huang, M-X., Hou, P., Wei, Q., Hu, Y. and Chen, F. (2008) A ribosome-inactivating protein (curcin 2) induced from *Jatropha curcas* can reduce viral and fungal infection in transgenic tobacco. *Plant Growth Regul., 54,* 115-123.

King, J.A., He, W., Cuevas, A.J., Freudenberger, M., Ramiaramanana, D. and Graham, A.I. (2009) Potential of *Jatropha curcas* as a source of renewable oil and animal feed. *Journal of Experimental Botany, 60(10),* 2897–2905.

Kittikajhon, S., Roytrakul, S., Wetprasit, N. and Ratanapo, S. (2009) Purification of curcin from physic nut (*Jatropha curcas* Linn) and identification of its antimicrobial effect. Proceedings for the 2nd. Biochemistry and Molecular Biology Conference.

Jaramillo-Quintero, L.P., Contis, M.A., Romero, R.A., Rojas-Hernández, S., Campos-Rodríguez, R. and Martínez-Ayala, A.L. (2015) Cytotoxic effect of the immunotoxin constructed of the ribosome-inactivating protein curcin and the monoclonal antibody against Her2 receptor on tumor cells. *Bioscience, Biotechnology, and Biochemistry, 79(6),* 896-906.

Lee-Huang, S., Huang, P.L., Huang, P.L., Bourinbaiar, A.S., Chen, H.C. and Kung, H.F. (1995) Inhibition of the integrase of human immunodeficiency virus (HIV) type 1 by anti-HIV plant proteins MAP30 and GAP31. *Proc. Natl. Acad. Sci. USA, 92,* 8818–8822.

Li, X.D., Chen, W.F., Liu, W.Y. and Wang, G.H. (1997) Large-scale preparation of two new ribosome-inactivating proteins-cinnamomin and camphorin from the seeds of *Cinnamomum camphora*. *Protein Expr. Purif. 10,* 27–31.

Lin, J., Chen, Y., Xu, Y., Yan, F., Tang, L. and Chen, F. (2003a) Cloning and expression of curcin, a ribosome-inactivating protein from the seeds of *Jatropha curcas*. *Acta Botanica Sinica, 45(7),* 858-863.

Lin, J., Yan, F., Tang, L. and Chen, F. (2003b) Antitumor effects of curcin from seeds of *Jatropha curcas*. *Acta Pharmacol. Sin., 24 (3),* 241-246.

Lin, J., Zhou, X., Wang, J., Jiang, P. and Tang, K. (2010) Purification and characterization of curcin, a toxin lectin from the seed of *Jatropha curcas*. *Preparative Biochemistry and Biotechnology, 40(2),* 107-118.

Lombard, S., Helmy, M.E. and Pieroni, G. (2001) Lipolytic activity of ricin from *Ricinus sanguineus* and *Ricinus communis* on neutral lipids. *Biochem. J., 358,* 773–781.

Luo, M., Yang, X., Liu, W., Xu, Y., Huang, P., Yan, F. and Chen, F. (2006) Expression, purification and anti-tumor activity of curcin. *Acta Biochimica et Biophysica Sinica, 38(9),* 663-668.

Makkar, H.P.S. and Becker, K. (2009) Challenges and opportunities for using byproducts from the production of biodiesel from Jatropha oil as livestock feed. *Proceedings of Animal Nutrition Association World Conference.* (pp. 168-170) New Delhi, India.

Makkar, H.P.S., Becker, K., Sporer, F. and Wink, M. (1997) Studies on nutritive potential and toxic constituents of different provenances of *Jatropha curcas*. *J. Agric. Food Chem., 45,* 3152-3157.

Makkar, H.P.S., Martínez-Herrera, J. and Becker, K. (2008) Variations in seed number per fruit, seed physical parameters and contents of oil, protein and phorbol ester in toxic and non-toxic genotypes of *Jatropha curcas*. *Journal of Plant Sciences, 3 (4)*, 260–265.

Martinez-Ayala, A.L., Sánchez Rivera, M., Martínez Herrera, J. and Rodríguez Ambriz, S.L. (2002) Seed storage proteins from mexican piñoncillo. In Food Chemistry: Proteins; Annual Meeting and Food Expo, Anaheim, CA; available at http://ift.confex.com/ift/2002/techprogram/paper_14098.htm.

Martínez, H.J., Jimenez, M.C., Martínez, A.A., Garduño, S.L., Mora, E.R., Davila, O.G., Chamorro, C.G., Makkar, H., Francis, G. and Becker, K. (2012) Evaluation of the nutritional quality of nontoxic kernel flour from *Jatropha curcas* L. in rats. *J. Food Quality, 35,* 152-158.

Martínez-Herrera, J., Siddhuraju, P., Francis, G., Dávila-Ortíz, G. and Becker, K. (2006) Chemical composition, toxic/antimetabolic constituents, and effects of different treatments on their levels, in four provenances of *Jatropha curcas* L. From Mexico. *Food chem., 96,* 80–89.

Mohamed, M.S., Veeranarayanan, S., Minegishi, H., Sakamoto, Y., Shimane, Y., Nagaoka, Y., Aki, A., Poulose, A.C., Echigo, A., Yoshida, Y., Maekawa, T. and Kumar, D.S. (2014a) Cytological and subcellular response of cells exposed to the type-1 RIP curcin and its hemocompatibility analysis. *Scientific Reports, 4,* 5747 DOI: 10.1038/srep05747.

Mohamed, M.S., Veeranarayanan, S., Poulose, A., Nagaoka, Y., Minegishi, H., Yoshida, Y., Maekawa, T. and Kumar, D.S. (2014b) Type 1 ribotoxin-curcin conjugated biogenic gold nanoparticles for a multimodal therapeutic approach towards brain cancer. *Biochimica et Biophysica acta (BBA), 1840(6),* 1657-1669.

Morlon-Guyot, J., Helmy, M., Lombard-Frasca, S., Pignol, D., Pieroni, G. and Beaumelle, B. (2003) Identification of the ricin lipase site and implication in cytotoxicity. *J. Biol. Chem., 278,* 17006–17011.

Peralta-Flores, L., Gallegos-Tintoré, S., Solorza-Feria, J., Dávila-Ortíz, G., Chel-Guerrero, L. and Martínez-Ayala, A. (2012) Biochemical evaluation of protein fractions from physic nut (*Jatropha curcas* L.). *Grasas y Aceites, 63(3),* 253-259.

Pieters, L.A.C., Bruyne, T.E.D. and Vlietinck, A.J. (1999) Low molecular weight compounds with complement activity. In Immunomodulatory Agents from Plants; Series: Progress in Inflammation Research; Wagner, Ed.; Birkh€auser Verlag: Basel, Switzerland, 137-160.

Puri, M., Kaur, I., Perugini, M.A. and Gupta, R.C. (2012) Ribosome-inactivating proteins: current status and biomedical applications. *Drug Discovery Today, 17,* 774-783.

Qin, W., Ming-Xing, H., M., Ying, X., Xin-Shen, Z. and Fang, C. (2005) Expression of a ribosome inactivating protein (curcin 2) in *Jatropha curcas* is induced by stress. *J. Biosci, 30,* 351-357.

Qin, X., Shao, C., Hou, P., Gao, J., Lei, N., Jiang, L., Ye, S., Gou, C., Luo, S., Zheng, X., Gu, X., Zhu, X., Xu, Y. and Chen, F. (2010) Different functions and expression profiles of curcin and curcin-L in *Jatropha curcas* L. *Z Naturforsch C., 65(5-6),* 355-362.

Rudi, P.L. (1993) Immunomodulatory compounds. In Bioactive Natural Products: Detection, Isolation, And Structure Determination; Steven, M.C., Russel, J.M., Eds.; CRC Press: London, U.K., 280-317.

Ruggiero, A., Chambery, A., Di Maro, A., Mastroianni, A., Parente, A. and Berisio, R. (2007) Crystallization and preliminary X-ray diffraction analysis of PD-L1, a highly glycosylated ribosome inactivating protein with DNase activity. *Protein Pept. Lett., 14*, 407–409.

Schrot, J., Weng, A. and Melzig, M.F. (2015) Ribosome-inactivating and related proteins. *Toxins, 7,* 1556-1615.

Schwarting, A.E. (1963) Poisonous seeds and fruits. *Prog. Chem. Toxicol.* 1, 385-401.

Selje-Assmann, N., Makkar, H.P.S., Hoffmann, E.M., Francis, G. and Becker, K. (2007) Quantitative and qualitative analyses of seed storage proteins from toxic and non-toxic varieties of *Jatropha curcas L.* International Symposium on Energy and Protein Metabolism, European Association for Animal Nutrition (Publication No. 124), Vichy, France; p 625, available at http://tinyurl.com/37wcfn4.

Shih, N., McDonald, K., Jackman, A., Girbés, T. and Iglesias, R. (1997) Bifunctional plant defense enzymes with chitinase and ribosome inactivating activities from *Trichosanthes kirilowii* cell cultures. *Plant Sci., 130*, 145–150.

Stirpe, F. (2004) Ribosome-inactivating proteins. *Toxicon, 44*, 371–383.

Stirpe, F., Barbieri, L., Gorini, P., Valbonesi, P., Bolognesi, A. and Polito, L. (1996) Activities associated with the presence of ribosome-inactivating proteins increase in senescent and stressed leaves. *FEBS Lett., 382*, 309–312.

Stirpe, F., Pession-Brizzi, A., Lorenzoni, E., Strocchi, P., Montanaro, L. and Sperti, S. (1976) Studies on the protein from seeds of *Croton tiglium* and of *Jatropha curcas. Biochem J., 156*, 1-6.

Vioque, J., Sánchez-Vioque, R., Clemente, A., Pedroche, J., Yust, M.M. and Millán, F. (2000) Péptidos bioactivos en proteínas de reserva. *Grasas y Aceites, 51(5),* 361–365.

Zhao, Q., Wang. W., Wang, Y., Xu, Y. and Chen, F. (2012) The effect of curcin from *Jatropha curcas* on apoptosis of sarcoma-180 cells. *Fitoterapia, 83,* 849-852.

Zheng, Q., Xiong, Y., Su, Z., Zhang, Q., Dai, X., Li, L., Xiao, X. and Huang, Y. (2013) Expression of curcin-transferrin receptor binding peptide fusion protein and its anti-tumor activity. *Protein Expression and Purification, 89,* 181-188.

In: The Promising Future of *Jatropha curcas*
Editors: Maira R. Segura-Campos et al.
ISBN: 978-1-63484-989-0
© 2016 Nova Science Publishers, Inc.

Chapter 8

MORPHOANATOMY AND ONTOGENY OF THE FRUIT AND SEEDS OF *JATROPHA CURCAS* L.: A PROMISING BIOFUEL PLANT

Natália Corte-Real[1], Laurício Endres[2], Karolyne P. O. Santos[2], Regina C. B. Figueirêdo[3], Emília C. P. Arruda[4], Cláudia Ulisses[5] and Marcelo. F. Pompelli[1,*]

[1]Plant Ecophysiology Laboratory,
Federal University of Pernambuco, Recife, Brazil
[2]Plant Physiology Laboratory,
Federal University of Alagoas, Maceió, Brazil
[3]Aggeu Magalhães Research Center/FIOCRUZ,
Department of Microbiology, Recife, Brazil
[4]Plant Anatomy Laboratory,
Federal University of Pernambuco, Recife, Brazil
[5]Plant Tissue Culture Laboratory,
Federal Rural University of Pernambuco, Recife, Brazil

ABSTRACT

The ripening physiology of *Jatropha curcas* fruits involves the production of photoassimilates in its formation and development, until the oil body becomes present in its seeds. The anatomy of fruits and seeds aids in understanding the main structures and compounds produced during ontogeny of this species. Thus, the purpose of this chapter is to describe the evolution of the development and main structures formed by *J. curcas* seeds and fruits four days after anthesis (DAA) to the completely ripe fruit at 58 DAA and to provide evidence that the seeds of this species are not waterproof and therefore dormant, as has widely described in the literature.

---

* Corresponding author: M. F. Pompelli. Tel: +55 81 2126-8844, fax: +55 81 2126-7803. E-mail address: mfpompelli@gmail.com.

In in this chapter, we describe that *J. curcas* fruits reach their maximum volume approximately 47 DAA and then reduce this volume from 55 DAA when they become dried and darkened. There is a major presence of lactiferous ducts over the fruit mesocarp, which may be or not branched or articulated. The seeds become fully ripe at 13 DAA, and the presence of malformation or abortion at one or even two ovules is relatively common, leading to a reduction of seeds in the fully ripened *J. curcas* fruit. At 18 DAA, the endosperm begins to be consumed as a consequence of increased embryo size, which is fully developed after 40 DAA with evidence of a hypocotyl-radicle axis, in addition to the cotyledons, primordial stem and root. The seed integument is composed of testa and tegmen, and each is clearly subdivided. The outer layer of testa has macrosclereids, fibers and lignified cells. However, the existence of some cells without lignification was observed along the entire seed surface. The macrosclereids and fibers present in the seminal integument are responsible for the high rigidity found in *J. curcas* seeds; however it seems that these characteristics do not affect the processes of seed imbibition and germination. Thus, from a reproductive point of view, these aspects along with those found in the literature, are essential to the adaptive and competitive success of this species in the different environmental conditions in which *J. curcas* is native and widely cultivated.

INTRODUCTION

In the last few decades, *Jatropha curcas* L. has become popular thanks to its broad capabilities and plethora of uses, especially in biodiesel production (Fairless, 2007). The seed oil is composed of approximately 73% unsaturated fatty acids (Pramanik, 2003) and is of commercial importance in *J. curcas*. The amount of oil in the seeds varies widely ranging from 17.5 to 41.6% (Pompelli et al., 2010, Sunil et al., 2013), resulting in a biodiesel yield up to 86% and is preferred by biodiesel industries. This oil can be easily extracted and used both for biodiesel production and as cooking/lighting fuel, medicine, a bio-pesticide, and for soap production. Additionally, the seed cake, a by-product of oil extraction by-product, can be used as organic fertilizer, as combustible fuel, or for biogas production (Arroyo et al., 2014). Along with these multiple uses, the expectations surrounding *J. curcas* are derived from the amazing adaptability of this succulent perennial tree, which is a drought avoidant that can thrive even on low-nutrient soils and under arid and semiarid conditions (Achten et al., 2010). The capability to grow on poor quality soils allows *J. curcas* not directly compete against food crops (Pompelli et al., 2011). Furthermore, this plant itself offers an ecological advantage of mitigating soil degradation and reclaiming marginal land or abandoned farmland; it can also be used as live-fencing as it acts like a livestock/fire barrier to protect fields (Kumar and Sharma, 2008).

Because seeds are the main commercial product of *J. curcas*, the formation of this structure should be studied from a botanical, biochemical and anatomical point of view. The heterogeneity of the Euphorbiaceae species makes it highly diversified, mainly due to the great variability in their vegetative and floral characteristics. As a result, the taxonomy of the family should not be based only on morphological characteristics, they often change in response to environmental stresses. It is known that the fruit and its quality are strongly influenced by photosynthetic activity, and these structures, along with the developing seeds, have a great ability to influence the production of assimilates during photosynthesis (Jones and Miller, 1992). In this sense, anatomical and embryological studies are very important tools for systematic ecophysiological studies of the Euphorbiaceae species.

Generally, the characterization of fruit and seeds are based on several morphological and anatomical characteristics, such as the consistency and the number of integuments; the location of the mechanical layer; the chemical composition; the classification of endosperm; the presence of appendices and other structures; and the size, color, shape and curvature of the embryo. Because most of these features cannot be easily observed in fruits and seeds when ripe, ontogenetic studies are necessary to complete the morphoanatomical description of the fruit and seeds. Ontogenetic and histochemical studies are commonly used to study morphological and anatomical patterns among taxa, to indicate organs specialization, and to provide a secure foundation for the understanding of important structures and compounds involved in species evolution. Furthermore, anatomical studies can provide information about chemical composition and can define integument characteristics involved in seed dormancy. Because the seed imbibition is the first and most critical step to successful reproduction of the species, the knowledge about the above mentioned characteristic can guide phytotechnical practices for *J.curcas* handling.

In this context, our aim in this chapter was to describe the main structures formed on *J. curcas* fruit and seeds during ontogeny, to define the major events that occur in each development step, and to identify the main chemical composition of the mature seeds. Moreover, we attempted to discuss the widespread misconception in the literature that describes *J. curcas* as waterproof, a phenomena that prevents seed germination and is characterized as a tegumentar dormancy.

FLOWER CHARACTERIZATION

The purging nut (*J. curcas*) presents unisexual and pentamerous flowers (Figure 1) that are arranged in terminal and monoic-type inflorescences, which are greenish-yellow when completely formed. The female flowers (Figure 1A and 1C) have long and isolated stems with the superior ovary containing the nectary in its base. The female flowers present three carpels, each containing one forked stigmas and locule, with a fertile ovule in each locule. The male flowers (Figure 1B and 1D) are diplostemonous with 10 homodinamous staminal filaments, united only by the base. The anthers are isolated from each other, with longitudinal dehiscence.

FRUIT ONTOGENY

In general, information about the anatomy of the reproductive organs of Euphorbiaceae is rare in the literature. This shortage is mainly due to the difficulty of obtaining plant material for analysis, in addition to the large presence of latex, a characteristics of this family, making it difficult to process. For this reason, many studies are based on external morphological characteristics, which can cause inaccurate or even erroneous information about the species. In this sense, we describe herein the full development of fruits of *J. curcas* during its ontogeny and maturation and highlighting its major events.

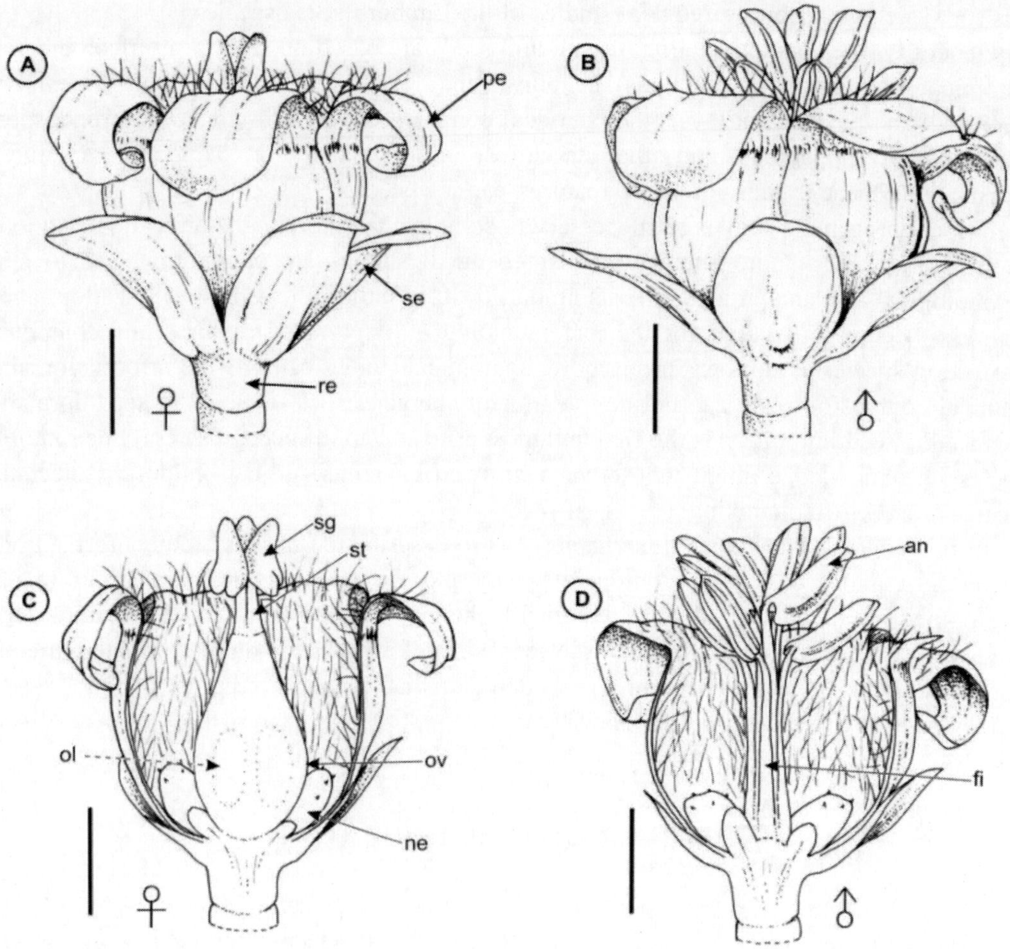

Figure 1. External and internal view of *Jatropha curcas* flowers. A and C – female flower; B and D – male flower. an = anther; sg = stigma; st = style; fi = filament; ne = nectary; ol = ovule; ov = ovary; pe = petal; se = sepal; re = receptacle. Bars 2 mm.

The fruits become easily identifiable from 4 days after anthesis (DAA) when they reach a polar and equatorial diameter of approximately 2.9 mm and 1.9 mm, respectively. The maturation process is completed at 47 DAA when the polar diameter is 33.9 ± 1.3 mm and the equatorial diameter is 27.6 ± 0.9 mm. These measurements show slight reductions accompanying the dehydration of fruits from 54 DAA when the fruit becomes completely dry and withered. Regardless of the stage of development, the fruits have an oval shape (Figure 2), with epicarp that is light green (4 to 27 DAA), citrus (40 to 47 DAA) or dark brown (from 54 DAA). The fruits of *J. curcas* present a tricarpellate and trilocular ovary, with one ovule in each locule (Figure 3; Figure 4A and 4B.). However, abortion and malformation of one (Figure 3D; Figure 4C and 4D) or even two (Figure 3E) ovules is commonly reported because at least 32% of the fruit has just two seeds and at least 8% show only one seed in each mature fruit. According to Mayhew (2006), abortion is usually reported in plants that have genetic incompatibility between the ovule and pollen tubes. In this sense, Bressan et al. (2013) described that pollination with pollen grains with low genetic variability in relation to the female flower increases the chances of miscarriage. This phenomenon can explain the loss of productivity in this type of seeds. It is worth noting that *J. curcas* presents a mixed system of reproduction by combining self-pollination, apomixis, and asexual reproduction (Bressan et al., 2013). These kind of combinations could lead to abortion.

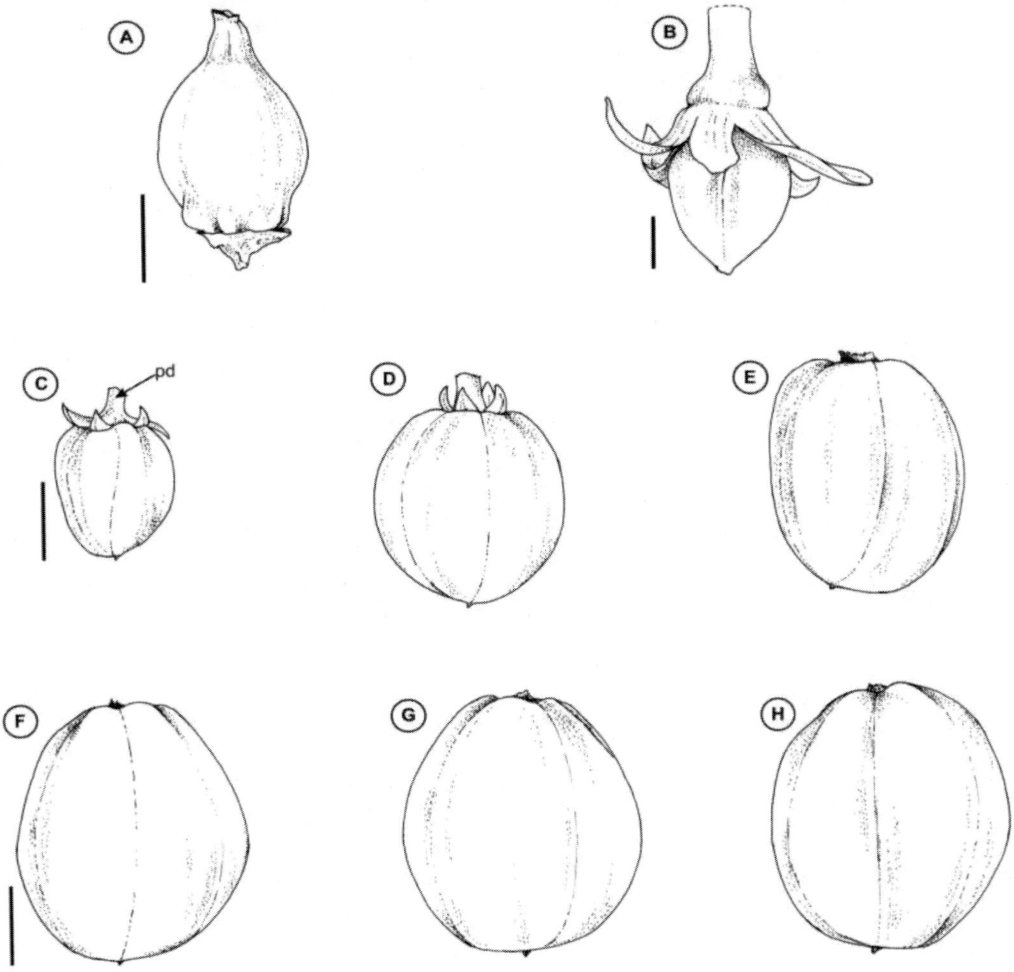

Figure 2. External view of *Jatropha curcas* fruit in different development stages. The fruit 13 (A), 18 (B), 22 (C), 27 (D), 40 (E), 47 (F), 54 (G) and 58 (H) days after anthesis (DAA). Bars: 2 mm (A, B) and 10 mm (C-H).

The central portion of the fruit has three vascular bundles (Figure 4C). Based on the dehiscence type and the persistent carpophore (Barroso et al., 1999), the *J. curcas* fruit should be classified as a schizocarp. The *J. curcas* fruits presents two dehiscent regions: a dorsal region and another region along the carpel sutures (Figure 4A, 4B, 4C and 5A). When mature, the dried fruit presents a dehiscence mechanism in which the structure and arrangement of the pericarp tissues are altered, as evidenced by a tissue suture from exocarp to endocarp.

In the early pericarp development stages, the exocarp (outer fruit layer) has a single layer of cells (Figure 5B), a thin cuticle, and stomata on the same level as the other epidermal cells. From 13 DAA, it is possible to observe mitotic divisions parallel to the surface (periclinal) with the formation of a new layer in the epicarp (inner layer), which then becomes bi-stratified (Figure 5B).

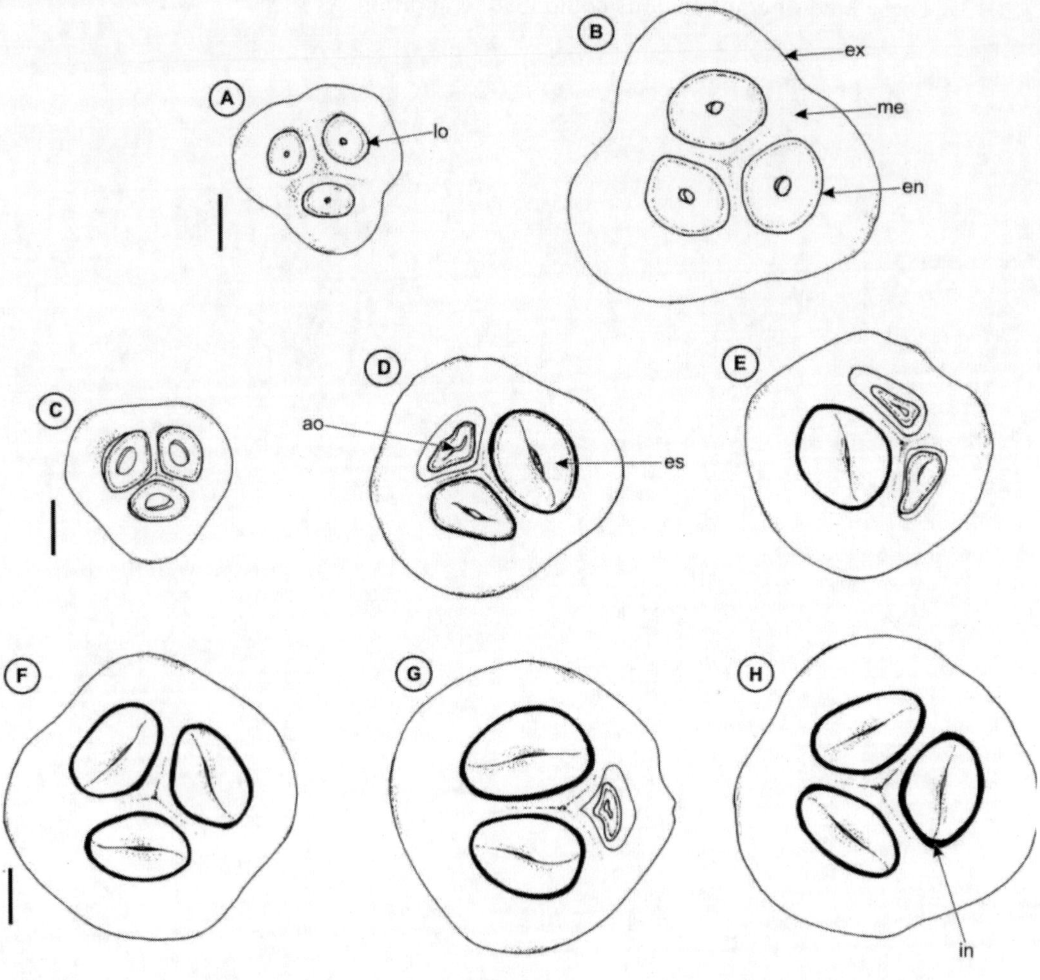

Figure 3. Internal view of *Jatropha curcas* fruit in different development stages. The fruit have 13 (A), 18 (B), 22 (C), 27 (D), 40 (E), 47 (F), 54 (G) and 58 (H) days after anthesis (DAA). en = endocarp; ex = exocarp; es = endosperm; lo = locule; me = mesocarp; ao = aborted ovule; in = integument. Bars: 1 mm (A, B) and 5 mm (C-H).

A multi-layered mesocarp (Figure 5A-F) is the result of cell divisions in multiple planes, consisting of parenchymal and isodiametric cells containing idioblasts in which druses (Figure 5C) and a large amount of latex are observed. Moreover, we observed the presence of collateral vascular bundles (Figure 5E-F). The laticiferous are dispersed throughout the mesocarp and may or may not be branched (Figure 5D) or articulated. Apparently, younger fruits present more laticiferous than mature fruits, probably due to the mesocarp having a larger area in more developed fruit.

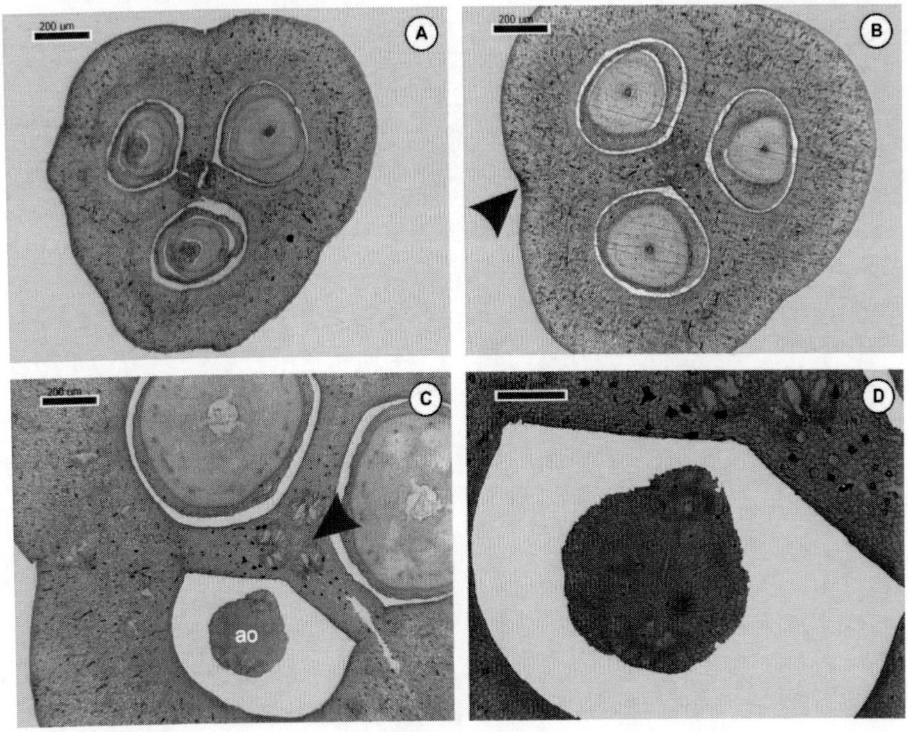

Figure 4. Cross-sections of *Jatropha curcas* fruits. A – overview of 4 DAA fruit showing the three locules; B – fruit at 7 DAA, highlighting the region that separates each mericarp (arrow); C – fruit at 13 DAA, highlighting the vascular bundle in the central portion (arrow); D – detail of aborted ovule. ao = aborted ovule. Bars = 100 μm (D) and 200 μm (A, B and C).

Approximately 27 days after anthesis (DAA) it is possible to observe collateral vascular bundles located in the border (outer and inner region) of the mesocarp, which form arcs (U-shape) (Figure 6A, fruit with 40 DAA). It is also possible to observe that from in 54 DAA, the mesocarp becomes thinner than fruit at earlier stages due to the collapse of some cells (Figure 6C and 6D) caused by fruit wilting when fully ripe. It is possible that fibers and macrosclereids are arranged in the mesocarp so that the cells of one layer cross the cells of another layer. It is probable that this crossed arrangement of cells is related to the movement of water lost during the dehiscence mechanism, which leads to an increase in the internal pressure of the pericarp resulting in fruit opening. It is possible that the dehydration that occurs in the ripening process leaves the brown-ripened fruit smaller than the green fruit, as was described in another Euphorbiaceae species (Gagliardi et al., 2012).

Generally, two dehiscence mechanisms are observed in fruits: hygroscopic and turgor. The hygroscopic-type is associated with shrinkage or distention of the dead cell walls, whereas the turgor-type mechanism depends on living cells. We described that the dehiscence of *J. curcas* fruit is hygroscopic-type, termed xerochasy. This observation has been reported in another species of the Euphorbiaceae family (Gagliardi et al., 2012).

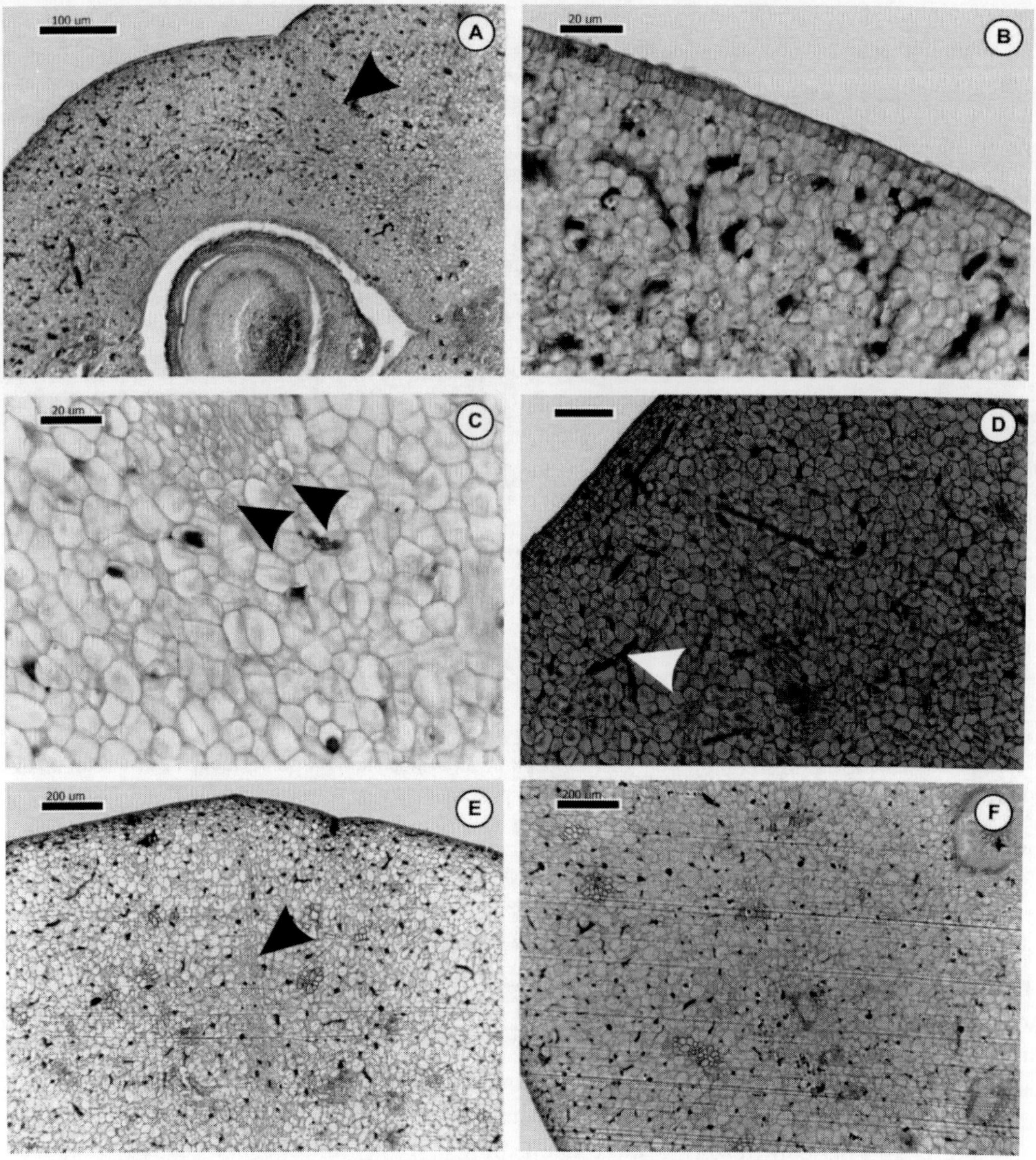

Figure 5. Cross-sections of the pericarp of *Jatropha curcas* fruit. A – Fruit and seed at 4 DAA, showing the region that separates each mericarp (arrow); B – 7 DAA fruit highlighting the laticiferous; C – 13 DAA fruit, highlighting the presence of druses (arrow); D –18 DAA fruit highlighting the branched laticiferous (arrow); E – 22 DAA fruit showing separation tissue (arrow) placed among two mericarps; F – mesocarp of the 27 DAA fruit showing vascular bundle. Bars = 20 µm (B and C), 100 µm (A and D) and 200 µm (E and F).

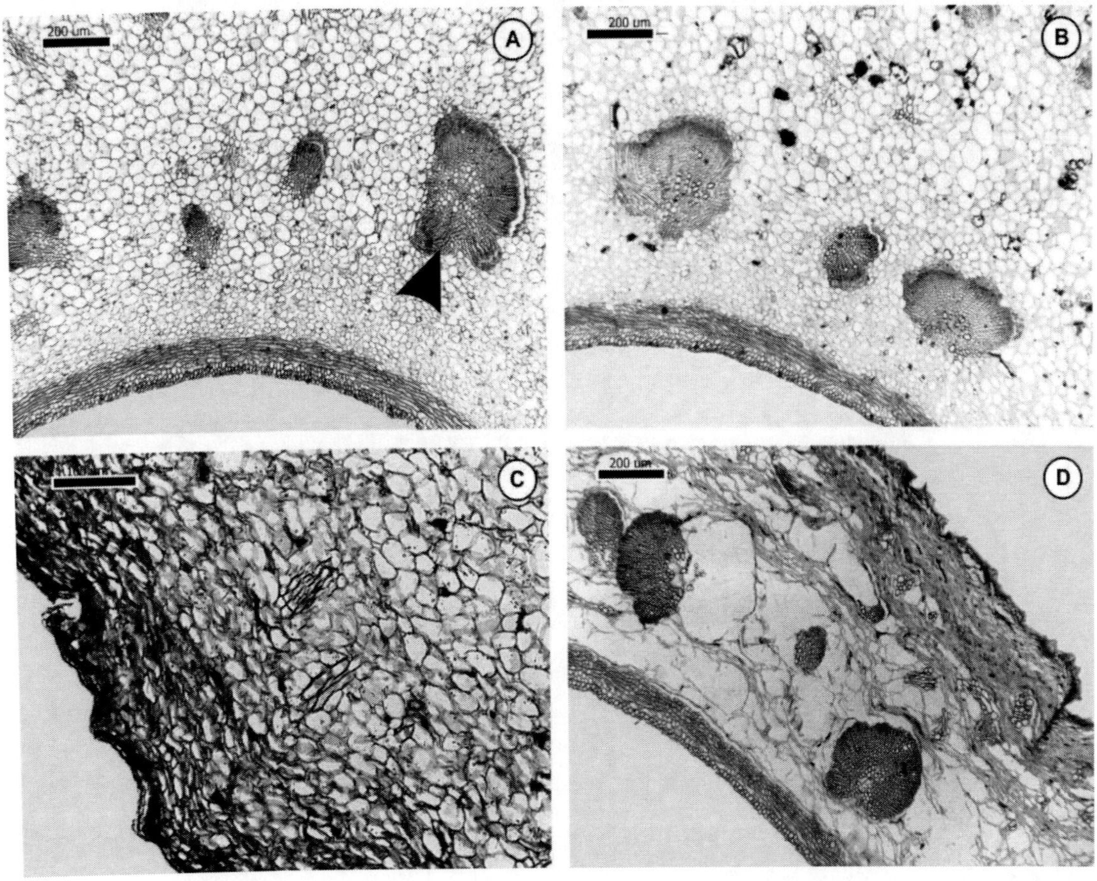

Figure 6. Cross-sections of *Jatropha curcas* pericarp. A – mesocarp and endocarp of the 40 DAA *J. curcas* fruit, highlighting vascular bundle forming arcs (arrow) in the inner region; B – detail of the vascular bundle located in the mesocarp in the nearest endocarp portion of the 47 DAA fruit; C – epicarp and mesocarp of the 54 DAA *J. curcas* fruits showing fully dehydrated fruit in the outer layers; D – pericarp presenting cell lysis at 58 DAA fully ripened *J. curcas* fruit. Bars = 100 μm (B and C) and 200 μm (A and D).

The endocarp of the *J. curcas* fruit is characterized by the presence of three to five layers of cells. However, with ripening, this layer shows a noticeable reduction to two cell layers, an outer layer, whose cells become elongated tangentially and obliquely, and an inner layer, whose cells assume an isodiametric form (Figure 6B). In both layers, the cells become completely thickened from 27 DAA (Figure 6A, 6B and 6D). The endocarp lignification, which occurs in late-stages of the development, contributes to the protection and growth of the seeds inside the fruit and helps in fruit dehiscence. This processes is promoted by the presence of a separation tissue in the mesocarp that facilitates fruit opening, which remains connected only by the carpophorus, a common structure in schizocarps fruits (Gagliardi et al., 2012). Even after complete dehydration, the *J. curcas* fruits keep their seeds inside, a mechanism described as serotiny, which is very common in regions with extreme seasonal climates, where water shortage is very frequent (Lamont et al., 1991) as well as in naturally fire-affected environments (e.g., Brazilian savanna-like areas in the South America and the chaparral biome in the Western region of the United States), where plants are frequently affected by fire (Simon and Pennington, 2012).

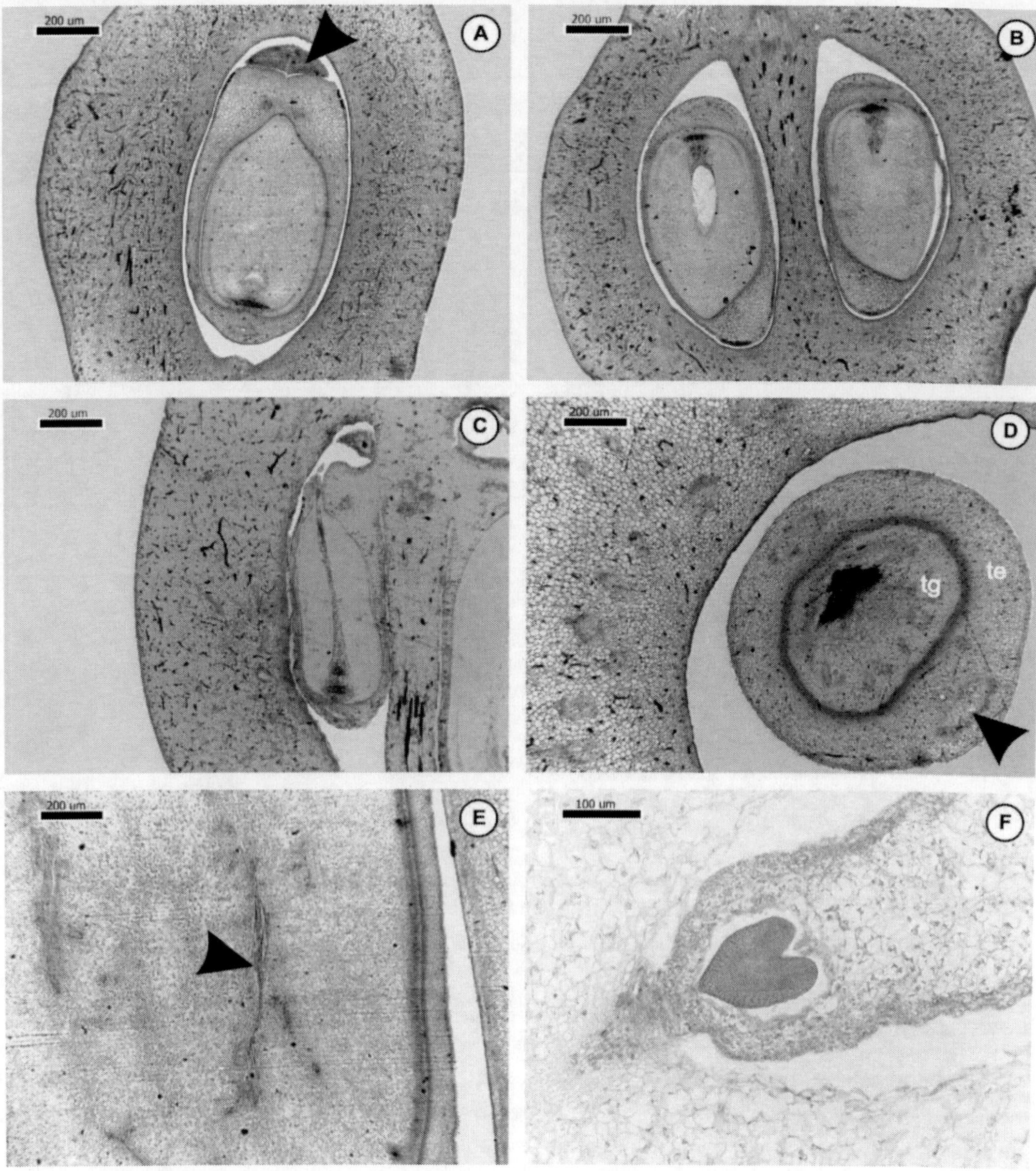

Figure 7. *Jatropha curcas* fruit in different developmental stages. A – longitudinal section of the 4 DAA *J. curcas* fruit showing the caruncle in the seed (arrow); B – longitudinal section of the 7 DAA fruit and seed, highlighting two locules; C – longitudinal section of the 13 DAA fruit and seed, showing anatropous fruit; D – cross-section of 18 DAA fruit and seed showing a vascular bundle in the raphe region (arrow); E – longitudinal section of the 22 DAA seed highlighting the cotyledon and its axis-cotyledon (arrow); F – embryo body in detail showing the cordiform embryo in 27 DAA fruit and seed. te = testa; tg = tegmen. Bars = 100 μm (F) and 200 μm (A, B, C, D and E).

SEED ONTOGENY

Seed development, including the endosperm and embryo at various stages, is outlined by Figures 7 to 10. Generally, seeds of Euphorbiaceae species have a very similar conformation. They all have a fleshy well-developed rich endosperm where oil bodies are very common as a storage compound (Barroso et al., 1999). The ovule of the *J. curcas* fruit are anatropous-type (Figure 7C); the micropyle and hilum are near each other as a result of unilateral growth. Moreover, the ovule is pendulous and crassinucellate; the archesporial cell cuts off a parietal cell and its derivatives make the megaspore, a deep-seated mother cell. The seeds have a whitish and well-developed caruncle covering the micropyle region formed by isodiametric parenchyma cells that accumulate oil or protein compounds (Figure 6A). *J. curcas* seeds have a well pronounced raphe and caruncle (i.e., a fleshy and edible seed region that attracts ants for seed dispersal, also known as myrmecochory). This description is very common in many Euphorbiaceae species such as *Ricinus communis* (Singh, 1954), *Croton bonplandianum* (Singh and Chopra, 1970), *Manihot caerulescens* and *Manihot tripartite* (Oliveira and Oliveira 2009) or *Dalechampia stipulacea* (Silva and Souza, 2009).

In Euphorbiaceae species, the caruncle can induce dormancy and can aid the seeds in soil water uptake (Lisci et al., 1996). The presence of a well-developed caruncle rich in oil compounds is a characteristic of *J. curcas* seeds. The presence of a caruncle was previously reported in other Euphorbiaceae species such as *Manihot caerulescens* and *M. tripartite* (Oliveira and Oliveira, 2009). However, for these *Manihot* species, the caruncle is cutinized. The form of the caruncle in *J. curcas* is devoid of cutine. In *Riccinus communis* seeds, the caruncle appears to be involved in the process of dehydration, rehydration and seed germination (Bianchini and Pacini, 1996). Here, we described that *J. curcas* seeds remains attached to the fruit by the funicle (Figure 7C and 7D) up to 7 days after anthesis, and this structure completely disappears by 13 DAA when the seed has fully ripened.

Seed Integument

The *J. curcas* seeds are bitegmented and show well-defined testa and tegmen (Figure 8B). The testa is multi-layered and has approximately 8 to 10 cell-thick layers with a distinct division between the exotesta, mesotesta and endotesta (Figure 8B and Figure 8D). The exotesta is unilayered and is characterized by the presence of epidermal cells, which are polyhedral until to 22 DAA, when they become elongated and sclerified by 27 DAA. However, some exotesta cells are not lignified throughout the seed (Figure 8B and 8D). We described here the macrosclereids-type in the exotesta (Figure 9A) as well as fibers that can be viewed as isolated (Figure 9B), arranged in a star pattern (Figure 9C) or arranged longitudinally (Figure 9D).

At first glance, testa is the integument that shows more changes over the course of development, mainly due to the collapse of cells and the thinning of the mesotesta layers, which incorporates a mechanism that promotes embryo communication with the external environment after the seeds are fully ripe. Some authors have suggested that this structure can facilitate the water, minerals and oxygen acquisition during the germination (Miranda et al., 2011). The mesotesta is multi-layered with isodiametric and non-lignified cells, where laticiferous and druses are very common. The endotesta is unilayered with short and columnar cells that exhibit thin walls (Figure 8B and 8D). The tegmen is multi-layered and thicker than the testa, either in the number of layers or in thickness and can reach up to 20 to 30 cell layers. In the same manner as described in the testa, the tegmen can be clearly divided into the exotegmen, mesotegmen and endotegmen (Figure 8B and 8D). The exotegmen is unilayered and has cell walls that are initially columnar and short and present as non-thickened; however, from 22 DAA, the cells become anticlinal elongated with thickened cell walls. The mesotegmen consists of large and non-lignified parenchymatous cells showing high mitotic activity during seed development, where we found collateral vascular bundles forming a ring surrounding the embryo. These vascular bundles are arranged tangentially to the mesotegmen and reach the chalaza region, which collapses during seed development. The endotegmen (*i.e.*, inner portion of the tegmen) is unilayered and is formed by short and non-thickened cells. From 27 DAA, it is possible to identify the presence of lipid bodies and phenolic compounds in the seed coat, as well as the presence of sclereids in the mesotegmen. From 40 DAA, the exotegmen, which was formed by a single layer of cells in previous stages (before 27 DAA), presents with a multi-layered arrangement (8B and 8D).

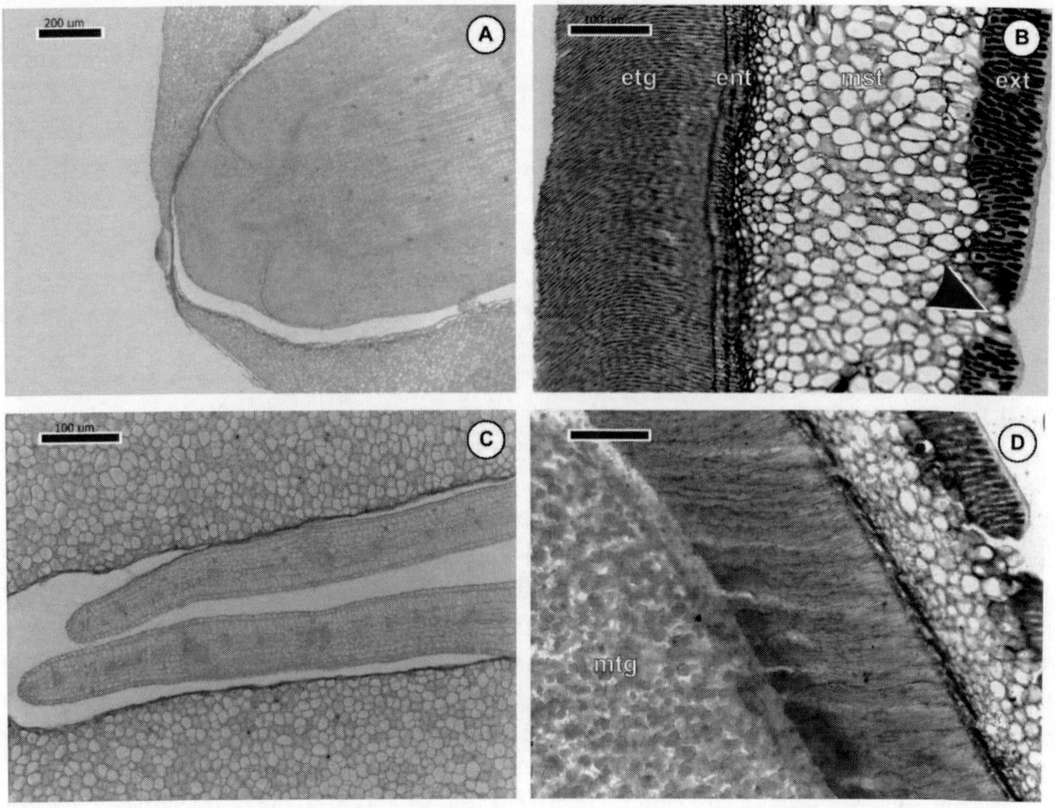

Figure 8. *Jatropha curcas* seeds in different development stages. A – 40 DAA embryo highlighting the radicular meristem; B – 40 DAA seed showing the testa surface without lignification (arrow); C – detail of the cotyledons in a 54 DAA embryo; D – seed integument in full ripened fruit stained with Sudan III, showing the reduction of the mesotesta. ext = exotesta; mst = mesotesta; ent = endotesta; etg = exotegmen; mtg = mesotegmen. Bars = 100 µm (B, C and D) and 200 µm (A).

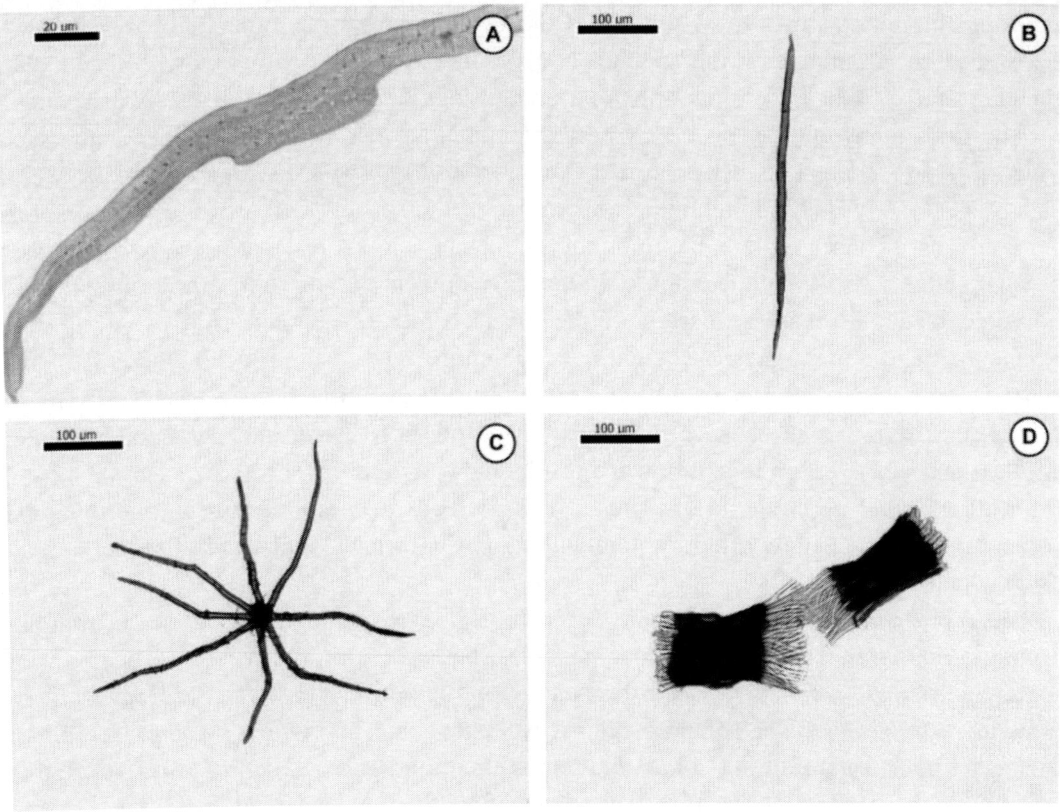

Figure 9. Structure presented in *J. curcas* seed integument stained with safranin. A – macrosclereids; B – isolated fibers; C – fibers arranged in a star conformation; D – fibers arranged longitudinally. Bars = 20 µm (A) and 100 µm (B, C and D).

Integument versus Waterproof

The inner portion of the *J. curcas* seed coat houses numerous macrosclereids which are closely related to the seed rigidity and waterproofing (Gopinathan and Babu, 1985) and are the cause of seed dormancy (Islam et al., 2009). These authors assumed that the presence of sclerified cells as well as the macrosclereids present in the exotesta can prevent the passage of water and gases to the embryo during development (Oguchi, 2005), resulting in seed dormancy. Although macrosclereids can impose seed tegumentar dormancy in many species, we hypothesized that in *J. curcas*, the macrosclereids present in the exotesta do not impose this physical barrier to the passage of water and gases because *J. curcas* are orthodox (*i.e.*, seeds will survive drying and/or freezing during *ex situ* conservation) and have a high water content (15 to 20%) when freshly harvested (Pompelli et al., 2010). This allows rapid and continuous water imbibition as a consequence of the matric potential of the dry seed tissues. Dried *J. curcas* seeds present a water potential of -63.1 ± 0.5 MPa, which quickly increases to -0.7 ± 0.2 MPa in the first 12 hours of water imbibition. In this lag phase, the seed water content is increased up to 600% compared to that of dried *J. curcas* seeds. Moreover, in recent tests, we found that the hardness of the seed coat is significantly reduced with increasing imbibition time. It is likely that the differential permeability shown here in *J. curcas* seeds is promoted by the spacing between the exotesta cells (Figure 8B and 8D), which are similar to the fissure line (pleurograma) of legume seeds, or even as a hilar channel, as found in in many another species. These structures are described as regions at the seed surface with greater weakness (Kursar and Coley, 1999, Lima, 1985, Margarido, 1977, Melo-Pinna et al., 1999). This highlights the importance of the spacing between the exotesta cells (*i.e.,* pores) as it causes remarkable thinning and collapse in exotesta cells, which facilitates the water and gas uptake by the embryo during development and promotes the resumption of embryonic metabolism and emission of the primary root, a characteristic event of seed germination. These pores are distributed over the entire seed coat surface (Figure 10A and 10B) and range from 20 μm to 120 μm (Figure 10C and 10D). The mesotesta remains exposed, thus reducing the barrier formed by the exotesta to water and gas uptake (Figure 10D, 10E and 10F). Taken together, these results indicate that the seeds of *J. curcas* have a hard seed coat, but not waterproof. Thus *J. curcas* seeds does not require pre-germination treatments, as previously reported by Pompelli et al. (2010) and Moncaleano-Escandon et al. (2013).

The *J. curcas* fruit and seeds showed a high concentration of idioblasts containing phenolic compounds. These phenolic compounds, when present in the seed coat, can prevent the passage of gases leading to seed dormancy (Murugesan et al., 2009). Moreover, these compounds can act as plant defenses against pathogens and herbivores (War et al., 2012), which is a fundamental trait in the arid and semiarid regions where *J. curcas* is native (Rao et al., 2008). In these areas, *J. curcas* is well-adapted to herbivores, water shortage, poor quality soils and other abiotic and biotic stressors. It is worth nothing that such phenolic compounds are soluble in water and may be leached out of *J. curcas* seeds during the imbibition process before seed germination.

Endosperm and Embryo Development

In the inner portion of the *J. curcas* seed coat, we found that the endosperm was merged in this structure and layered in the embryo, which up to 22 DAA was formed by a simple and undifferentiated cells that exhibited radial symmetry, which characterizes the globular stage. From 27 DAA, this embryo become differentiated, with rapid cell division in the two regions on either side of the future shoot apical meristem to form the cotyledon primordia, giving the embryo bilateral symmetry and characterizing the heart stage (Figure 7F). The embryo is developed at the same time as the endosperm, which is composed of large parenchymal cells, used for the storage of the oil body and protein compounds. From 18 DAA, the endosperm begins to be consumed by the embryo; however the *J. curcas* endosperm is too large to be exhausted by the embryo and remains in ripe seeds; thus, *J. curcas* seeds are classified as endospermic seeds. From 40 DAA, the embryo presented a well-developed hypocotyl-radicle axis (Figure 8A), a plumule, which will become the future shoot and two large, thin, flat and sessile cotyledons (Figure 7E and 8C).

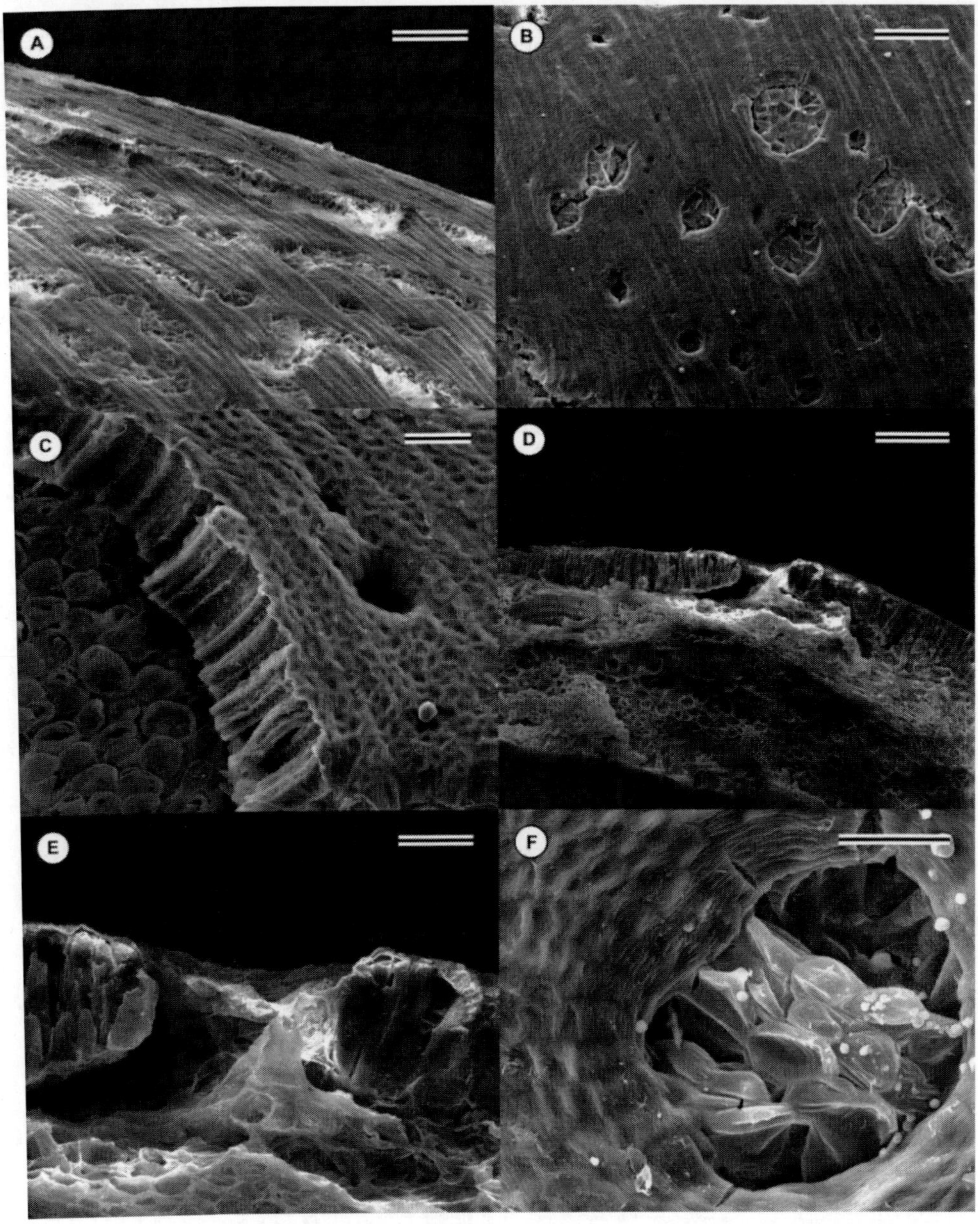

Figure 10. Electron scanning micrograph of *Jatropha curcas* seeds showing external integument and pores on the surface. A – general image of the external integument showing innumerous pore on the surface (B); C – oblique and transverse (D) vision of the the pore through which water penetrates into the integument measuring approximately 55 µm; E – Zoom of the pore showing mesotesta in the inner portion. Bars. 200 µm in A, B and D and 50 µm in C, E and F.

CONCLUSION

Taken together, these results indicate that ontogeny aids in the study of plant morphoanatomy. Here, we outlined *Jatropha curcas* fruit and seed ontogeny. The laticiferous is very common in *J. curcas* fruit. The seeds have a prominent caruncle and a well-developed endosperm that occupies the major portion of the seed. The presence of a rigid seed coat and a large endosperm, rich in oil bodies and protein compounds, guarantees the seed survival against biotic and abiotic stressors and also allows late development of the embryo leading to germination. The rigidity of the *J. curcas* seed coat is caused by the presence of fibers and macrosclereids, which are certainly involved in the germination process. However, innumerous pores on the seed surface can be overwhelm the waterproofing imposed by macrosclereids. This knowledge could lead to a better understanding of the amazing adaptability and competitive success of *J. curcas* in many conditions, particularly arid and semiarid environments, leading to oil production in this promising biofuel plant.

References

Achten W.M.J., Nielsen L.R., Aerts R., Lengkeek A.G., Kjær E.D., Trabucco A., Hansen J.K., Maes W.H., Graudal L., Akinnifesi F.K., Muys B. (2010). Towards domestication of *Jatropha curcas*. *Biofuels*, 1, 91-107.

Arroyo B.J., Vanegas Y.O., Pompelli M.F., Garrido C.G., Bezerra Neto E., Orozco A.J. (2014). Detoxification of *Jatropha curcas* L. seed meal as possible alternative livestock feed in the Colombian Caribbean. *Rev Actualidad and Divulgación Científica*, 17, 171-178.

Barroso M.B., Morin N.P., Peixoto A.L., Ichaso C.L.F. (1999). *Frutos e sementes. Morfologia aplicada à sistemática de dicotiledôneas*. Viçosa: Editora da UFV.

Bianchini M., Pacini E. (1996). The caruncle of *Ricinus communis* L. Its development and role in seed dehydration, rehydration and germination. *Int J Plant Sci*, 157, 40-48.

Bressan E.A., Sebbenn A.M., Ferreira R.R., Gerald Lee T.S., Figueira A. (2013). *Jatropha curcas* L. (Euphorbiaceae) exhibits a mixed mating system, high correlated mating and apomixis. *Tree Genet Genomes*, 9, 1089-1097.

Fairless D. (2007). The little shrub that could - maybe. *Nature*, 449, 652-655.

Gagliardi K.B., Rosado A., Souza L.A., Moscheta I.S., Albiero A.L.M. (2012). Structure and development of fruits and seeds of weed species of Euphorbiaceae. *Acta Bot Bras*, 26, 38-45.

Gopinathan M.C., Babu C.R. (1985). Structural Diversity and its Adaptive Significancein Seeds of *Vigna minima* (Roxb.) Ohwi and Ohashi and its Allies (Leguminosae-Papilionoideae). *Ann Bot*, 56, 723-732.

Islam A.K.M.A., Anuar N., Yaakob Z. (2009). Effect of genotypes and pre-sowing treatmentson seed germination behavior of Jatropha. *Asian J Plant Sci*, 8, 433-439.

Kumar A., Sharma S. (2008). An evaluation of multipurpose oil seed crop for industrial uses (*Jatropha curcas* L.): A review. *Ind Crops Prod*, 28, 1-10.

Kursar T.A., Coley P.D. (1999). Contrasting modes of light acclimation in two species of the rainforest understory. *Oecologia*, 121, 489-498.

Lamont B.B., Lemaitre D.C., Cowling R.M., Enright N.J. (1991). Canopy seed storage in woody-plants. *Bot Rev*, 57, 277-317.

Lima M.P.M. (1985). Morfologia dos frutos e sementes dos gêneros da tribo Mimoseae (Leguminosae-Mimosoideae) aplicada à Sistemática. *Rodriguesia*, 37, 53-78.

Lisci M., Bianchini M., Pacini E. (1996). Structure and function of elaiosome in some Angiosperm species. *Flora*, 191, 131-141.

Margarido J.M., Aspectos anatômicos do desenvolvimento do fruto e da semente de *Dimorphandra mollis* Benth.(Leguminosae), in, Universidade Federal de São Paulo, São Paulo, 1977.

Mayhew P. (2006). *Discovering Evolutionary Ecology: Bringing together ecology and evolution*. Oxford: Oxford University Press.

Melo-Pinna G.F.A., Neiva M.S.M., Barbosa D.C.A. (1999). Estrutura do tegumento seminal de quatro espécies de Leguminosae (Caesalpinioideae), ocorrentes numa área de caatinga (PE Brasil). *Rev bras bot*, 22, 375-379.

Miranda R.Q., Oliveira M.T.P., Correia R.M., Almeida-Cortez J.S., Pompelli M.F. (2011). Germination of *Prosopis juliflora* (Sw) DC seeds after scarification treatments. *Plant Species Biol*, 26, 186-192.

Moncaleano-Escandon J., Silva B.C.F., Silva S.R.S., Granja J.A., Alves M.C.J.L., Pompelli M.F. (2013). Germination responses of *Jatropha curcas* L. seeds to storage and aging. *Ind Crops Prod*, 44, 684-690.

Murugesan A., Umarani C., Chinnusamy T.R., Krishnan M., Subramanian R., Neduzchezhain N. (2009). Production and analysis of bio-diesel from non-edible oils - A review. *Renew Sust Energy Rev*, 13, 825-834.

Oguchi R., Hikosaka K., Hirose T. (2005). Leaf anatomy as a constraint for photosynthetic acclimation: differential responses in leaf anatomy to increasing growth irradiance among three deciduous trees. *Plant Cell Environ*, 28, 916-927.

Oliveira J.H.G., Oliveira D.M.T. (2009). Morfoanatomia e ontogênese fo pericarpo de *Manihot caerulescens* Pohl e *M. tripartita* Müll. Arg. (Euphorbiaceae). *Rev Bras Bot*, 32, 117-129.

Pompelli M.F., Ferreira D.T.R.G., Cavalcante P.P.G.S., Salvador T.L., Hsie B.S., Endres L. (2010). Environmental influence on the physico-chemical and physiological properties of *Jatropha curcas* L. seeds. *Aust J Bot*, 58, 421-427.

Pompelli M.F., Orozco A.J., Oliveira M.T.O., Rodrigues B.M., Barbosa M.O., Santos M.G., Oliveira A.F.M., Almeida-Cortez J.S. (2011). Crise energética mundial e o papel do Brasil na problemática de biocombustíveis. *Revista Agronomía Colombiana*, 29, 361-371.

Pramanik K. (2003). Properties and use of *Jatropha curcas* oil and diesel fuel blends in compression ignition engine. *Renew Energ*, 28, 239-248.

Rao G.R., Korwar G.R., Shanker A.K., Ramakrishna Y.S. (2008). Genetic associations, variability and diversity in seed characters, growth, reproductive phenology and yield in *Jatropha curcas* (L.) accessions. *Trees Struct Funct*, 22, 697-709.

Silva A.C., Souza L.A. (2009). Morphology and anatomy of the developing fruit and seed of *Dalechampia stipulacea* Müll. Arg. (Euphorbiaceae). *Acta Sci Biol Sci*, 35, 425-432.

Simon M.F., Pennington T. (2012). Evidence for adaptation to fire regimes in the tropical savannas of the Brazilian Cerrado. *Int J Plant Sci*, 173, 711-723.

Singh R.P. (1954). Structure and development of seeds in Euphorbiaceae: *Ricinus communis* L. *Phytomorphology*, 4, 118-123.

Singh R.P., Chopra S. (1970). Structure and development of seeds in *Croton bonplandianum*. *Phytomorphology*, 20, 83-87.

Sunil N., Kumar V., Sujatha M., Rao G.R., Varaprasad K.S. (2013). Minimal descriptors for characterization and evaluation of *Jatropha curcas* L. germplasm for utilization in crop improvement. *Biomass Bioenerg*, 48, 239-249.

War A.R., Paulraj M.G., Ahmad T., Buhroo A.A., Hussain B., Ignacimuthu S., Sharma H.C. (2012). Mechanisms of plant defense against insect herbivores. *Plant Signaling Behavior*, 7, 1306-1320.

In: The Promising Future of *Jatropha curcas*
Editors: Maira R. Segura-Campos et al.

ISBN: 978-1-63484-989-0
© 2016 Nova Science Publishers, Inc.

Chapter 9

JATROPHA CURCAS: A SUSTAINABLE FEEDSTOCK FOR BIODIESEL SYNTHESIS

Nurudeen Ishola Mohammed[*] and Nassereldeen Ahmed Kabbashi[†]
Bioenvironmental Engineering Research Centre (BERC),
Department of Biotechnology Engineering, Faculty of Engineering,
International Islamic University Malaysia, Jalan Gombak, Kuala Lumpur, Malaysia

ABSTRACT

The development of biodiesel as a renewable fuel is primarily dependent on two major impediments encountered in the fossil fuel consumption. These constraints include dwindling oil reserves leading to probable dry-out of the resource in time and pollution related impact due to release of obnoxious gases when combusted. Oil seed price accounts for about 75% of the production cost on average, thereby, effort to abate this high cost adopts the use of low-grade resources such as waste cooking oil (yellow grease), animals fats (tallow) and trap grease (brown grease). Nevertheless, uncertainty surrounding the availability of these materials calls for utilization of non-edible oil crop of which *Jatropha* stands prominent. Owning to the chemical composition of *Jatropha*, its properties and health related issues, a number of operations had been considered for a holistic biofuel development recording an unprecedented success relative to other non-food materials. In this chapter, we presented the significance of *Jatropha curcas* oil and the dependence of the feedstock to production route selection. The use of various catalytic approaches was explored and the trends in the production capacities in different locations of the world. Findings revealed that, of more than 350 oil seed crops identified as potential feedstock for biodiesel development, the relevance of *Jatropha curcas* stands out due to the advantage of its oil content, physical and chemical features, emission characteristics and it propensity to be used in all routes for biodiesel production. *Jatropha* biodiesel showed unrivalled conformity with standard specifications, hence it use in

[*] Corresponding author: N. I., Mohammed: Tel: + 2348123825929, Fax: + 60 361964442, Email: sholanourou @yahoo.com.
[†] E-mail: nasreldin@iium.edu.my.

compression ignition engine pose no engine malfunction and undesirable byproduct as against other oil seed crop. Thus, it is a sustainable resource of immense benefit and relevance in biodiesel industry.

Keywords: *Jatropha curcas*, biodiesel, properties, production, emission

INTRODUCTION

An alternative fuel of similar properties and performance to conventional or fossil fuel is referred to as biodiesel. It is basically produced from base materials of organic origin such as vegetable oil, animal oil or fats as well as waste cooking/frying oil. Biodiesel is a promising substitute as an alternative fuel. It has gained significant attention due to predicted depletion of the fossil fuel reserves and environmental degradation caused by the combustion of the fossil fuel resulting in the release of obnoxious greenhouse gases and other pollutants (Raja et al., 2011).

Renewable organic sources such as vegetable oils and animal fats from which biodiesel is made are esters of glycol possessing different chain length and degree of saturation owing to the source and nature of the materials. It has recently attracted attention throughout the world as a result of its availability, renewability, non-toxicity, low-pollutant emission and biodegradability. The conversion process is primarily by reaction of the vegetable oil or animal fat with short-chain alcohol usually methanol and ethanol in the presence of chemical catalyst to produce mono-alkyl esters and glycerin. Consisting of long chain fatty acid esters (Abreu et al., 2004), biodiesel is majorly produced by transesterification reaction process of vegetable oils with short chain alcohols (Noureddini et al., 1998).

The interest on alternative energy source is because of growing concern of political and environmental issues of fossil fuel. The fossil fuels play an important role in industrial growth, transportation, agriculture and to meet many other human energy needs. However, these fuels are limited and depleting day by day as consumption increases rapidly due to the world population boom and technological advancement. Biodiesel remains one of the best available options to provide the world energy demand.

Besides, emission of greenhouse gases from the combustion of fossil fuels poses serious environmental threat and health related problems to the society. This calls for inquiry into renewable energy sources that can function as though conventional fuel without engine modification and as well be economically viable. Thus, researchers are developing means of energy provision through which alternative energy can be utilized. Currently, production rate of biodiesel requires an incentive to compete with petroleum-based fuels (Raja et al., 2011).

The production of biodiesel from oil origin has utilized feedstocks from available biomass throughout the world producing different capacities. For instance, in USA, the combined vegetable oil and animal fat production amounts to about 35.3 billion pounds per year (Pearl, 2002). Similarly, in Brazil, since the advent of the first biodiesel plant in 2005, Brazilian production has witnessed predominant increase with about 2 million tons/year in 2009 utilizing about 60 plant oils (Aranda et al., 2009).

Moreover, as at 2002, of the world production of biodiesel about 84% was from rapeseed oil, 13% from sunflower while palm oil and soybean biodiesel both record 1% (Korbiz, 2002), subject to the availability of these seed oil and the energy needs in the various locations of the world. Majority of the feedstock highlighted and used commonly for biodiesel production like soybean, rapeseed/canola, palm, corn, sunflower, cottonseed, peanut, and coconut oils are food materials. Moreover, many of these oilseed crops are usually utilized as a result of their relative availability and abundance in various location of the world. Hence, the food-energy feud of these oil seeds.

Biodiesel synthesis and usage was primarily assisted by benefits of subsidy for oil crop cultivation. Moreover, biodiesel exemption from the oil tax was another measure to encourage the biodiesel development. The Common Agricultural Policy Reform of 1992 stipulated a crop-specific settlement for every hectare as a compensation for the abatement of institutional prices. The earmarked reference share is currently 10%, but the applied earmarked rates have been adapted periodically, taking into account market forecasts. Furthermore, farmers are flexible in the management of their set-aside obligations (Demirbas, 2008).

In the past, conventional oilseed utilized for biodiesel synthesis sourced for feedstocks primarily from available resource such as soybean, rapeseed/canola, palm, corn, sunflower, cottonseed, peanut, and coconut oils. Although, research is tailored towards improving oil content of these resources through genetic modification and molecular breeding, however, despite the potential of this practice to raise the throughput, inquiry into other biodiesel feedstocks remains significantly important. Predominantly, majority of these oil resources are researched as biodiesel feedstock relative to their abundance in many locations of the world. Moreover, the cost of edible vegetable oils, like canola oil, palm oil and soybean oil, are higher than that of diesel fuel. Hence, waste restaurant or frying oils and non-edible crude vegetable oils such as karanj, *Jatropha* and neem oils are considered as potential cheap biodiesel feedstock (Pinto et al., 2005).

Consequently, regions with deficient oil resource from the many usual feedstocks explore the potential of non-edible vegetable oils such as *Jatropha curcas*, waste cooking oil and animals fats in an effort to reduce food-energy competitiveness of the major oilseed and as part of effort to reduce reliance on imported petroleum as sole energy provision. For Example, due in part to veracity of its properties compared to other alternative feedstock, non-edible *Jatropha curcas* (*Jatropha*) oil has in later days beckon commensurate research interest as a feedstock for biodiesel production in India, Sudan and other climatically parallel regions of the world (Mohibbeazam et al., 2005; Kumartiwari et al., 2007; Kalbande et al., 2008).

The *Jatropha* tree is a perennial shrub (growing only up to about 5 m), belonging to the Euphorbiaceae family and have extractable oil content above 30 wt. % (Mohibbeazam et al., 2005). High oil content of *Jatropha* has high free fatty acids (FFA) implying a relatively high percentage of saturated fatty acids of about 34 wt. % on average. The corresponding biodiesel synthesized from it is characterized with poor low-temperature operability, as a measure of its low Pour Point value of 2°C (Kumartiwari et al., 2007).

Jatropha plant can grow in diverse climatic conditions of rainfall range between 250 to 3000mm (Vaknin et al., 2011). It is considered drought resistant because of its ability to adapt to arid and semi-arid conditions with seed potential of about 40% by mass (Rodriguez et al., 2011). The primary use of *Jatropha* plant was to control erosion, land reclamation, live fence and utilization as raw material in soap industries and lamp fuel (Openshaw, 2000). Concerned with the propensity of *Jatropha* in biodiesel synthesis, it has become area of research consideration and as such local and global business groups are exploring cultivation of the plant in many countries around the world (FAO, 2008). Besides, *Jatropha* oil cannot be utilized for nutritional purposes without detoxification (Akbar et al., 2009), this makes it promising as energy source.

The seed oil of *Jatropha* was utilized as a diesel fuel alternative during World War II and can as well serve as blends with conventional diesel (Foidl et al., 1996; Gubitz et al., 1999). Hence, the use of *Jatropha curcas* indicates a more suitable feedstock for synthesis of renewable fuel such as biodiesel (Meher et al., 2006a; Meher et al., 2006b). The oil content of *Jatropha* is very high reaching up to 30-50% (Foidl et al., 1996; Azam et al., 2005).

Recently *Jatropha curcas* has become the subject of many research and breeding programs worldwide due primarily to its non-edible property and its potential as an oil crop for biodiesel production (Jongschaap et al., 2007; King et al., 2009; Divakara et al., 2010). *Jatropha* plant has a long productive period with effective yield up to the 50 years and yielding large returns annually. The production of the seeds is about 0.8kg/m^2 per year (Banapurmath et al., 2008). The oil content of *Jatropha* seed ranges from 30% to 50% by weight and the kernel ranges from 45% to 60% (Pramanik, 2003). In large areas of Belize, Brazil, China, Egypt, Ethiopia, Gambia, Honduras, India, Indonesia, Mozambique, Myanmar, Philippines, Senegal and the United Republic of Tanzania, *Jatropha curcas* plantation has experienced improved cultivation with international and national investors rushing to its establishment (FAO, 2008). The fatty acid composition of *Jatropha* classifies it as a linoleic or oleic acid type, which are unsaturated fatty acids. Fresh *Jatropha* is a slow drying, odourless and colourless oil and become yellow after aging. The average production of *Jatropha* oil on average soil can yield about 1.6 million tones/ha of oil (Sharma et al., 2008).

Jatropha oil is a non-edible vegetable oil. Its methyl esters properties are close to diesel fuel and its utilization in diesel engine reduce CO_2 emission to the atmosphere (Martinez-Herrera et al., 2006; Yong et al., 2006; Hameed et al., 2009). However, the combustion of *Jatropha* oil in diesel engine faces many problems associated with the viscosity which is about ten times higher than the viscosity of diesel fuel. This high viscosity of the oil is as a result of the large molecular weight and chemical structure (Tippayawong et al., 2002). The problems generated as a result of viscosity of the oil include poor atomization, incomplete combustion, coking of fuel injectors, ring carbonization and fuel accumulation in the lubricating oil. As a result, reduction of the oil viscosity so as to make it suitable as alternative fuel is of prime importance. Several approaches have been devised over time; these methods include dilution, micro-emulsification, pyrolysis and transesterification (Tippayawong et al., 2002; Demirbas, 2009).

Jatropha is thus a potential feedstock for biodiesel production and it has attracted much attention when compared to other edible and non-edible oil resource such as palm oil and algae which are reported feedstocks to produce better yield in biodiesel synthesis (Azam et al., 2005). *Jatropha curcas* oil is used as oil feedstock for biodiesel production due to its non-edible characteristics and can thus avoid food-fuel competitiveness. Moreover, the fatty acid composition of *Jatropha curcas* oil which is similar to edible oil such as palm oil, it positive life cycle energy balance and satisfactory engine performance test has made it a promising candidate to replace edible oils as feedstock for biodiesel production (Ting et al., 2008; Yee et al., 2011).

Economic Considerations in Biodiesel Development

Industrial biodiesel synthesis started in the 1990s. Biodiesel of rapeseed oil origin is the most common biodiesel available in Europe. Soybean oil and yellow grease are the primary feedstocks utilized in biodiesel synthesis in the United State of America. Biodiesel stands prominent in Europe and America when compared with other organic fuels. Blends of biodiesel performs much better than conventional diesel; nevertheless, relative abundance and high cost of pure feedstock which make up more than 70% of the production estimate impede the commercialization of the product (Demirbas, 2008).

Prices of feedstock, cost of production, selling price of petroleum fuels, and energy products taxation among others are the various factors inhibiting development of Biodiesel industry. The economic sustenance and advantages of a biodiesel enterprise depend majorly on the value added to the raw material, increasing rural employment opportunities, high income taxation and low health care expenditure resulting from improved air quality and reduction in greenhouse gas emission (Demirbas, 2008).

In 2005, about 89% of biodiesel synthesized globally was from Europe. Biodiesel consumption in the United States reached about 18%. This makes U.S. the largest biodiesel market globally. Economic benefits of biodiesel include low greenhouse gas emission, helps to reduce a country's dependence on fossil fuel imports, and improves agricultural practice through employment of new labour and revenue generation for domestic crops. Furthermore, it augments lubrication and is widely endorsed by vehicle manufacturers (Demirbas, 2008).

The feedstock constitutes highest cost in biodiesel synthesis amounting to about 80% of the operating cost. Some other determining factors are labour, alcohol used, and catalyst. The cost of biodiesel fuels depends on the feedstock purity, geographical location, diversity in periodic crop cultivation, the price of fossil fuel, and some other factors. Biodiesel costs more than doubled the price of conventional diesel. However, effort to reduce the production cost informs the use of low cost feedstocks in biodiesel synthesis.

Biodiesel beckons increasing consideration from institutions and individual desiring home brewed biodiesel production. It is economically feasible in majority of oilseed-producing regions of the world. It is a technologically feasible substitute to petro-diesel, but its selling price doubled that of fossil diesel in most advance countries. Although, biodiesel is still currently produced in relatively small scale compared to fossil fuel, present market price is not competitive. Hence, biodiesel at the present economic situation does not satisfactorily rivaled petro-diesel (Demirbas, 2008).

Apart from the price of the feedstock material, costs of adopted conversion approach also determines biodiesel competing strength. The recent rise in the potential use of biodiesel has been majorly attributed to booming production plants but also to the sophisticated facilities used in the process. The overwhelming improvement in the biodiesel enterprise is imperative to reducing the price of biodiesel source materials. A prior appraisal of the probable feedstocks for biodiesel by Hanna et al. (2005) established an anticipated price pressures on biodiesel feedstocks.

Budgetary motivations and financial bait like feedstock and processing costs reduction as well as taxation waive will be a vital ingredient for supporting biodiesel consumption as an alternative fuel. Biodiesel offers many benefits due to some characteristics such as lower emission of obnoxious exhaust (Dunn, 2001). Other advantages of biodiesel must be explored beyond its current sole utilization for agriculture transport, and energy needs.

SIGNIFICANCE OF *JATROPHA CURCAS* OIL PROPERTIES

The shift in utilization of feedstock in energy generation as renewable fuel beckons enormous 'technical know-how' on the requirement of the fuel ranging from its characteristics, chemistry of formation and reaction mechanism as well as the storage requirements. This is required for informed social awareness and technical knowledge related to the fuel. Several Researchers have investigated the characteristics of *Jatropha curcas* oils and reported comparatively the variations in properties either due to different climatic region of propagation or the pre-treatment operation adopted for the feedstock.

In a comparative study carried out by Rodriguez et al. (2011) they characterized two species of *Jatropha curcas* grown at different climatic regions of Cuba. In their finding, they reported a variance in some of the properties of the fuels produced from the two species of *Jatropha* oils. It was concluded that the variation in properties of the two samples was due to effect of the different climatic region and seed processing operations.

Moreover, a similar study carried out by Salimon and Ahmed in 2012, investigated the properties of three *Jatropha curcas* species sourced from Malaysia, Indonesia and India. Their finding revealed that *J. curcas* grown on Malaysia soil had higher oil content, higher iodine value, lower acid value and some other variance in other properties when compared to those from Indonesian and Indian plantations. The Table 1 presents comparatively the properties that were investigated.

Table 1. Physical and Chemical Properties of Tropical *J. curcas* oil

Properties	Malaysian *Jatropha. curcas* oil	Indonesian *Jatropha. curcas* oil	Indian *Jatropha. curcas* oil
Oil content	33.7%	32.7%	30.3%
% Moisture	4.1%	3.3%	4.2%
Spe. Gravity at 28°C	0.8750	0.9190	0.8895
Refractive Index at 25°C	1.472	1.463	1.465
Viscosity at 28°C	58 ± 1	38 ± 1	63 ± 1
Colour at 28°C	3.2Y - 0.5R	20Y,2R	20Y,0.5R
Iodine value	103.2 ± 0.6	99.8 ± 0.5	97 ± 0.7
Acid value mg NaOH/g	2.4 ± 0.2	9.9 ± 0.3	7.6 ± 0.1
Free Fatty Acids as Oleic %	1.68 ± 0.02	6.99 ± 0.03	5.35 ± 0.05
Saponification value mg/g	197 ± 0.1	183 ± 0.2	156.2 ± 0.2
Unsaponification matter %	1.99 ± 0.15	2.15 ± 0.06	2.35 ± 0.08

Source: Salimon and Ahmed, 2012.

The data summarized in Table 1 shows the diversity in properties of *J. curcas* oil grown in different climatic region. This implies that there is need to take into cognizance the source of *J. curcas* oil relative to the climate of where it is cultivated before consideration as feedstock in biodiesel synthesis. *Jatropha curcas* oil content is relatively higher than soybean oil which is 18.35% (Akbar et al., 2009). Soybean oil is a major feedstock for biodiesel development, however, the oil content of *Jatropha* makes it a more suitable as feedstock in biodiesel industries. Iodine value indicates the degree of unsaturation in oil resource. The European EN 14214 specifies iodine value of 120 for biodiesel feedstock, lower value of *J. Curcas* oil is due to polymerization of glycerides of higher unsaturated fatty acids resulting in formation of deposits or engine malfunction (Mittelbach, 1996). Furthermore, *Jatropha curcas* oil of relatively low peroxide value compared to some other seed crops (Ahmad et al., 2014), inform an oxidative stability of the oil. This indicates a better property of the semidrying oil purposes.

Besides, pretreatment operation is also a probable factor for *Jatropha* oil property diversity. In a similar research carried out by our group, we studied the trend in property variation of three species of *J. curcas* oil and juxtaposed the significance of the properties in selecting production route in biodiesel production. Biodiesel is produced through different production approaches and the approach that is use may be highly dependent upon the nature of the feedstock and its properties.

We procured two crude *Jatropha curcas* oil sample from two different sources in Malaysia and another one sample oil from Sudan. We characterized all the oil samples and compared their properties. Findings in our study enabled us to suggest the best approach that could be employed in producing biodiesel from these different feedstocks that were considered relative to their properties. Below is the result of our investigation as presented in Table 2.

The summary of the result presented in the Table 2 above shows that *Jatropha curcas* oil has high free fatty acids (FFA) as a measure of the acid value in comparison with some other feedstocks reported in literature (Ahmad et al., 2014). Moreover, Saponification value indicates the nature of fatty acids available in triacylglycerol. The longer the carbon chain of the fat hydrolyzed, the reduced the quantity of acid liberated per gram of sample and hence the reduced the saponification value of such oil sample. The high saponification value of *Jatropha curcas* shows its propensity of soap formation with alkaline transesterification. Hence, it significance as raw material in oleo-chemical industries.

Table 2. Physical and Chemical Properties of Selected *J. curcas* oils

Properties	Units	BD	UK	SD
Moisture content	%	0.06	0.23	0.17
Density	g/cm^3	3.40	3.60	3.60
Velocity	Pa.s	0.053	0.078	0.054
Saponification value	$mgKOH^{-1}$	220.01	215.99	218.79
Peroxide value	mMol/Kg	31.84	53.84	24.64
Acid Value	$KOHg^{-1}$	14.60	8.42	17.20
Phase at room temperature (26°C)		Liquid	Liquid	Liquid

Source: Mohammed et al., 2014b.

CATALYTIC PATHWAY TO THE DEVELOPMENT OF *JATROPHA* BIODIESEL

A catalyst is fundamental in reactions to generate biodiesel, but the catalyst used differs from one biodiesel processing plant to another. Caustic soda (NaOH), Caustic potash (KOH) and sulfuric acid (H_2SO_4) are three commonly utilized as chemical catalysts in biodiesel processing. Other studies had reported the use of enzymes (biological catalyst) (Akoh et al., 2007; Ghaly et al., 2010) in biodiesel industries. Du et al. reported a 20,000 tons of biodiesel plant in China utilizing biological catalyst. Aside this, biodiesel production is most successful with the use of basic chemical catalyst.

Jatropha curcas oil has been utilized in myriad of biodiesel production approaches. Our group explored the various approaches used in producing biodiesel from *Jatropha curcas* oil and the economic importance of the *Jatropha* plant to Malaysia biodiesel industry. In our investigation, we reported comparatively the throughput achieved from the various routes. The Table 3 below shows the summary of the methods adopted as investigated by our group. This table illustrates the diversity of utilization of *J. curcas* oil as feedstock in biodiesel production.

Table 3. Production routes in the development of *Jatropha* biodiesel and comparison of throughput

Production route	Product yield	Advantages	Disadvantages	References
Acid catalyzed reaction	90%	No soap formation	High temp. Requirement Longer reaction time	Kay and Yasir 2012
Alkaline catalyzed reaction	90-95%	High yield Low reaction time and temp.	Soap formation Product contamination	Tiwari et al., 2007
Acid/alkaline catalyzed reaction	90-95%	High yield Low temp requirement	Long reaction time	Kamil et al., 2011; Berchmans and Hirata, 2008
Heterogeneous solid base catalyst	84%	Low temp. Catalyst recycle	Long reaction time Excess alcohol usage	Zhu et al., 2006; Vyas et al., 2009
Super critical alcohol condition	80%, 60-70%	No catalyst is required	High energy requirement	Rathore, 2007
Continuous microwave process	96.5%	Reaction rate increase High yield	High power requirement	Tippayawong Et al., 2002; Liao and Chung, 2011
Reactive extraction	80%	No danger of toxicity	Difficult to commercialize	Kasim and Harvey, 2011
Lipase catalyzed process	98%	Product purity Low reaction temp.	Longer reaction time	Shah and Gupta, 2007
Ultrasonic transesterification	96.4%	Short reaction time Effective in product yield	High power requirement	Deng et al., 2010

Source: Mohammed et al., 2014a.

PERFORMANCE OF *JATROPHA* BIODIESEL IN COMPRESSION IGNITION ENGINES (CIES)

Combustion of fossil fuel in cars, power plants and industries has resulted in indiscriminate release of greenhouse gases. The release of these poisonous gases has altered the climatic conditions of the world over. The devastating effect of which is largely linked with serious health problems, environmental degradation and acid rain (Zhang et al., 2002; Zhu et al., 2010; Qi et al., 2011; Saravanan et al., 2012). Measures including the use of renewable fuel have been put in place to rescue the mother earth from the major effect of global warming due to release of the gaseous pollutants.

Agarwar and Agarwar (2007), investigated the properties of emissions emanating from the utilization of *Jatropha* oil blends in a direct injection compression ignition engine. The proportion of the *Jatropha* biodiesel in the oil blend was found to influence the emissions when benchmarked with pure biodiesel. The effect of injection timing and high injection pressure on the emission characteristics was determined. It was recorded that emission reduction of HC, CO and smoke was achieved with *Jatropha* biodiesel as fuel in diesel engine. It however offered a diverse property when utilized in CIE.

Besides, utilization of biodiesel in diesel engine has been discovered to reduce the release of pollutants such as carbon monoxide, carbon dioxide, hydrocarbon and sulphur (Canakci, 2005, Anand et al., 2011). Nevertheless, a constraint limiting the complete use of this resource is the increased amount of nitrous oxides (NOx) and particulate matter (PM) released in biodiesel in comparison with fossil diesel (Tesfa et al., 2012), consequently, a more stringent regulation had been put on these two pollutants. This possess a stumbling constraint to the diesel fuel substitute especially if biodiesel industry succeeded in revolutionizing the diesel provision of the world.

Many researchers have adopted different approach in an effort to reduce NOx in biodiesel. The main technologies expected to reduce NOx are classified into two categories (Shin et al., 2011). The two approaches are basically geared toward prevention of formation of NOx and conversion of the formed NOx into less harmful constituents. Among the effort put in place was the EGR and injection timing on combustion and emission characteristics of emission produced from a split injection DI-diesel engine utilizing biodiesel (Qi et al., 2011). In the report, NOx was observed to be significantly reduced though there were some complexities in the measure of soot emission, which was not altered.

Furthermore, other studies revealed that to ameliorate the exhaust emissions of biodiesel and its blends in petroleum diesel in various engine, measures such as post treatment techniques like selective catalytic reduction (SCR), exhaust gas recirculation (EGR), diesel oxidation catalysts are to be adopted, (Knothe et al., 2006; Moser, 2009). Jindal et al. (2010) investigated the emissions from pure *Jatropha* oil, diesel fuel and *Jatropha* biodiesel. The study revealed that when the injection pressure and compression ratio were increased, emission of HC, NOx, smoke opacity and temperature of exhaust were reduced in *Jatropha* oil in comparison with conventional diesel. However, *Jatropha* biodiesel was observed to produce increased HC and CO, crude *Jatropha* produced more smoke and particulate matter (PM).

Banapurmath et al. (2008) studied the emission characteristics of DI compression ignition engine run on *Jatropha* biodiesel. The smoke opacity, HC and CO for the *Jatropha* biodiesel was observed to be higher in relation to conventional diesel fuel, while NOx is lower in the *Jatropha* biodiesel compared to diesel as a result of low cylinder peak pressure and lower heat proportion generated. Chauhan et al. (2010) reported the effect of fuel inlet temperature on engine emission utilizing dual fuel engine test rig with tube heat exchanger. High NOx emission was recorded for *Jatropha* oil. Although, pretreatment of the oil resulted in high emission of CO_2 while other emissions such as CO, HC, and smoke opacity were reduced significantly. Predeep and Sharma recommended the use of hot exhaust gas recirculation (EGR) for control of NOx in diesel engine utilizing *Jatropha* oil. Result shows that effective reduction of NOx was achieved. However 15% of EGR was optimized to reduce NOx emission without pronounced effect on the performance, smoke and other emission.

Conclusion

Among the more than three hundred oil seed crops identified, the prospect of *Jatropha curcas* cannot be underestimated. It is a plant that is capable of growth in arid and semi-arid regions with relatively low rainfall. It has high oil content and possess some other value added product that can be utilized for many other purposes. Its advantages as a non-food oil has propelled research in utilizing *Jatropha curcas* oil as feedstock in myriad of production processes for optimum yield at minimum resource utilization.

This chapter reported some comparative findings of researchers and their supposition relative to physical and chemical properties of *Jatropha curcas* oil samples sourced from different locations. Findings informed that *Jatropha curcas* oil in biodiesel production is a prominent resource and the prospect of which cannot be under estimated as being reported in different production approaches. Moreover, the property of *Jatropha curcas* is required for route selection in biodiesel synthesis.

Besides, the use of *Jatropha* biodiesel in blend or pure form has comparatively lowered emission except for NOx, which has been reported to be lowered in some report and raised in some others literature. However, the characteristics emission of *Jatropha* biodiesel as well as the biodiesel properties is observed to be comparative in performance with conventional diesel fuel. Thus it use in compression ignition engine pose no engine malfunction to the engine. Hence, *Jatropha curcas* oil from propagation to utilization is considered a sustainable feedstock among the myriad of oil crop resource for biodiesel synthesis.

References

Abreu, F. R., Lima, D. G., Hamu, C. W. and Suarz, P. A. Z. (2004). Utilization of metal complexes as catalysts in the transesterification of Brazilian vegetable oils with different alcohols. *Journal of Molecular Catalysis A: Chemical*. 209(1): 29-33.

Agarwal, D. and A. K. Agarwal, (2007). Performance and emissions characteristics of *Jatropha* oil (preheated and blends) in a direct injection compression ignition engine. *Applied Thermal Engineering*. 27(13): 2314-23.

Ahmad, J., Yusup, S., Bokhari, A. and Mohd. Kamil, R. (2014). Study of fuel properties of rubber seed oil based biodiesel. *Energy Conversion and Management*. 78: 266-275.

Akbar, E., Yaakob, Z., Kamarudin, S. K., Ismail, M. and Salimon, A. (2009). Characteristic and composition of *Jatropha curcas* oil seed from Malaysia and its potential as biodiesel feedstock. *European journal of scientific research*. 29(3): 396-403.

Akoh, C. C., Chang, S. W., Lee, G. C. and Shae, J. F. (2007). Enzymatic approach to biodiesel production. *Journal of Agricultural and Food Chemistry*.55: 8995-9005.

Anand, K. Sharma, R. P. and Mehta, P. S. (2011). Experimental investigations on combustion, performance and emissions characteristics of neat karanji biodiesel and its methanol blend in a diesel engine. *Biomass and Bioenergy*. 35: 533-541.

Aranda, D. A. G., da Silva, C. C. C. and Detoni, C. (2009). Current processes in Brazilian biodiesel production. *International review of chemical Engineering* (I.RE.CH.E). 1(6): 603-608.

Azam, M. M., Waris, A. and Nahar, N. M. (2005). Prospects and potential of fatty acid methyl esters of some non-traditional seed oils for use as biodiesel in India. *Biomass and Bioenergy*. 29(4):293-302.

Banapurmath, N. R., Tewari, P. G. and Hosmath, R. S. (2008). Performance emissions characteristics of a DI compression ignition engine operated on Honge, *Jatropha*, and sesame oil methyl esters. *Renew Eng*. 33:1982-1988.

Berchmans, H. J. and S. Hirata, 2008. Biodiesel production from crude *Jatropha curcas* L. seed oil with a high content of free fatty acids, *Bioresour. Technol.*, 99(6): 1716-1721.

Canakci, M. (2005). Performance and emission characteristics of biodiesel from Soybean oil. *P Mech Eng. D. J Aut*. 219:1-8.

Chauhan, B. S., Kumar, N. Jun, Y. D. and Lee, K. B. 2010. Performance and emission study of preheated *Jatropha* oil on medium capacity diesel engine. *Energy*, 35(6): 2484-92.

Demirbas, A. (2008). *Biodiesel: A realistic Fuel Alternative for Diesel Engine*. DOI 10.1007/978-1-84628-995-8. Springer. ISBN-13: 9781846289941.

Demirbas, A. (2009) Progress and recent trends in biodiesel fuels. *Energy Conversion and Management*. 50(1):14-34.

Deng, X., Z. Fang and Y-h. Liu, 2010. Ultrasonic transesterification of *Jatropha curcas* L. oil to biodiesel by a two-step process. *Energy Conversion and Management*. 51: 2802-7.

Divakara, B. N., Upadhyaya, H. D., Wani, S. P. and Gowda, C. L. L. (2010). Biology and Genetic improvement of *Jatropha curcas* L.: a review. *Applied Energy*. 87(3):732-742.

Du, W., Li, W., Sun, T., Chen, X. and Liu, D. (2008). Perspectives for biotechnological production of biodiesel and impacts. *Applied Microbiol. Biotechnol*. 79:331-337.

Du, W., Xu, Y., Liu, D. and Zeng, J. (2004). Comparative study on lipase-catalyzed transformation of soybean oil for biodiesel production with different acyl acceptors. *J. Mol. Catal. B: Enz*. 30:125 129.

Dunn, R. O. (2001). Alternative jet fuels from vegetable-oils. *Trans ASAE*. 44:1151-1757.

FAO - Food and Agricultural Organization of the United Nations, (2008). *The State of Food and Agriculture*. Biofuels: Prospects, Risks and Opportunities. FAO, Rome.

FAO (Food and Agricultural Organization). (2012). *Oilseed Market Summary, Trade and Market Division*. Food Outlook.

Foidl, N., Foidl, G., Sanchez, M., Mittelbach, M. and Hackel, S. (1996). *Jatropha curcas* L. As a source for the production of biofuel in Nicaragua. *Bioresour Technol*. 58:77-82.

Ghaly, A. E., Dave, D., Brooks, M. S. and Budge, S. (2010): production of biodiesel by enzymatic transesterification: review. *American journal of biotechnology and biotechnology*. 6(2): 54-76.

Gubitz, G. M., Mittelbach, M. and Trabi, M. (1999). Exploitation of the tropical seed plant *Jatropha Curcas* L. *Bioresour Technol*. 67:73-82.

Hameed, B. H., Lai, L. F., Chin, L. H., (2009). Production of biodiesel from palm oil (*Elaeis guineensis*) using heterogeneous catalyst: an optimized process. *Fuel Process Technol.* 90, 606-610.

Hanna, M. A., Isom, L. and Campbell, J. (2005). Biodiesel: current perspectives and future. *J Sci Ind Res*. 64:854-857.

Jindal, S., B. P. Nandwana, N. S. Rathore and V. Vashistha, 2010. Experimental investigation of the effect of compression ratio and injection pressure in a direct injection diesel engine running on *Jatropha* methyl ester. *Applied Thermal Engineering*, 30(5): 442-8.

Jongschaap, R. E. E., Corré, W. J., Bindraban, P. S. and Brandenburg, W. A. (2007). Claim and facts on *Jatropha curcas* L. Wageningen: *Plant Research International.* B.V.

Kalbande, S. R., More, G. R. and Nadre, R. G. (2008). Biodiesel production from non-edible oils of *Jatropha* and karanj for utilization in electrical generator. *Bioenerg. Res.* 1:170-178.

Kamil, M., Nik, R., Yusup, S. and Umer, R. 2011. Optimization of polyol ester production by tranesterfication of *Jatropha*-based methyl ester with trimethylolpropane using Taguchi design experiment. *Fuel*. 90: 2343-2345.

Kasim, F. H. and Harvey, A. P. (2011). Influence of various parameters on reactive extraction of *Jatropha curcas* L. for biodiesel production. *Chemical engineering Journa.* 171: 1373-1378.

Kay, K. H. and Yasir, S. M. (2012). Biodiesel Production from Low Quality Crude *Jatropha* Oil Using Heterogeneous Catalyst. *APCBEE Procedia*. 3: 23 – 27.

King, A., He, Cuevas, J., Freudenberger. M., Ramiaramanana, D. and Graham, I. (2009). Potential of *Jatropha curcas* as a source of renewable oil and animal feed. *J Exp Bot.* 60(10):2897-2905.

Knothe, G., Sharp, C. A. and Ryan, T. W. (2006). Exhaust emissions of biodiesel, petro-diesel, neat methyl esters, and alkanes in a new technology engine. *Energy Fuels.* 20:403-408.

Korbitz, W. (2002). *New trends in developing biodiesel world-wide*. Asia Bio-Fuels: Evaluating and Exploiting the Commercial Uses of Ethanol, Fuel Alcohol and Biodiesel. 22-23 April 2002, Singapore.

Kumartiwari, A. K., Kumar, A. and Raheman, H. (2007). Biodiesel production from *Jatropha* oil (*Jatropha curcas*) with high free fatty acids: An optimized process. *Biomass. Bioenerg*, 31:569-575.

Liao, C. C., and T. W. Chung, 2011. Analysis of parameters of the microwave-assisted continuous transesterification process of *Jatropha* oil using response surface methodology. *Chemical engineering research and design.* 89 (12): 2575-2581.

Martinez-Herrera J, Siddhuraju, P., Francis, G., Davila-Ortız, G., Becker, K. (2006). Chemical composition, toxic/antimetabolic constituents, and effects of different treatments on their levels, in four provenances of *Jatropha curcas* L. from Mexico. *Food Chem*; 96:80e9.

Meher, L. C., Dharmagadda, V. S. S. and Naik, S. N. (2006b). Optimization of alkali-catalyzed transesterification of Pongamia pinnata oil for production of biodiesel. *Bioresour Technol*. 97:1392-1397.

Meher, L. C., Saga, D. V. and Naik, S. N. (2006a). Technical aspects of biodiesel production by transesterification-a review. *Renew. Sust. Engerg Rev.* 10: 248-268.

Mohammed, N.I., Kabbashi, N.A., Alam, M.Z. and Mirghani, M.E. (2014a). Approaches in the utilization of *Jatropha curcas* oil as feedstock for biodiesel synthesis: A review. *Advances in Environmental Biology*. 8(3): 626-632

Mohammed, N.I., Kabbashi, N.A., Alam, M.Z. and Mirghani, M.E. (2104b). *Jatropha Curcas* Oil Characterization and Its Significance for Feedstock Selection in Biodiesel Production. *IPCBEE*. 65(12): 57-62.

Mohibbeazam, M. M., Waris, A. and Nahar, N. M. (2005). Prospects and potential of fatty acid methyl esters of some non-traditional seed oils for use as biodiesel in India. *Biomass. Bioenerg*. 29:293-302.

Moser, B. R. (2009). Biodiesel production, properties and feedstocks. *In vitro cell. Dev. Biol-Plant*. 45:229-266.

Noureddini, H., Gao, V. and Philkana, R. S. (2005). Immobilized *Pseudomonas cepacia* lipase for biodiesel fuel production from soybean oil. *Bioresour. Technol*. 96: 769-777.

Openshaw, K., (2000). A review of *Jatropha curcas*: an oil plant of unfulfilled promise. *Biomass and Bioenergy*. 19(1):1-15.

Pearl, G. G. (2002). *Animal fat potential for bioenergy use*. In: Bioenergy. The Biennial Bioenergy Conference, Boise, ID, September 22-26.

Pinto, A. C., Guarieiro, L. L. N., Rezende, M. J. C., Ribeiro, N. M., Torres, E. A., Lopes, W. A., Pereira, P. A. P. and Andrade, J. B. (2005). Biodiesel: an overview. *J Brazil Chem Soc*. 16:1313-1330.

Pradeep, V. and Sharma, R. P. (2007). Use of Hot EGR for NOx control in a compression ignition engine fueled with bio-diesel from *Jatropha* oil. *Renewable Energy*. 32:1136-1154.

Pramanik, K. (2003). Properties and use of *Jatropha curcas* oil and diesel fuel blends in compression ignition engine. *Renewable Energy*. 28:239-248.

Qi, D., Leick, M., Liu, Y. and Lee, C. F. (2011). Effect of EGR and injection timing on combustion and emission characteristics of split injection strategy DI-diesel engine fueled with biodiesel. *Fuel*. 90: 1884-1891.

Rathore, V. G., 2007. Synthesis of biodiesel from edible and non-edible Q5 oils in supercritical alcohols and enzymatic synthesis in supercritical carbon dioxide. *Fuel,* 86: 2650-9.

Rodriguez, R. P., Perez, L. G., Alfonso, M., Duarte, M., Caro, R., Galle, J., Sierens, R., Verhelst, S. (2011) Characterization of *Jatropha curcas* oils and their derived fatty acid ethyl esters obtained from two different plantation in Cuba. *Biomass and Bioenergy*. 35(9): 4092-4098.

Salimon, J. and Ahmed, W. A. (2012). Physicochemical Characteristics of tropical *Jatropha curcas* seed Oil. *Sains Malaysiana* 41(3): 313-317.

Saravanan, S., Nagarajan, G., Anand, S. and Sampath, S. (2012). Correlation for thermal NOx formation in compression ignition (CI) engine fuelled with diesel and biodiesel. *Energy*. 42: 401-410.

Shah, S. and M. N. Gupta, 2007. Lipase catalyzed preparation of biodiesel from *Jatropha* oil in a solvent free system. *Process Biochemistry*, 42(3): 409-14.

Sharma Y. C., Singh B., Upadhyay, S. N. (2008) Advancements in development and characterization of biodiesel: a review. *Fuel*; 87:2355-73.

Shin, B., Cho, Y., Han D., Soonho, S., and Chun, K. C. (2011). Hydrogen effects on NOx emissions and brake thermal efficiency in a diesel engine under low-temperature and heavy-EGR conditions. *International journal of hydrogen energy*. 36:6281:6291.

Tesfa, B., Mishra, R., Gu, F., Ball, A. D. (2012). Water injection effects on the performance and emission characteristics of a CI engine operating with biodiesel. *Renewable Energy*. 37: 333-344.

Ting, W.-J., Huang, C.-M., Nair, G., Wu, W.-T., (2008). An enzymatic/acid-catalyzed hybrid process for biodiesel production from soybean oil. *J. Chin. Inst. Chem. Eng*. 39, 203-210.

Tippayawong, N., T. Wongsiriamnuay and W. Jompakdee, (2002). Performance and emissions of a small agricultural diesel engine fueled with 100% vegetable oil: effects of fuel type and elevated inlet temperature. *Asian Journal on Energy and Environment*, 3:139-58.

Tiwari, K. A., A. Kumar and H. Raheman, (2007). Biodiesel production from *Jatropha* oil (*Jatropha curcas*) with high free fatty acids: an optimized process. *Biomass Bioenergy*, 31(8): 569-575.

Vaknin, Y., Ghanim, M., Samra, S., Dvash, L., Hendelsman, D. E. and Samocha, Y. (2011). Predicting *Jatropha curcas* seed-oil content, oil composition and protein content using near-infrared spectroscopy - A quick and non-destructive method. *Industrial crops and products*. 34: 1029-1034.

Vyas, A. P., N. Subrahmayam and P. A. Patel, 2009. Production of biodiesel through transesterification of *Jatropha* oil using KNO3/Al2O3 solid catalyst. *Fuel*, 88(4): 625-8.

Yee, K. F., J. C. S. Wu and K. T. Lee, 2011. A green catalyst for biodiesel production from *Jatropha* oil: optimization study. *Biomass and Bioenergy*, 35: 1739-1746.

Yee, K. F., Lee, K. T., Ceccato, R. and Abdullah, A Z. (2011). Production of bio-diesel from *Jatropha curcas* L. oil catalyzed by SO2_4/ZrO_2 catalyst: Effect of interaction between process variables. *Bioresource Technology*. 102: 4285-4289.

Yong, W., Ou, S., Liu, P.-H., Feng, X., Tang, S., (2006). Comparison of two different processes to synthesize biodiesel by waste cooking oil. *J. Mol. Catal. A: Chem*. 252, 107-112.

Zhang, J. J., Hu, W., Wei, F., Wu, G., Korn, L. R. and Chapman, R. S. (2002). Children's Respiratory Morbidity Prevalence in Relation to Air Pollution in Four Chinese Cities *Environmental Health Perspectives* Volume 110|Number 9|September 2002.

Zhu, H., Z. Wu, Y. Chen, P. Zhang, S. Duan and X. Liu, 2006. Preparation of biodiesel catalyzed by solid super base of calcium oxide and its refining process. *Chinese Journal of Catalysis*, 27: 391-6.

Zhu, L. Cheung, C. S. Zhang, W. G. and Huang, Z. (2010). Emissions characteristics of a diesel engine operating on biodiesel and biodiesel blended with ethanol and methanol. *Science of the Total Environment*. 408: 914-921.

In: The Promising Future of *Jatropha curcas*
Editors: Maira R. Segura-Campos et al.
ISBN: 978-1-63484-989-0
© 2016 Nova Science Publishers, Inc.

Chapter 10

JATROPHA CURCAS SEED OIL AND BIOFUELS

***J. García-Davila*[1,*] *and A. L. Martinez-Ayala*[2,†]**
[1]Universidad Politécnica de Tlaxcala, San Pedro Xalcaltzinco, Tepeyanco, Tlaxcala, México
[2]Instituto Politécnico Nacional, Centro de Desarrollo de Productos Bióticos, Col. San Isidro, Yautepec, Morelos, México

ABSTRACT

Jatropha curcas is a potential source of hydrocarbons that have created interest all over the world for the use of its seed oil as a commercial source of fuel. The oil contains mainly oleic (37-63%) as principal fatty acid unlike other *Jatropha* species, which show predominance of linoleic acid. *Jatropha* can supply raw materials for biodiesel production and can chance marginal lands into valuable economic lands. Bio-hydrogenated diesel is called the second generation bio-diesel and it is produced by the hydrotreatment of triglycerides can convert vegetables oil to biofuels liquids products, such as green naphtha (C_5-C_{10}), green jet fuel (C_{11}-C_{13}), and green diesel (C_{14}-C_{20}). During hydroprocessing of triglycerides, the type of catalyst is one of the most important factors to determine the yield and composition for all unsaturated C = C double bonds have been hydrogenated and the oxygen atoms have been removed in the hydrotreatment process. Tree types of catalysts have been reported for the hydrotreatment process: noble metal catalysts (Pd, Pt, etc.), desulfurized catalysts (Ni-Mo, Co-Mo, Ni-W, Co-W, etc.) and solid acid catalysts (such as H-ZSM-5, SO_4/ZrO_2, etc.) Although C_{18}, C_{17}, C_{16}, and C_{15} are the main products over all catalysts, the amount of light hydrocarbons ($\leq C_{14}$) formed increased when the reaction temperature increased from 300 to 400°C. At these conditions, occur a cracking of normal paraffins caused the increase of light hydrocarbons ($\leq C_{14}$) at high reaction temperatures. Recently, the production of bio-hydrogenated diesel by the hydrotreatment of *Jatropha curcas* seed oil over Ni-Mo based catalysts in a high pressure fixed bed flow reactor system at 350°C under 4 MPa of hydrogen. *Jatropha curcas* L. oil was converted into paraffins by one-step

* Universidad Politécnica de Tlaxcala Av. Universidad Politécnica No. 1, San Pedro Xalcaltzinco, Tepeyanco, Tlaxcala, 90180, México. E-mail: jorge.garcia@uptlax.edu.mx.
† Instituto Politécnico Nacional. Centro de Desarrollo de Productos Bióticos, Carretera Yautepec-Jojutla, Km. 6, calle Ceprobi No. 8, Col. San Isidro, Yautepec, Morelos 62731, México. E-mail: alayala@ipn.mx.

hydrotreatment process. Ni-Mo/SiO_2 catalyst favored the production of C_{18}, to C_{15}. These long normal hydrocarbons have a high melting point and thus the liquid hydrocarbon product has poor cold properties. The use of Ni-Mo supported on zeolites, i.e., Ni-Mo/H-Y or Ni-Mo/H-ZSM5, USY/Al_2O_3 resulted in the production of a large amount of gasoline hydrocarbons.

Considering the problem of biofuels competing with food sources, non-edible plant oil as *Jatropha curcas* seed oil have the potential to be used as feedstock to produce green diesel for the transportation fuel.

Keywords: biodiesel, hydroconversion, fatty acid, transesterification

INTRODUCTION

The limited resources of fossil fuels and the increasing demand for clean hydrocarbon, the increase price of crude oil and environmental fears have been the diverse motives for exploring the production of renewable materials-based in the use of vegetable oils as alternative fuels.

The use of biofuels has promoted in the recent years to contribute to decreasing of CO_2 emissions partial reduction of the dependence of fossil energy resources, mainly petroleum and natural gas. Biofuels play an important role in the fuel quality improvement, as they can be denoted clean fuels since they do not contain aromatics and sulfur compounds. The use of vegetable oils is considered important raw material in the production of biofuels due to their capacity to store large amounts of energy, this capacity is related to their chemical structure.

Different source of triglycerides has been treated to produce green fuels such as palm, sunflower, *Jatropha*, rapeseed and soybean oils. Waste cooking oil has also been used to produce green fuels. Most of the vegetable oils contain linear paraffinic chains between 15 to 18 carbon atoms, mainly monounsaturated chains (C_{16}:1 to C_{18}:1).

The conventional ways to produce biofuels from biomass are by transesterification to obtain biodiesel, by hydroprocessing to produce green diesel, both can be used as additives to diesel fossil. The technology for producing it well established, during the last decade it has been an increasing interest in research and development for their optimal operations conditions. The main drawbacks seem to be the availability of raw material and create the best catalyst to favor de process.

Hydrotreating process are widely used in the petrochemical industry to improve fuel characteristics such as the octane, remove contaminants such as, sulfur, nitrogen, heavy metals, aromatic, oxygenated groups and others. All this in order to get lighter fractions of oil clean, which meet environmental standards quality. The hydroconversion process is similar to the hydrotreating process. The hydroconversion process used as initial material vegetable oil of different qualities. The process resides in the processing of free fatty acids of vegetable oils to get linear hydrocarbon chain length fractions of gasoline and diesel type renewable. This can be generated by renewable fuels that meet the specifications of the International Standards of fossil fuels.

The seed kernel of *Jatropha curcas* have reached 60% weight of oil, this performance depends of many factors such as genotype, soil type, climate and growing conditions The characteristics of *Jatropha curcas* make it are a viable and sustainable for the generation of alternative biofuel. Furthermore, the physicochemical properties of *Jatropha curcas* L. seed oil are recommended for the production of biofuels because of their excellent flow properties at low temperature and high calorific value is slightly less than the petrodiesel. This oil allows for easy incorporation into existing hydrotreating processes, obtaining a mixture similar to fossil fuel also called bio-crude oil.

The commercial incorporation of the culture of *Jatropha curcas*, like renewable source of energy contributes to the economic, social and environmental development, the generation of employments and the reduction in the emission of pollutant gases.

JATROPHA CURCAS SEED OIL

Jatropha curcas is an oilseed plant, is an inedible species endemic with origins in Central America and México. In addition, these are two toxic and non-toxic varieties. The seed kernel of *Jatropha curcas* have reached 60% weight of oil, this performance depends of many factors such as genotype, soil type, climate and growing conditions. *Jatropha curcas* is resistant to drought and can adapt in dry soil, so his crop can be established on land not suitable for other crops, so it does not displace food crops (Openshaw, 2000).

Jatropha curcas plant is resistant and easily adapt to adverse environmental conditions. Therefore, this plant could be cultivated in more extended areas without competing for arable land. (Mendes, 2015) However, this is a toxic plant, and some local varieties in Mexico have low concentrations of toxic compounds and antinutritional agents, such phorbol esters, which allows for the generation of by-products in the extraction process as a protein-rich paste for use in animal feedstocks as well for human consumption.

The characteristics of *Jatropha curcas* make it are viable and sustainable for the generation of alternative biofuel. Furthermore, the physicochemical properties *of Jatropha curcas* seed oil are recommended for the production of biofuels because of their excellent flow properties at low temperature and high calorific value is slightly less than the petrodiesel. This oil allows for easy incorporation into existing hydrotreating processes, obtaining a mixture similar to fossil fuel also called bio-crude oil. Renewable fuels that meet the specifications of the International Standards of fossil fuels can generate this.

Jatropha curcas was introduced to Africa a few centuries ago. It is currently widely distributed throughout the wide ranges of products are made from *Jatropha* (Demirbas, 2003). The by-product from extracting oil from the hulls, and the shells are made into briquettes used to generate biogas and/or applied as organic manure to cultivated areas (Mkoma and Mabiki, 2012). Reveal that the press cake is an excellent fertilizer and made from selling *Jatropha* for bioenergy, household, medicinal and agricultural byproducts improves food security (Figure 1).

Jatropha curcas has the benefit on other plants due to its drought hardiness, rapid growth, easy propagation, low cost of seeds, high oil content (27.0 to 48.5% seed oil content), small gestation period, wide adaptation, production on good and degraded soils and the optimum plant size that makes the seed collection more appropriate (Kumar and Sharma, 2005, Becker and Makkar, 2008). *Jatropha curcas* was identified as potential biodiesel source and comparing with other sources, which has advantages as rapid growth, higher seed productivity, suitable for tropical and subtropical regions in the world. *Jatropha curcas* oil and fossil diesel share similar features that make the former an option to combine or replace the fossil fuel (Becker and Makkar, 2008) (Table 1).

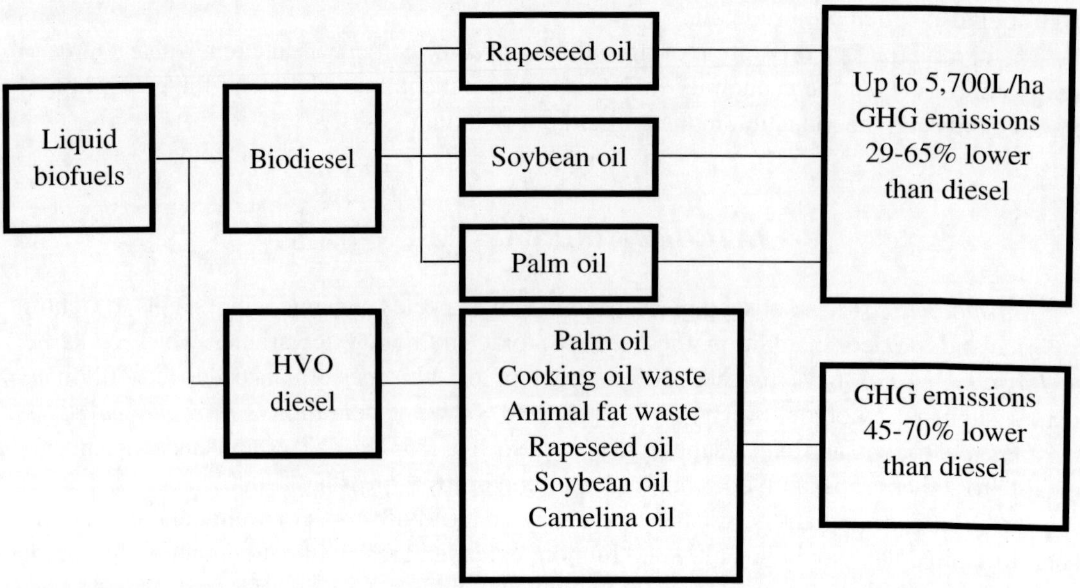

Figure 1. Current feedstock and biofuels in industrial scale. HVO (Hydrotreated Vegetable Oil), GHG (Greenhouse Gases).

Jatropha have interest all over the world for the use of its seed oil as a commercial source of fuel (Mkoma and Mabiki, 2012). The oil contains mainly oleic (37-63%) as principal fatty acid unlike other *Jatropha* species, which show predominance of linoleic acid (Corral, et al., 2015) Oleic acid is the most abundant (44.8%) followed by linoleic acid (34%), palmitic acid (12.8%), and stearic acid (7.3%). While composition of the oil is similar to other oils, which are used for edible purposes, the presence of some antinutritional factors such as toxic phorbol esters reduces this oil inappropriate for use in cooking. Thus, it is a good choice as the starting oil for production of biodiesel.

Oil quality and consistency are important for producing biodiesel. The physical and chemical content of *Jatropha curcas* oil can be extremely variable. Oil characteristics appear to be influenced by environment and genetic interaction, as the seed size, weight and oil content. The mature of the fruits also can affect the fatty acid composition of the oil, and processing and storage.

Jatropha curcas has been promoted as a productive tropical oil seed that can be grown on marginal land (Demirbas, 2003). However, there is little reviewed literature to support these claims. While 10 ton. ha^{-1} has been reported on good soils, more typical yields are round 2 ton. ha^{-1}, this would still be a very small amount of oil per hectare. There are further practical problems. Fruits on the plant mature at very different times and the mixture of unripe, ripe, and overripe fruits borne at the same time excludes mechanized harvesting. The oil of most cultivars is highly carcinogenic, raising serious handling issues. The crop also has a harvest index of just 10% and oil production decrease on acid soils, eliminating the crop from much marginal land (Kant and Wu, 2011). However, it should not be ignored that established crops might have had several parallel problems in their domestication.

Table 1. Typical physical and chemical properties of *Jatropha curcas* seed oil

Properties	*Jatropha* seed oil
Calorofic value	37.8 MJ/Kg
Appearance	Light yellow liquid
Specific gravity at 30°C	0.92
Acid value	1.24
Sulfur	0.13%
Carbone residue	0.024
Kinematic viscosity	50.73 cs
Flash point °C	110-240
Cetane number	51.0
Sulphur %	0.13

Adapted from: cooperation project Daimler Chrysler, CSMCRI and Hohenheim (2008).

Therefore, *Jatropha curcas* seed oil production costs are promising (Becker and Makkar, 2008) as well the physicochemical properties of seed oil (Makkar et al., 2008) allow an easy feeding industrial catalytic hydroprocessing of vegetable oils (CHVO) processes.

BIODIESEL PRODUCTION

Today's diesel engines require a clean burning, stable fuel that performs well under the variety of operating conditions. Biodiesel is the only alternative fuel that can be use directly in any existing unmodified diesel engine. Because it has similar properties to diesel fuel, biodiesel can be blended at any ratio with diesel fuel.

Currently there are a variety of methods and processes for the production of biofuels from vegetable oils. They will mention in Figure 2 only those processes applied *to Jatropha curcas* seed oil for the different renewable liquid biofuels.

TRANSESTERIFICATION

Transesterification is a chemical process in which alcoholysis of oil by methyl or ethyl alcohol in the presence of a strong acid or base produces biodiesel and glycerol. In the transesterification process, C-O cleavages are achieved using catalysts and an alcohol as co-reactant to produce fatty acid methyl esters (biodiesel) (Ramesh et al., 2006), but this product results to be also an oxygenated biofuel (fuel quality directive 2009/30/EC has limited the use of biodiesel to maximum 7 vol. %). Additionally, the coproduction of impure glycerol is a disadvantage of this technology (Chiappero, 2011).

The transesterification catalyzed by a base is much faster and less corrosive than the catalytic acid reaction. Thus, hydroxides are the most commonly used catalysts However, if the feedstock has a high free fatty acid (FFA) content, the alkali causes loss of the free fatty acids as their insoluble soaps. As an alternative, one can conduct an acid-catalyzed reaction that requires higher reaction temperatures (100°C) and longer reaction times than alkali-catalyzed transesterification. (Ramesh et al., 2006) have evaluated *Jatropha curcas* as a source for the production of biofuel in Nicaragua using base- and acid-catalyzed for transesterification.

Have reported very positive results obtained in respect of biodiesel by transesterification compared to fossil diesel in most of their physicochemical properties as summarized in Table 2. However, there are several problems when is used in diesel engine, like coking of injectors on piston and head of engine, carbon deposits on piston and excessive engine wear. These drawbacks are due to high viscosity of raw oil, incomplete combustion of fuel and poor combustion at part load with raw oil (Mikulec et al., 2010).

Natural fats and vegetable oils are potential feedstock for biofuel production. Currently, edible and non-edible vegetable oils are the main feedstock for biodiesel production. These oils, consists of fatty acid methyl esters (FAME) that are produced by transesterification of triglycerides with methanol (Romero et al., 2015) However, the FAME can develop undesirable properties derived from natural oil including low cetane number, high viscosity and low flow properties that are determined by the source oil composition (Murata et al., 2010). These are increasing requirements on the feedstock quality resulting in higher prices.

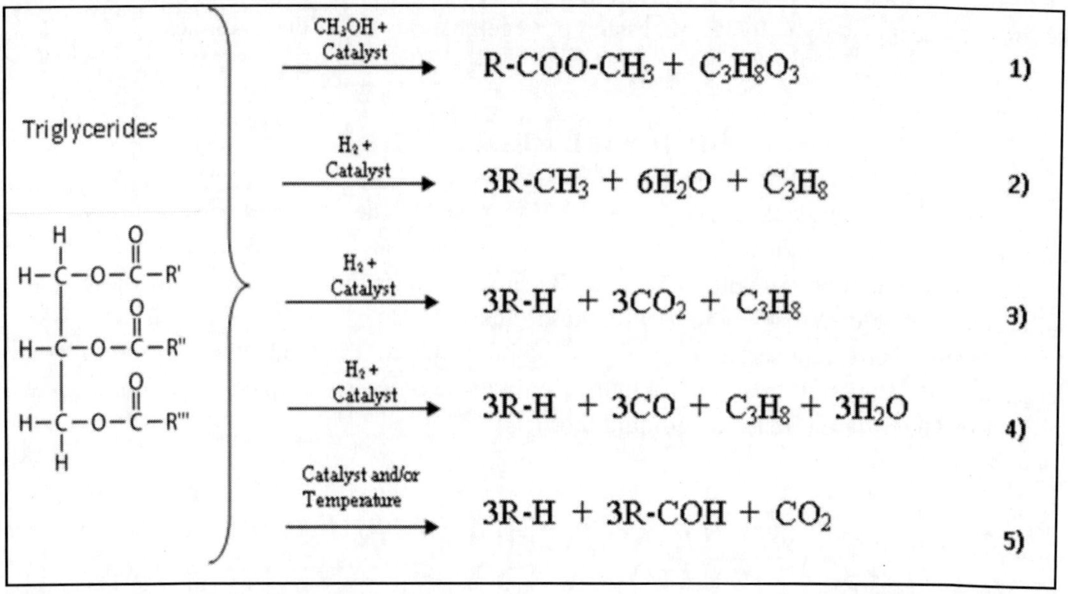

Figure 2. Main reaction paths in transformation of *Jatropha curcas* seed oil into biofuels 1) transesterification, 2) hydrodeoxygenation, 3) hydrodecarboxylayion, 4) decarbonylation, 5) pyrolysis. Process 1), 2) and 3) called hydroconversion (Sotelo et al., 2012).

TRANSESTERIFICATION BY ENZIMES

The possibility of using enzymes in organic solvents has opened up several applications for biotransformations. One strategy has been the use of lipase for transesterification reactions. *Jatropha curcas* seeds themselves are reported to contain lipase activity, which could also catalyze transesterification reactions (Shweta et al., 2004).

Table 2. Fuel properties of biodiesel from *Jatropha curcas* and fossil diesel (Sotelo et al., 2012)

Properties	*Jatropha* Biodiesel	Diesel
Density, g/ml	0.865	0.841
Viscosity @ 40°C, Cst	5.2	4.5
Calorific value, MJ/Kg	39.2	42.0
Flash point, °C	175	50
Cloud point, °C	13	9
Cetane number	47.8	51.0

Many reports have been done about lipase application as catalyze in the transesterification of triglyceride (Köse et al., 2002). In the case of biocatalysts in no aqueous media, immobilization is also reported to result in better activity. *Chromobacterium viscosum lipases* in both free and immobilized were tested for transesterification activity for conversion of *Jatropha curcas* seed oil into biodiesel. The lipases from *Chromobacterium viscosum* are capable of transesterification with a yield of 62% for the ethyl ester after 8 h (Shweta et al., 2004).

Maximum yield was obtained when the amount of buffer salt was 0.2 mmol/g of enzyme with 50 mg of enzyme with immobilized lipase from *Chromobacterium viscosum* under the improved conditions getting a yield of 89% in 10 hr. These results show the capability of using lipases for obtaining biodiesel from *Jatropha curcas* seed oil. Making a comparison between the yields by chemical transesterification (93%) and enzymatic transesterification (92%) were closely the same.

PYROLISIS

Pyrolysis is thermal decomposition of biomass in the absence of oxygen that produces a solid (charcoal), a liquid (pyrolysis oil or bio-oil), and a gas product (Bridgwater et al., 2003). The relative amounts of the three co-products depend on the operating temperature and the residence time used in the process. High heating rates (fast pyrolysis) of the biomass feedstocks at moderate temperatures (450°C to 550°C) result in oxygenated oils as the major products (60% to 75%). With the residue split between a biochar and gases. Slow pyrolysis, also known as carbonization, produces about 35% of charcoal, 30% of tars/oils, and gas. At medium heating rates, about 50% of pyrolysis oil formed has lower oxygen content than those from fast pyrolysis, and about 20% char and gases (Laird et al., 2009).

Pyrolysis of vegetable oils at intermediate temperatures permits the production of hydrocarbon biofuels (Elliott et al., 2012a). Pyrolysis of different raw material under atmospheric pressure and temperatures extending between 350°C and 550°C were reported. Was studied the effects of oil type on products obtained in the presence of zeolite catalysts. Additionally, the use of acidic catalytic materials such as protonic zeolites has been the object of several experiments. Although the use of minor acidic materials as catalysts seems suitable to produce liquid fuels in the diesel range.

Conventional pyrolysis is associated with the promotion of C-O cleavages and poor selectivity due to uncontrollable reactions such as cracking or polymerization of the hydrocarbons, resulting as product a type of biofuel with a high amount of oxygenated compounds as unreacted carboxylic acids, acrolein and ketones (Corral et al., 2015). On the other hand, similar tests for *Jatropha curcas* seed oil deoxygenation were performed in a mechanically stirred 500 mL reactor operating in batch mode at high temperatures and pressures (Elliott et al., 2012a). In a typical experiment, approximately 180 g of *Jatropha* oil and 3% of catalysts were added to the reactor.

Jatropha curcas oil deoxygenation was tested also without catalyst (controlled pyrolysis). Before each experiment, the reactor was purged with nitrogen for 3 min to remove the remaining oxygen and pressurized to around 3 bars. The temperature was increase up to 350°C to 400°C, maintaining isothermal conditions for 3 or 6 hr. At the end of the process, the pressure reaches around 40 and 110 bars at 350°C and 400°C, respectively.

HYDROPROCESSING OF TRIGLYCERIDES

Hydrotreating process are widely used in the petrochemical industry to improve fuel characteristics such as the octane, remove contaminants such as, sulfur, nitrogen, heavy metals, aromatic, oxygenated groups and others. All this in order to get lighter fractions of oil clean, which meet environmental standards quality. The hydroconversion process is similar to the hydrotreating process. The hydroconversion process used as initial material vegetable oil of different qualities. The process resides in the processing of free fatty acids of vegetable oils to get linear hydrocarbon chain length of gasoline and diesel type renewable.

Current thermochemical technologies include hydroprocessing and hydroisomerization technology used in petroleum refineries, although various biochemical strategies are under investigation. The products consist predominantly of isoparaffins with some residual normal paraffins and proportions can be adjusted for diesel fractions or jet fuel fractions. Technoeconomic analyses and size or production are different depending on the feedstock and co-products (Arbogast et al., 2013).

Renewable diesel is a commercial hydrocarbon biofuel introduced in 2007 that reached 10% of biodiesel production by 2013. It is also referred to as “green diesel” and includes Hydro-processed esters and fatty acids (HEFA), Hydrogenated Vegetable Oil (HVO) produced from fatty acids (fats, oils, and greases) or vegetable oils, or tail oil from trees.

As an alternative renewable raw material, the triglycerides present in fats and vegetables oils can be industrially hydroconverted at the petroleum refineries. Bio-hydrogenated diesel, (BHD) Figure 2 is called the second generation bio-diesel and it is produced by the hydrotreatment of triglycerides can convert vegetables oil to biofuels liquids products, such as green naphtha (C_5-C_{10}), green jet fuel (C_{11}-C_{13}), and green diesel (C_{14}-C_{20}) (Liu et al., 2011; García et al., 2014). During hydroprocessing of triglycerides, the type of catalyst is the most important factor to define the yield and composition for all unsaturated C = C double bonds have been hydrogenated and the oxygen atoms have been removed in the hydrotreatment process.

The hydroprocessing of triglycerides involves the hydrogenation of the double bonds of the side chains and the removal of oxygen on the metal sites of the catalysts. The hydrotreating of most vegetable oils leads to the production of C_{15}-C_{18} hydrocarbons, a liquid mixture within the boiling point range of diesel which is normally called "green diesel", "renewable diesel" or "bio-hydrogenated diesel", and therefore with the same chemical nature than petroleum-derived diesel (Murata, K. et al., 2010).

Conventional hydrotreating involves removal of heteroatoms, such as sulfur and nitrogen. The hydrocracking process involves saturation and breakage of C-C bonds to produce high quality gasoline and diesel fuels (Romero et al., 2015). The CHVO process results in better flow properties compared to biodiesel obtained by transesterification (Sotelo et al., 2012) reported an estimated cost processing that represents 50% of transesterification process related costs.

Vegetable oil hydrotreating and its mixtures with heavy vacuum gas oil have been explored for biofuel production by, employing hydrotreating catalysts based on Ni-Mo/Al_2O_3 and Co-Mo/Al_2O_3 (Becker and Makkar, 2008) under HDS nominal operating conventional conditions. Several hydrotreated vegetable oils have been studied including sunflower (Huber, G. W. et al., 2007], cottonseed (Sebos et al., 2009) palm oil (Bezergianni et al., 2009) *Jatropha curcas L* oil (García et al., 2014) and residual oils (Liu et al., 2011). The use of these raw materials on the availability, cost and environment in each country.

In comparison with the process to produce fatty acid methyl esters (FAME) the hydroprocessing of vegetable oils for production green diesel has the advantages that the product is compatible with existing engines, flexibility with the feedstock, e.g., the content of free fatty acids in the vegetable oil does not matter, higher cetane number, higher energy density, it does not increase the emissions of NOx (Sotelo et al., 2012)

Recently, the catalytic hydrotreating process is carried out under similar conditions to those normally used for hydrotreating in oil refineries were used to obtain free-oxygen biofuels from vegetable oils. Followed by mild hydrocracking to improve cold properties, the biofuel produced using this technology was demonstrating to have nearly identical properties to normal petrochemical diesel (Murata et al., 2010).

Three types of catalysts have been reported for the hydrotreatment process: noble metal catalysts (Pd, Pt, etc.), desulfurized catalysts (Ni-Mo, Co-Mo, Ni-W, Co-W, etc.) and solid acid catalysts (such as H-ZSM-5, SO_4/ZrO_2, and Zeolites, etc.) Although C_{18}, C_{17}, C_{16}, and C_{15} are the main products, the amount of light hydrocarbons ($\leq C_{14}$) formed increased when the reaction temperature increased from 300 to 400°C. At these conditions, occur cracking reactions of normal paraffins caused the increase of light hydrocarbons ($\leq C_{14}$) at high reaction temperatures. Hence, the reaction temperature also can be used to regulate the composition of the Bio-Hydrogenated Diesel (BHD) product.

Works reports the reaction pathways of the catalytic hydrotreating of *Jatropha curcas* seed oil using different catalysts (Ni, Pd, CoMoSx, and NiMoSx USY, NiMo/Al_2O_3 and CoMo/Al_2O_3). Using several temperatures, hydrogen pressures, and reaction times were treated through this technology with optimum results. The temperature and pressure the most important parameters on the process, their influence on the intermediate and final products was studied The use of high amounts of hydrogen represents a significant production cost, consequently, needs to be reduce to guarantee the large-scale industrialization of advance biofuels.

Recently, the production of biodiesel by the hydrotreatment of *Jatropha curcas* seed oil over Ni-Mo based catalysts in a high-pressure fixed bed flow reactor system at 350°C under 4 MPa of hydrogen. *Jatropha curcas L.* oil was converted into paraffin's by one-step process. Ni-Mo/SiO_2 catalyst favored the production of n-C_{18}, to n-C_{15} (Kant P., and Wu, S. 2011). These long normal hydrocarbons have a high melting point and thus the liquid hydrocarbon product has poor cold properties. The use of Ni-Mo supported on zeolites, i.e., (nickel-molybdenum) Ni-Mo/H-Y or Ni-Mo/H-ZSM5 (Zeolite Socony Mobil-5), (ultrastable Y-type) USY/Al_2O_3 resulted in the production of a large amount of gasoline hydrocarbons.

Hydrocarbons selectivity (i.e., alkenes and alkanes) increased as temperature and weight hour space velocity (WHSV) increased. At temperatures of 310°C there are no hydrocarbons presence by Gas Chromatography/Mass Spectroscopy Detector (GC/MSD), only large amounts of oxygenates and free fatty acids compounds. Hydrocarbons selectivity increase up to 31% of product at 390°C. The triglycerides degradation only by pressure and temperature effect is slight; the hydroconversion process to hydrocarbons formation is due to catalyst effect (Lui et al., 2011) used NiMo/Al_2O_3-SiO_2 catalyst to hydrogenate *Jatropha* seed oil to produce C_{15}-C_{18} paraffins, which were further cracked to generate -C_{15}-C_{18} iso-paraffins. With a yield of 83.5% at pressure of 4 Mpa, H_2/oil ratio 800 Nm^3/m^3 and Liquid Hourly Space Velocity (LHSV) 7.6 hr^{-1} these parameters are higher values compared with those used in this research. H_2/oil ratio 400 (Normal cubic meter per cubic meter) Nm^3/m^3 pressure of 2.8 Mpa (Murata et al., 2010) report conversion of 83.8% and yield of hydrocarbons of 67.7% with similar conditions only with a mayor pressure of 6.5 Mpa.

Increasing the temperature in the CHVO process of 350°C to 390°C favors the formation of short chain hydrocarbon due to cracking effect of the original fatty acids. These compounds haves the same number of carbon atoms in the molecule as the corresponding fatty acid chain, from which it was derived. However, these products have one less carbon atom than the original fatty acid chain (Figure 3). The presence of heptadecane and pentadecane do suggest that decarboxylation is also occurs.

The liquid yield from the CHVO process was 95 wt %. The gas effluent content consisted primarily of CO_2, CO, H_2O, methane and propane. The liquid effluent was analyzed by Fourier transform infrared spectroscopy (FTIR). The intensity variations in the absorption bands located at 1160 cm^{-1} and 1743 cm^{-1} wave numbers and characteristic of the vibrational modes of carboxylic groups (i.e., C-O and C = O bonds, respectively) were investigated in *J. curcas* seed crude oil (Morgan, T. et al., 2010). The changes in the intensities corresponding to the bands associated with carboxyl groups, especially for the C-O and C = O bonds. The signal reduction was quantified and correlated with the Hydrodeoxygenation (HDO) conversion.

Similar results have been reported by (Chiappero et al., 2011 and Liu et al., 2011) who reported higher conversions of 80% and 90% respectively at higher WHSV values compared with those used by (García et al., 2014). However, the extent of these reactions depends on the catalyst type and hydroprocessing reaction conditions. Alcohols and esters production have also been reported to take place during hydroprocessing of triglycerides (Romero et al., 2015). The use of sulfided catalysts during hydrodeoxygenation of triglycerides under Hydrodesulfurization (HSD) conditions produces carboxylic acids as intermediates (Satyarthi and Srinivas, 2011). Charusiri et al., (2006) report a hydrocarbon yield of 80% with cuts of naphtha, kerosene and paraffin's in diesel range. Using a batch reactor with a HZSM-5 catalyst and a reaction time of 1.5 hr (Huber et al., 2007) work with a similar Ni-Mo catalyst at temperatures from 300 to 450°C and pressure of 5 MPa. With a hydrogen/feed oil ratio of 1600 N m^3/m^3 and obtained linear alkanes in the range of C_{15}-C_{18} as the main products with a yield of 75%.

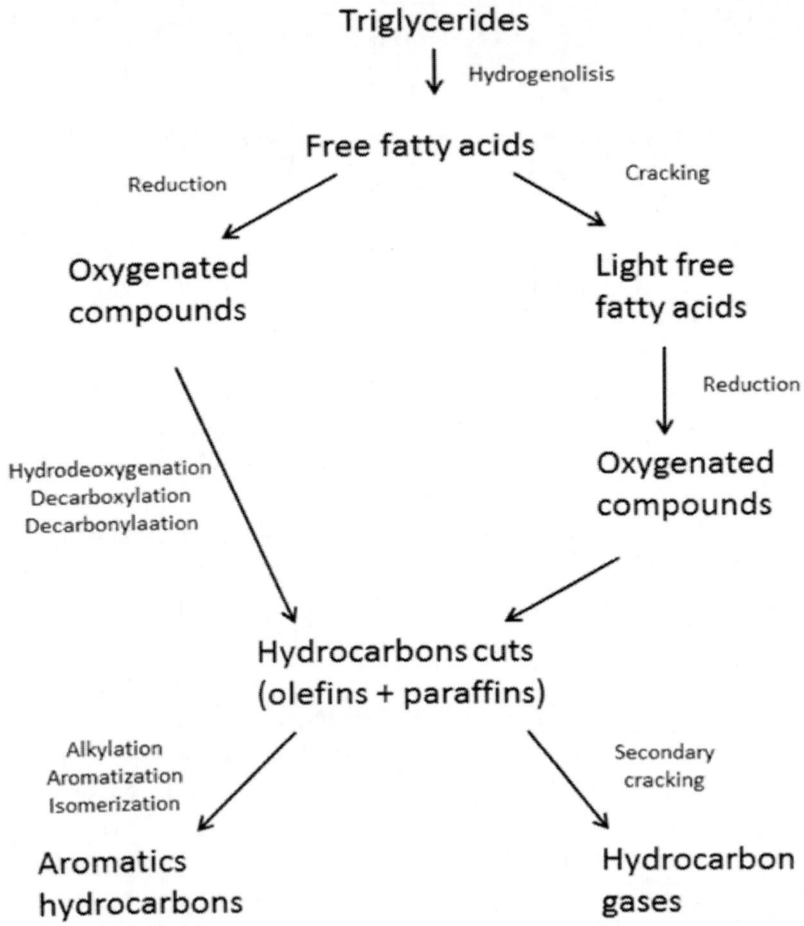

Figure 3. Main reaction pathways in the CHVO process using *Jatropha curcas* seed oil. (García et al., 2014).

Parallel results were achieved by (Satyarthi and Srinivas, 2011) present the following fractions: 3-4% of larger hydrocarbons up to 18 carbons and 1% of hydrocarbons with less than 15 carbons, using a commercial NiMo/C-Al_2O_3 catalyst at 340°C and 40 bar, hydrogen flow was set at 50 mL/min. On the other hand, (Liu, Y. et al., 2011) performed the hydrotreatment of residual oil at similar conditions. (temperature 350°C, hydrogen pressure of 2 MPa, LHSV of 15.2 h^{-1} with (ruthenium) Ru/HY type and Ru/Al_{13}-Mont catalysts, with a tubular reactor with internal diameter of 1 cm, a length of 50 cm and a hydrogen/oil ratio of 400 ml/ml). Established in the liquid phase yields of 82.1-84% of hydrocarbons about 42.5% gasoline and 56.5% for hydrocarbon fractions including diesel and jet fuel was reported.

We observe that the CHVO process is feasible at a temperature of 390°C, a pressure of 2.8 Mpa and a hydrogen/feed oil ratio of 400 N m3/m3 to obtain a hydrocarbon base for the production of biofuels. Comparatively, (Huber et al., 2007) reported temperatures of 300-450°C and pressures of 5 MPa with a hydrogen/oil feed ratio of 1600 N m^3/m^3.

Considering the problem of biofuels competing with food sources, non-edible plant oil as *Jatropha curcas* seed oil have the potential to be used as feedstock to produce green diesel for the transportation fuel.

ALTERNATIVE RESEARCH FOR *JATROPHA CURCAS* OIL AND BIODIESEL

One of the means of increasing the diesel fuel production lies in the utilization of waste and excess triacylglycerols (TAG) of vegetable and animal origin in a catalytic process. To the advantages of the process, the utilization of hydrorefining catalysts and excellent emission profile of the products can be included. As a disadvantage is, the high unit investment costs mentioned. Such disadvantages could be eliminating via connecting the process of hydrogenation desulphurization of gas oil with TAG hydrodeoxygenation by means of technology modification and relatively simple adjustment of the corresponding equipment (Smagala et al., 2012).

The catalytic hydroprocessing of vegetable oils (CHVO) is a promising alternative for biofuel production (Mendes et al., 2015), which could employ the existing infrastructure of petroleum refineries (Shweta et al., 2004).

CONCLUSION

Jatropha curcas is a potential source to be used as feedstock to produce energy, which can be effectively obtained with the highest technology. Green diesel is currently not a substitute of fossil fuels, but it is useful as additive to boost the properties of their corresponding fossil fuels.

The transform technology will probably continue to develop in the present years, but there are some factors that need to take into consideration. The catalyst plays an important part in the reaction process to optimize. Is necessary the use of kinetics models, to understand the chemistry in the hydrocracking, hydroisomerization and hydrodeoxygenation pathways reactions in the hydroconversion process.

REFERENCES

Abreu, F. *Alternative by-products from Jatropha.* (2008) Available from: http://www.ifad.org/events/jatropha/harvest/F_Abreu.ppt.

Arbogast, S., Bellman, D., Paynter, J. D. and Wykowski, J. (2013). Advanced biofuels from pyrolysis oil. Opportunities for cost reduction. *Fuel Processing Technology, 106*, 518-525.

Becker, K. and Makkar, H. P. S. (2008). *Jatropha curcas*: A potential source for tomorrow's oil and biodiesel. *Lipid Technology, 20(5)*, 104-107.

Bezergianni, S. and Kalogianni, A. (2009). Hydrocracking of used cooking oil for biofuels production. *Bioresource Technology, 100*, 3927-3932.

Bridgwater, A., Czemik, S., Diebold, J., Meier, D., Oasmaa, A., Piskorz, J. and Radlein, D. (2003). *Fast Pyrolysis of Biomass: A Handbook*, Kingdom United, CPL Press, Berks.

Charusiri, W., Yongchareon, W. and Vitidsant, T. (2006). Conversion of used vegetable oils to liquid fuels and chemicals over HZSM-5, sulfated zirconia and hybrid catalysts. *Korean Journal of Chemical Engineering, 23(3),* 349-55.

Chiappero, M., Do, P. T. M., Crossley, S., Lobban, L. L. and Resasco, D. E. (2011). Direct conversion of triglycerides to olefins and paraffins over noble metal supported catalysts. *Fuel, 90*, 1155-1165.

Corral, S., Legna, D. and Romero D. (2015). Integrated assessment of biofuel production in arid lands: *Jatropha* cultivation on the island of Fuerteventura. *Renewable and Sustainable Energy Reviews, 52*, 41-53.

Demirbas, A. (2003) Biodiesel fuels from vegetable oils via catalytic and non-catalytic supercritical alcohol transesterifications and other methods: a survey. *Energy Conversion and Management, 44*, 2093-109.

Egharevba, H. O. and Kunle O. F., (2013) Broad Spectrum Antimicrobial Activity of Extracts of *Jatropha curcas. Journal of Applied Pharmaceutical Science, 3(4)*, 83-87.

Elliott, D. C., Hart, T. R., Neuenschwander, G. G., Rotness, L. J., Olarte, M. V., Zacher, A. H. and Solantausta, Y. (2012a). Catalytic Hydroprocessing of Fast Pyrolysis Bio-oil from Pine Sawdust. *Energy and Fuels, 26*, 3891-3896.

Huber, G. W., O'Connor, P. and Corma, A. (2207). Processing biomass in conventional oil refineries: production of high quality diesel by hydrotreating vegetable oils in heavy vacuum oil mixtures. *Applied Catalysis A: General, 329*: 120-9.

García, J., Ocaranza, E., Rojas, M., Muñoz., J. A., Ramírez, J. and Martínez, A. L. (2014). *Jatropha curcas* L. oil hydroconversion over hydrodesulfurization catalysts for biofuel production. *Fuel, 135*, 380-386.

Kant P. and Wu, S. (2011). The Extraordinary Collapse of Jatropha as a Global Biofuel. *Environmental Science and Technology, 45*, 7114-7115.

Köse, Ö., Tüter, M. and Aksoy, H. A. (2002). Immobilized Candida antárctica lipase-catalyzed alcoholysis of cotton seed oil in a solvent-free médium. *Bioresource Technology, 83,* 125-129.

Kubickova, I. and Kubicka, D. (2010). Utilization of Triglycerides and Related Feedstocks for Production of Clean Hydrocarbon Fuels and petrochemicals: A Review. *Waste Biomass Valor, 1*, 293-308.

Kumar, N. and Sharma, P. B. (2005). *Jatropha curcas* - A sustainable source for production of biodiesel. *Journal of Scientific and Industrial Research, 64*, 883-889.

Laird, D. A., Brown R. C., Amonette, J. E. and Lehmann, J. (2009). Review of the pyrolysis platform for coproducing bio-oil and biochar Biofuels. *Bioproducts and Biorefining, 3*, 547-562.

Liu, Y., Sotelo, R., Murata, K., Minowa, T. and Sakanishi, K. (2011). Hydrotreatment of vegetable oils to produce bio-hydrogenated diesel and liquefied petroleum gas fuel over catalysts containing sulfided Ni-Mo and solid acids. *Energy and Fuels, 25*, 4675-4685.

Maher, K. D. and Bressler, D. C. (2007). Pyrolysis of triglyceride materials for the production of renewable fuels and chemicals review. *Bioresource Technology, 98*, 2351-2368.

Makkar, H. P. S, Martínez, J. and Becker, K. (2008). Variation in seed number per fruit, seed physical parameters and contents of oil, protein and phorbol ester in toxic and no-toxic genotypes of *Jatropha curcas. Journal of plant Sciences, 3*, 260-265.

Mendes, G., Reynaldo, V., Joly, C. A. and Verdade, L. M. (2015) *Bioenergy and Sustainability: Bridging the gaps.* Scientific Committee on Problems of the Enviroment (SCOPE), Sao Paulo.

Mikulec, J., Cvengros, J., Joríkova, L., Banica, M. and Kleinova, A. (2010). Second generation diesel fuel from renewable sources. *Journal of Cleaner Production, 18*, 917-926.

Miller, P., Sultana, A. and Kumar, A. (2012). Optimum scale of feedstock processing for renewable diesel production. *Biofuel, Bioproducts and Biorefining, 6*, 188-204.

Mkoma, S. L. and Mabiki, F. P. (2012). Jatropha as energy potential biofuel in Tanzania, *International Journal of Environmental Sciences, 2(3),* 1553-1563.

Morgan, T., Grubb, D., Santillan, E. and Crocker, M. (2010). Conversion of triglycerides to hydrocarbons over supported metal catalysts. *Topics in Catalysis, 53,* 820-829.

Murata, K., Liu, Y., Inaba, M. and Takahara, I. (2010). Production of Synthetic Diesel by Hydrotreatment of *Jatropha* Oils Using Pt-Re/H-ZSM-5 Catalyts, *Energy and Fuels, 24*, 2404-2409.

Openshaw K. (2000). A review of *Jatropha curcas*: an oil plant of unfulfilled promise. *Biomass Bioenergy, 19,* 1-15.

Ramesh, D., Samapathrajan, A. and Venkatachalam, P. (2006). Production of biodiesel from *Jatropha curcas* oil by using pilot biodiesel plan. *Agrl. Engg. College and Research Institute*, *1*, 1-6.

Romero, M., Pizzi, A., Toscano, G., Casazza, A. A., Busca, G., Bosio, B. and Arato, E. (2015). Preliminary experimental study on biofuel production by deoxygenation of *Jatropha* oil, *Fuel Processing Technology 137*, 31-37.

Satyarthi, J., K. and Srinivas, D., (2011). Fourier transform infrared spectroscopic method for monitoring hydroprocessing of vegetable oils to produce hydrocarbon-based biofuel. *Energy Fuels, 25,* 3318-22.

Sebos, I., Matsoukas, A., Apostolopoulos, V. and Papayannakos, N. (2009). Catalytic hydroprocessing of cottonseed oil in petroleum diesel mixtures for production of renewable diesel. *Fuel, 88,* 145-149.

Shweta S., Shweta S. and Gupta M. N. (2004). Biodiesel Preparation by Lipase-Catalyzed Transesterification of *Jatropha* Oil, *Energy and Fuels, 18,* 154-159.

Smagala, T. G., Christensen, E., Chistison, K. M., Mohler, R. E., Gjersing, E. and McCormick, R. L. (2012). Hydrocarbon Renewable and Synthetic Diesel Fuel Blendstocks: Composition and Properties. *Energy and fuels, 27,* 237-246.

Sotelo, R., Trejo, F. and Hernández, F. J. (2012). *Hydrogenation: Hydroconversion of Triglycerides into Green Liquid Fuels* Prof. Iyad Karamé (Ed.), ISBN: 978-953-51-0785-9, InTech, DOI: 10.5772/48710. Available from http://www.intechopen.com/books/hydrogenation/hydro-conversion-of-triglycerides-into-green-liquid-fuels.

In: The Promising Future of *Jatropha curcas*
Editors: Maira R. Segura-Campos et al.

ISBN: 978-1-63484-989-0
© 2016 Nova Science Publishers, Inc.

Chapter 11

PERSPECTIVE OF BIOENERGY CROP *JATROPHA CURCAS* IMPROVING AGRICULTURE AND MITIGATION OF GLOBAL CLIMATE CHANGE

Santosh Ranjan Mohanty** and Bharati Kollah
Indian Institute of Soil Science,
Indian Council of Agricultural Research, Nabibagh, Bhopal, India

ABSTRACT

Climate change has caused irreversible changes to our environment. Greenhouse gasses (GHG) emitted from human activities as well as from natural ecosystem are the main cause for climate change. Rise in atmospheric GHG like CH_4, N_2O, and CO_2 are the main cause of global change. Renewable bioenergy holds promise to resolve energy crisis and countries across the globe are driven exploring different sources of bioenergy. One such is bioenergy from *Jatropha curcas*, which is widely recognized as source for biodiesel. Intensive cultivation of Jatropha is under serious consideration globally, but there is need of better understanding of how this plant can be used not only for biodiesel but also for agriculture and climate mitigation as well. Manuscript addresses perspective of *J. curcas* towards improving soil biological health, carbon sequestration and greenhouse gas emission reduction. *J. curcas* harbors many important microbial species in its rhizosphere including N fixers, P solubilizers, PGPRs, mycorrhiza fungi etc. Many important arbuscular mycorrhizae belonging to *Acaulospora sp., Gigaspora sp., Glomus sp., Sclerocystis sp.* and *Scutellospora sp*. Bacterial species like *Azotobacter, Rhizobium, Pleomorphomonas diazotrophica*, *Bacillus megaterium, Bacillus thuringiensis* has been found associated with *J. curcas* rhizosphere. These microbial groups are the key to improve soil function important for agriculture and maintaining soil health. Manuscript highlights different bioactive molecules like organic acids, amyrin, flavonoids, phytates, etc present in different parts of the *J. curcas* and perspective improving agriculture. *J. curcas* cultivated in one hectare of land can absorb as much as 25 tonnes of CO_2 yr^{-1}. Due to high photosynthesis rate, *J. curcas* accumulates significant amounts of C to soil. Overall, a 3 – 5 year old plantation may add around 4000 kg plant biomass equivalent yr

* Email: mohantywisc@gmail.com.

[1]. Manuscript highlights C sequestration potential of *J. curcas* in different ecological scenarios and usefulness rehabilitating degraded lands. *J. curcas* biomass significantly stimulates CH_4 and CO_2 production from soil and inhibits CH_4 consumption activity. Amendment of *J. curcas* biomass at 1% level increased CH_4 production potential of soil to nearly 50-fold than un-amended soil. Soil CO_2 production also doubled when the J. curcas biomass was added. Data reveal optimum level of biomass incorporation to manage its biomass and at the same time mitigate GHG emission. Manuscript concluded with recommendation of *J. curcas* cultivation and approaches managing agriculture, ecosystem and global change issues.

Keywords: *J. curcas*, rhizosphere, microorganism, agriculture, greenhouse gas, climate change

INTRODUCTION

Jatropha curcas (physic nut) is a multipurpose shrub that has gained popularity as a potential biofuel crop. It is member of the Euphorbiaceae family, native to tropical America and is widely adapted to tropical and subtropical Asia and Africa. However, it can also grow in temperate conditions and arid and semi-arid soils (Jongschaap et al., 2007). Its seeds contain up to 34% oil that is easily converted to biodiesel, which has stimulated a growing interest in its use as a renewable energy source. Jatropha can tolerate high temperatures and low fertility; adopted well under arid and semi-arid conditions (Augustus et al., 2002). Establishment of Jatropha plantations can put marginal land into productive utilization by way of biofuel production. Additionally, when marginal lands are planted with Jatropha, it reclaims them by exploring the soil with an adequate deep root system that results in recycling of nutrients from deeper soil layers. It is assumed that marginal lands might have restored by increase of soil organic matter. While having positive effect on the soil quality Jatropha could enhance the productivity of rural poor farmers for livelihood. Approximately 15% of Earth's land area has been degraded through anthropogenic activities, water erosion, wind erosion, chemical degradation and physical degradation (Becker and Francis, 2000; Francis et al., 2005). Hence, *Jatropha* can be one of the potential crops to restore soil fertility of normal and marginal lands (Becker and Francis, 2000). Because of these attributes, Jatropha cultivation has been promoted across the globe as a crop to enable enhance socio economic status of developing countries and reduce the dependency on fossil fuels, and as a climate change mitigation strategy.

Its rhizosphere harbors myriad of microbial community and aid in nutrient uptake. Phosphorus is one of the major limiting factors for crop production on many tropical and subtropical soils (Norman et al., 1995). Inorganic fertilizer may affect soil biological properties and this practice may further reduce the soil function including microbial mediated nutrient transformations. To enhance productivity of *J. curcas,* amendment of organic fertilizers and biofertilizers can be the best practice (Dobson et al., 1997; Mbagwu, 1992). For instance, a study suggest that nitrogen and phosphorus fertilization effectively promotes the growth of *Jatropha* (Sinha et al., 2011).

Although there are apprehensions about negative impact of *J. curcas* on soil quality due to its allelopathic effect, but there is no scientific evidence to support it. Allelopathy generally refers to the beneficial or harmful effects of one plant on another plant, from the release of biochemicals. These chemicals (also known as allelochemicals) are released from plant parts by leaching, root exudation, volatilization, residue decomposition, and other processes in both natural and agricultural systems. Allelochemicals with negative allelopathic effects are an important part of plant defense against competitors and herbivores (Bertin et al., 2003). To exploit this crop to its full potential, there is need of better understanding on the plant and its interaction with ecosystem. Several studies have defined the potential of Jatropha bioenergy and its contribution to climate change mitigation by increasing soil carbon stocks. This is most effective when *J. curcas* is grown on marginal land. This chapter highlights microbial diversity associated in the rhizosphere of this crop. Different phytochemicals and their significance towards agriculture have been described. Biomass production, relation between plant diameter and above and below ground biomass has been discussed to define carbon sequestration potential of this crop.

Microbial Community Associated with *Jatropha curcas*

It is unclear how Jatropha is able to grow in limited nutrient or degraded soil. Scientists have explored the rhizosphere to find the clues for this query. Few groups have studied microbial community dynamics in the rhizosphere. In a study abundance of total heterotrophs, N_2 fixers, and P solubilizers estimated in bulk soil, rhizospheric soil, rhizosplane and endorhizosphere. Microbial groups irrespective of nutrient cycling activity were highly populated in the rhizosplane followed by rhizosphere and bulk soil (Mohanty et al., 2012). In addition soil enzymatic activity analyzed to characterize the differential microbial activity in bulk and rhizospheric soil. Microbial enzymatic activity comprising fluorescent diacetate assay (FDA), dehydrogenase (DHA) and phosphatase were more in the rhizospheric soil compared to the bulk soil irrespective of soil types. FDA ranged from 5 to 10 μg fluorescein released g^{-1} soil h^{-1}. Whereas, DHA varied in the range of 0.79 to 2.18 μg TPF released g^{-1} soil day^{-1}. Results confirmed high microbial metabolism in the rhizosphere. Rhizosphere of Jatropha has specific microbes that usually play relevant role during nutrient availability for plants. Prevalent microbial groups were arbuscular mycorrhizal fungi (AMF), actinomycetes, nitrogen-fixing bacteria, and plant growth-promoting rhizobacteria (PGPR). In a study AMF associated with rhizosphere of *J. curcas* L. from different ecological regions of India were identified by taxonomical key features of spores as described by (Pérez and Schenck, 1990). A total of 20 AMF species were recorded, which consisted of two species of *Acaulospora* and 18 species of *Glomus*. Among the AMF, the *Acaulospora* species were detected from the rhizosphere of *J. curcas* L. and *P. pinnata*. *Glomus* species were more abundant in the soil samples than Acaulospora and other AMF species (Kamalvanshi et al., 2010a). Similar screening of arbuscular mycorrhizal (AM) fungi on *Jatropha curcas* were carried out which identified altogether, 21 AM fungal spores. *Glomus* sp is the most predominant among the recorded AM fungal spores. A small number of *Sclerocystis species* were recorded in all rhizospheric samples. In *Jatropha curcas*, *Acaulospora* sp., were more abundant followed by *Gigaspora* sp, *Glomus* sp, *Sclerocystis* sp, *and Scutellospora* sp. In other hand soil samples from *Pongamia pinnata* were predominantly occupied by *Glomus* sp followed by

Acaulospora sp, *Gigaspora* sp, *Sclerocystis* sp, and *Scutellospora* sp. (Venkatesh et al., 2009).

Few papers have acknowledged the presence of nitrogen-fixing *Azospirillum* sp. in the roots of biomass biofuel crop *Miscanthus.* For instance, (Eckert et al., 2001) detected a new species, *Azospirillum doebereinerae* sp. nov. and (Kirchhof et al., 1997) found *A. lipoferum* and *Herbaspirillum* sp. Plant growth promoting gamma-proteobacterium, *Enterobacter cancerogenus* MSA2 were isolated from the rhizosphere of *Jatropha curcas L. E. cancerogenus* produced growth hormone ACC (1-amino-cyclopropane-1-carboxylate) deaminase, phytases, auxin IAA (indole acetic acid), siderophores, ammonia and it solubilized phosphate. On sowing of *Jatropha* seeds coated with bacterial inoculums in earthen pots with regularly watered sandy loamy soil, the *Enterobacter* was able to enhance various parts of *J. curcas* such as root length by 124.14%, fresh root mass by 81%, fresh shoot mass by 120.02%, dry root mass by 124%, dry shoot mass by 105.54%, leaf number by 30.72%, chlorophyll content by 50.41% and biomass by 87.20% as compared to control under experimental conditions (Jha et al., 2012).

BIOACTIVE MOLECULES FROM *JATROPHA CURCAS*

J. curcas contains numerous bioactive compounds that are of significance to industrial and medicinal values (Islam et al., 2011). Identification of these phytochemicals revealed that there is still untapped pool of bioactive chemicals. Allmost all parts of the plant are rich source of functional chemicals. Like other perennial plants, *J. curcas* has main stem, branch, and leaves. These aerial parts contain organic acids (Aiyelaagbe et al., 2011; Devappa et al., 2010), amyrin in stem bark (Islam et al., 2011; Misra and Misra, 2010), terpenes in leaves (Boateng and Kusi, 2008; Jummai and Okoli, 2014), curcins in latex (Falodun et al., 2014; Santos et al., 2014), phorbolesters in seeds (Agarwal et al., 2013; Bashir et al., 2013), saponins in kernel (Gámez-Meza et al., 2013; Pasha et al., 2013; Sharath et al., 2014), and sitosterol in roots (Abdelgadir and Van Staden, 2013; Khalil et al., 2013). Phytochemicals found in *J. curcas* can be explored for bio-prospecting new and novel bioactive compounds. Many studies demonstrate the efficacy of *J. curcas* against a wide array of bacteria (Paulillo et al., 2014; Sabandar et al., 2013; Setha et al., 2014) and fungi (Bashir et al., 2013; Colvin and Lambert, 2013). Anticancer and antitumor properties of *J. curcas* has been well acknowledged (Fernandes et al., 2013; Sabandar et al., 2013). Its leaves contained 1.99% N, 0.14% P, 1.08% K, 2.95% total carbohydrates, 12.46% protein, 2.47 mg/g total chlorophyll, 0.40 mg/g carotenoids, amino acids (0.0159 mg/g aspargine, 0.0306 mg/g proline, 0.0127 mg/g cystine and 0.0205 mg/g histidine), vitamins (0.064 mg/g B1, 0.121 mg/g B2, 0.058 mg/g pantothenic acid, 0.049 mg/g niacin, 0.229 mg/g inositol, 0.75 mg/g α-tocopherol, 0.18 mg. g^{-1} γ-tocopherol and 0.30 mg/g), 1.29 mg/g phenolic acid, 0.540 mg/g flavonoids, mg/g 0.870 tannins (Ahmed and Zabar, 2013). Chemical composition of *Jatropha* from different geographical locations were evaluated (Martinez-Herrera et al., 2006). *J. curcas* fruit contains shell and seed. Later contains husk and kernel (Singh et al., 2008). The seed kernels were rich in crude protein (31–34.5%), lipid (55–58%) and about 6% starch and total soluble sugars. The major fatty acids found in the oil samples were oleic (41.5–48.8%), linoleic (34.6–44.4%), palmitic (10.5–13.0%), stearic acids (2.3–2.8%), cis-11-eicosenoic acid (C20:1), and cis-11,14-eicosadienoic acid (C20:2) were found in the oil (Akbar et al., 2009; Caballero-Mellado et al., 2007). Phorbolesters were present in high concentrations in the kernels (3.85 mg/g dry meal) in few samples (Gübitz et al., 1999; Martinez-Herrera et al., 2006). Trypsin inhibitors, phytates (8.5–9.3% of dry meal as phytic acid equivalent), saponins (2.1–2.9% of dry meal) and lectins (0.35–1.46 mg/ml) were found in all the seed meals. In a study the chemical composition of seed were evaluated and found that it contains 66.4% oil present in the seed. Triacylglycerol is the dominant lipid species, mainly 1,2 Dioleoyl-3-linoleoyl-rac-glycerol. Linolenic acid was the dominant fatty acid in the oil. Ten sterols and thirteen triterpene were also identified in the unsaponifiable fraction of the oil (Adebowale and Adedire, 2006). Roots of the *J. curcas* exhibited bioactive compounds identified. Compounds were 5α-stigmastane-3, 6-dione, nobiletin, β-sitosterol, taraxerol, 2S-tetracosanoic acid glyceride-1,5-hydroxy-6,7-dimethoxycoumarin, jatropholone A, jatropholone B, 6-methoxy-7-hydroxycoumarin, caniojane, 3-hydroxy-4- methoxybenzaldehyde, 3-methoxy-4-hydroxybenzoic acid and daucosterol (Kong et al., 1996). Phytochemicals from *J. curcas* should be studied extensively. There is lot of potential utilizing these chemicals from *J. curcas* for beneficial use in agriculture and environmental problems.

CARBON SEQUESTRATION AND CLIMATE CHANGE MITIGATION BY *J. CURCAS*

Global warming is due to rise in atmospheric CO_2, which traps heat and warms the planet. Burning of fossil fuels like coal, oil and natural gas for energy or cutting down and burning forests causes accumulation of CO_2 in atmosphere. Certain waste management and agricultural practices aggravate the problem by releasing other potent global warming gases, such as CH_4 and N_2O. Carbon sequestration means capturing CO_2 from the atmosphere or capturing anthropogenic CO_2 from large-scale stationary sources like power plants before it is released to the atmosphere. Once captured, the CO_2 gas (or the carbon portion of the CO_2) is put into long-term storage. CO_2 sequestration has the potential to significantly reduce the level of carbon that occurs in the atmosphere as CO_2 and to reduce the release of CO_2 to the atmosphere from major stationary human sources, including power plants and refineries. Terrestrial (or biologic) sequestration means using plants to capture CO_2 from the atmosphere and then storing it as carbon in the stems and roots of the plants as well as in the soil. In photosynthesis, plants take in CO_2 and give off the O_2 to the atmosphere as a waste gas. The plants retain and use the carbon to make biomass. When the plant dies, part of the carbon from the plant is preserved in the soil. Terrestrial sequestration is a set of land management practices that maximizes the amount of carbon that remains stored in the soil and plant material for the long term. No-till farming, wetland management, rangeland management, and reforestation are examples of terrestrial sequestration practices that are already in use. Terrestrial sequestration does not store CO_2 as a gas but stores the carbon portion of the CO_2. If the soil is disturbed and the soil carbon exposed then soil carbon with atmospheric O_2 form CO_2 and released to atmosphere, reducing the amount of carbon in storage (Ciais et al., 2014).

CARBON SEQUESTRATION BY *J. CURCAS*

Role of *J. curcas* in improving soil quality and carbon sequestration potential has been studied in many parts of the world. High growth rate and ability to grow under limited environment makes this plant a potential agent to enhance soil carbon sequestration. So far, the carbon sequestration function of Jatropha trees has received much less attention than its oilseed production potential, but there is a growing interest in exploiting its potential carbon sequestration services. Moreover, the carbon sequestration capacity of Jatropha shrubs is an essential determinant of the overall GHG reduction performance of its biofuel (Bailis and Baka, 2010). Two aspects are crucial for reliable carbon sequestration estimations. First, we have to compare the biomass content of *J. curcas* with other representative trees, growing under same ecological circumstances. Second, we have to find out how much of its biomass actually constitutes carbon. Dry biomass accumulation by *Jatropha curcas* tree has been addressed by many workers (Benge, 2006). However, these early reports were based on rough estimations. Therefore destructive methods are recommended, where the weight of a Jatropha tree is determined by digging out and weighing the tree. This methodology gives accurate data of the biomass that is produced in a Jatropha tree.

An experiment predicted above ground variables (height and weight) from basal diameter and crown-depth (Ghezehei et al., 2009). Tree samples of different age/ height were cut off at ground level and all relevant variables were measured just before cutting. Samples represented representing leaves, branches and stem were oven dried and then weighed. Highly significant relationships were found between the above ground variables and the basal stem diameter and the crown depth, respectively. The allometric equations for the relationship between the basal diameter (BD) and the total above-ground dry mass (TAM), and with the woody above-ground dry mass (WAM), followed the relation as $TAM = 0.000907\ BD^{3.354}$, whereas $WAM = 0.000283BD^{3.529}$.

In a study, allometric relationships determined for young Jatropha seedlings, under three different levels of drought stress (Achten et al., 2010). Stem diameters data collected from *J. curcas* of 64, 78, 92, and 116 days. After this last measurement the weights of both above- and below- ground parts were determined. Dry masses were determined after oven drying. The results from these measurements were used to calculate the root-shoot ratios. They indicated that water stressed seedlings had significantly higher root-shoot ratios compared to wet treatment plants, but they are all around 0.50 for these young seedlings. The allometric equation of $TAM = 0.029BD^{2.328}$.

Potential of *J. curcas* towards carbon sequestration was studied by various researchers. A study determined the potential of *J. curcas* to sequester carbon of *J. curcas*. The study was executed at the agriculture park of the Putra University in Malaysia, where a total area of one hectare is planted with *Jatropha curcas*, at a planting distance of 2x3m. Three trees of different ages (10, 17, and 32 months) were explored in this study (Firdaus et al., 2010). Plant weights (both root and shoot) collected by destructive method. Dried biomass total carbon content, litterfall data of trees, photosynthesis rate, and soil carbon was analyzed for each plant. The measurements on soil and litterfall did not indicate any addition to the total carbon content of soil by the litterfall. Above ground plants carbon content varied from 46.2% to 46.9%, while the below ground biomass had carbon of 44.6% to 44.9%. Above ground dry biomass were in the range of 1.23 to 12.68% while the dry biomass below ground was 0.70 to

3.95%. The carbon data suggested that Jatropha's carbon fraction is well below the commonly assumed 0.5.

In an experiment the effect of cultivation of the biofuel crop *J. curcas* on soil quality in Burkina Faso was evaluated. Amelioration of *J. curcas* biomass to a severely declining cropland enhanced soil quality through additional organic matter input and erosion control. This experiment revealed that *J. curcas* potentially increased soil organic carbon (SOC) sequestration in traditional and newly introduced *J. curcas* production systems. Inter-planting with annual crops, intensely managed monoculture plantations, afforestation of abandoned cropland and traditional living fences were compared. Linear regression analysis did not reveal any significant clear trend of SOC dynamics over time for the short-term chronosequences. A 20 year chronosequence of *J. curcas* living fences exhibited a significant SOC increase of 62±23g $m^{-2}y^{-1}$. Taking into account of soil coverage of 1,200 m^2 ha^{-1} by living fences (400-m hedge with a width of 3m), 1.5 Mg SOC was sequestered. These findings on traditional *J. curcas* systems indicate the potential of this biofuel crop for improving cropland soil fertility in the long run (Wani et al., 2012).

In a study soil carbon sequestration and reclamation potential of *J. curcas* evaluated by growing in varying soils of central India. Plant biomass, litter turn over, microbial biomass, soil enzymes and carbon and nutrients stock of *J. curcas* plantations were estimated. Results indicated that irrespective of soil types, *J. curcas* plantations significantly enhanced ($\alpha = 5\%$; $p \leq 0.05$) the total organic carbon, total Kjeldahl nitrogen, available phosphorus and potassium content of the soil. During the fourth year of plantations, the plant biomass (both above and below ground biomass) of *J. curcas* in all sites increased from 15.20 ± 4.60 to 203.00 ± 40.60 t ha^{-1} $year^{-1}$ with a subsequent total biomass carbon content of 7.60 ± 2.30 to 101.50 ± 13.52 t ha^{-1} $year^{-1}$, respectively. Soil carbon stock of the *J. curcas* plantation sites varied from 20.59 to 50.45 Mg ha^{-1} $year^{-1}$. Microbial biomass carbon content of the sites was in the range of 132.64 ± 9.28 to 641.32 ± 38.48 μg g^{-1} soils (Anand et al., 2015). Study indicated that *J. curcas* plantations could significantly enhance the soil quality by improving soil carbon and microbial biomass. Thus, in addition to biofuel production, *J. curcas* can be used to increase soil carbon sequestration and soil reclamation.

It will be interesting to know how *J. curcas* can actually do carbon sequestration. In an experiment the effect Jatropha cultivation in degraded lands were evaluated on carbon sequestration and soil properties. Jatropha accumulated and added to soil significant amounts of C (305 kg ha^{-1} year–1) from the year one itself. Overall, a 3–5 year old plantation added up around 4000 kg plant biomass in a year. This biomass is equivalent to 1450 kg C ha^{-1}, consisting of 800 kg C as leaves, 150 kg C as pruned twigs, and 495 kg C as de-oiled Jatropha cake. Biodiesel C replacement in the fossil fuel was 230 kg ha^{-1}. Besides adding biomass to the soil, and C replacement in fossil fuel; the standing Jatropha rendered ecosystem service by fixing 5100–6100 kg ha^{-1} C as the shoot and root biomass. Carbon additions by Jatropha during 4 years increased soil C content by 19%, resulting in about 2500 kg ha^{-1} C sequestered. Huge C additions and high rhizospheric activity in Jatropha cultivated lands increased microbial population, respiration rate and microbial biomass C and N in soil. Along with C additions, 4000 kg ha^{-1} $year^{-1}$ plant biomass recycled into the soil 85.5 kg nitrogen, 7.67 kg phosphorus, 43.9 kg potassium, 5.20 kg sulphur, 0.11 kg boron, 0.12 kg zinc and other nutrients. The C additions improved water-holding capacity of the soil under Jatropha as compared with the adjacent control soil (Singh et al., 2013).

Conclusion

Jatropha curcas is a multi-facet plant which has potential to improve soil biological health, carbon sequestration and greenhouse gas mitigation. This plant harbors many important functional microbial groups including N fixers, P solubilizers, PGPRs, mycorrhiza fungi etc. Further research required to understand how the plant-microbial interaction play significant role in plants adaptation to climate change as well as adaptation to extreme environment. We also highlighted that this plant is a store house of important biomolecules. However, the information in this line are fragmented. Comprehensive research required to identify these bioactive molecules and strategy to use these biomolecules in agriculture. Manuscript concluded that *J. curcas* has promising future in agriculture, cosystem and global climate change.

Acknowledgments

This work was funded by the Department of Biotechnology, New Delhi, India. Authors acknowledge the research fellows and staffs for their support carrying out the experiments and drafting of manuscript.

References

Abdelgadir, H. A., Van Staden, J., 2013. Ethnobotany, ethnopharmacology and toxicity of <i> *Jatropha curcas*</i> L.(Euphorbiaceae): A review. *South African Journal of Botany* 88, 204–218.

Achten, W. M., Maes, Wh., Aerts, R., Verchot, L., Trabucco, A., Mathijs, E., Singh, Vp., Muys, B., 2010. Jatropha: from global hype to local opportunity. *Journal of Arid Environments* 74, 164–165.

Adebowale, K. O., Adedire, C. O., 2006. Chemical composition and insecticidal properties of the underutilized *Jatropha curcas* seed oil. *African Journal of Biotechnology* 5, 901–906.

Agarwal, P., Bhatt, V., Singh, R., Das, M., Sopory, S. K., Chikara, J., 2013. Pathogenesis-related gene, JcPR-10a from *Jatropha curcas* exhibit RNase and antifungal activity. *Molecular Biotechnology* 54, 412–425.

Ahmed, A., Zabar, F., 2013. Chemical composition and nutritional value of jatropha *Jatropha curcas* L. leaves. *Journal of Genetic and Environmental Resources Conservation* 1, 221–226.

Aiyelaagbe, O., Hamid, A., Fattorusso, E., Taglialatela-Scafati, O., Schröder, H. C., Müller, W., 2011. Cytotoxic activity of crude extracts as well as of pure components from Jatropha species, plants used extensively in African traditional medicine. *Evidence-Based Complementary and Alternative Medicine* 2011.

Akbar, E., Yaakob, Z., Kamarudin, S. K., Ismail, M., Salimon, J., 2009. Characteristic and composition of *Jatropha curcas* oil seed from Malaysia and its potential as biodiesel feedstock feedstock. *European Journal of Scientific Research* 29, 396–403.

Anand, K. V., Kubavat, D., Trivedi, K., Agarwal, P. K., Wheeler, C., Ghosh, A., 2015. Long-term application of Jatropha press cake promotes seed yield by enhanced soil organic carbon accumulation, microbial biomass and enzymatic activities in soils of semi-arid tropical wastelands. *European Journal of Soil Biology* 69, 57–65.

Augustus, G., Jayabalan, M., Seiler, G. J., 2002. Evaluation and bioinduction of energy components of *Jatropha curcas*. *Biomass and Bioenergy* 23, 161–164.

Bailis, R. E., Baka, J. E., 2010. Greenhouse gas emissions and land use change from *Jatropha curcas*-based jet fuel in Brazil. *Environmental Science and Technology* 44, 8684–8691.

Bashir, L. U., Gashua, I. B., Isa, M. A., Ali, A., 2013. The antifungal activity of aqueous and ethanol extracts of *Jatropha curcas* L. against Aspergillus Niger (Van Tieghem) that cause black mould rot of onion bulbs in Sokoto, Nigeria. *International Journal of Environment* 2, 83–90.

Becker, K., Francis, G., 2000. Bio-diesel from Jatropha plantations on degraded land, Multifunctional Plants- Food, Feeds and Industrial Products. Department of Aquaculture Systems and Animal Nutrition, University of Hohenheim, Stuttgart, Germany.

Benge, M., 2006. Assessment of the potential of *Jatropha curcas*,(biodiesel tree,) for energy production and other uses in developing countries. USAID Report.

Bertin, C., Yang, X., Weston, L. A., 2003. The role of root exudates and allelochemicals in the rhizosphere. *Plant and Soil* 256, 67–83.

Boateng, B. A., Kusi, F., 2008. Toxicity of jatropha seed oil to Callosobruchus maculatus (Coleoptera: Bruchidae) and its parasitoid, Dinarmus basalis (Hymenoptera: Pteromalidae). *Journal of Applied Sciences Research* 4, 945–951.

Caballero-Mellado, J., Onofre-Lemus, J., Estrada-de los Santos, P., Martinez-Aguilar, L., 2007. The Tomato Rhizosphere, an Environment Rich in Nitrogen-Fixing Burkholderia Species with Capabilities of Interest for Agriculture and Bioremediation. *Applied and Environmental Microbiology* 73, 5308–5319.

Ciais, P., Sabine, C., Bala, G., Bopp, L., Brovkin, V., Canadell, J., Chhabra, A., DeFries, R., Galloway, J., Heimann, M., 2014. Carbon and other biogeochemical cycles, in: Climate Change 2013: The Physical Science Basis. Contribution of Working Group I to the Fifth Assessment Report of the Intergovernmental Panel on Climate Change. Cambridge University Press, pp. 465–570.

Colvin, J., Lambert, J., 2013. Phytofungitoxic Effects of Medicinal Plants (*Jatropha curcas* and Mangifera indica) on Pathogenic Fungi of Cultivated Tomato (Lycopersicon esculentum Mill., syn. Solanum lycopersicon). *European Journal of Chemical Engineering and Chemistry* 11, 18–26.

Devappa, R. K., Makkar, H. P., Becker, K., 2010. Jatropha toxicity—a review. *Journal of Toxicology and Environmental Health, Part B* 13, 476–507.

Dobson, A. P., Bradshaw, A. D., Baker, A. J.., 1997. Hopes for the future: restoration ecology and conservation biology. *Science* 277, 515–522.

Eckert, B., Weber, O.B., Kirchhof, G., Halbritter, A., Stoffels, M., Hartmann, A., 2001. Azospirillum doebereinerae sp. nov., a nitrogen-fixing bacterium associated with the C4-grass Miscanthus. *International Journal of Systematic and Evolutionary Microbiology* 51, 17–26.

Einhellig, F. A., 1995. Allelopathy: current status and future goals.

Falodun, A., Udu-Cosi, A. A., Erharuyi, O., Imieje, V., Falodun, J. E., Agbonlahor, O., Hamann, M. T., 2014. Jatropha Tanjorensis-Review of Phytochemistry, Pharmacology, and Pharmacotherapy. Journal of Pharmaceutical and Allied Sciences 10.

Fernandes, E. de S., Rodrigues, F. A., Tófoli, D., Imamura, P. M., Carvalho, J. E. de, Ruiz, A. L., Foglio, M. A., Minguzzi, S., Silva, R. C., 2013. Isolation, structural identification and cytotoxic activity of hexanic extract, cyperenoic acid, and jatrophone terpenes from Jatropha ribifolia roots. *Revista Brasileira de Farmacognosia* 23, 441–446.

Firdaus, M. S., Hanif, H. M., Safiee, S., Ismail, R., 2010. Carbon sequestration potential in soil and biomass of *Jatropha curcas*, in: Proceedings of the 19th World Congress of Soil Science, Brisbane, Australia. pp. 1–6.

Francis, G., Edinger, R., Becker, K., 2005. A concept for simultaneous wasteland reclamation, fuel production, and socio-economic development in degraded areas in India: Need, potential and perspectives of Jatropha plantations. *Natural Resources Forum* 29, 12–24.

Gámez-Meza, N., Alday-Lara, P. P., Makkar, H. P., Becker, K., Medina-Juárez, L. A., 2013. Chemical characterisation of kernels, kernel meals and oils from Jatropha cordata and Jatropha cardiophylla seeds. *Journal of the Science of Food and Agriculture* 93, 1706–1710.

Ghezehei, S. B., Annandale, J. G., Everson, C. S., 2009. Shoot allometry of *Jatropha curcas*. *Southern Forests* 71, 279–286.

Gübitz, G. M., Mittelbach, M., Trabi, M., 1999. Exploitation of the tropical oil seed plant< i> *Jatropha curcas*</i> L. *Bioresource Technology* 67, 73–82.

Islam, A., Yaakob, Z., Anuar, N., 2011. Jatropha: A multipurpose plant with considerable potential for the tropics. *Scientific Research and Essays* 6, 2597–2605.

Jha, C. K., Patel, B., Saraf, M., 2012. Stimulation of the growth of *Jatropha curcas* by the plant growth promoting bacterium Enterobacter cancerogenus MSA2. *World Journal of Microbiology and Biotechnology* 28, 891–899.

Jongschaap, R. E.., Corre, W. J., Bindraban, P. S., Brandenburg, W. A., 2007. Claims and Facts on *Jatropha curcas* L.: Global *Jatropha curcas* evaluation, breeding and propagation programme (No. 158). Plant Research International B. V., Wageningen.

Jummai, A. T., Okoli, B. J., 2014. Curcin from *Jatropha curcas* seed as a potential anthelmintic. *Advancement in Medicinal Plant Research* 2, 47–49.

Kamalvanshi, M., Kumar, A., Jha, A., Shukla, A., 2010. Isolation of Acaulospora species from *Jatropha curcas* L. and Pongamia pinnata Pierre. *International Journal of Pharmaceutical Sciences and Research* 1, 74–78.

Khalil, H. P. S., Aprilia, N. A., Bhat, A. H., Jawaid, M., Paridah, M. T., Rudi, D., 2013. A< i> Jatropha</i> biomass as renewable materials for biocomposites and its applications. *Renewable and Sustainable Energy Reviews* 22, 667–685.

Kirchhof, G., Reis, V. M., Baldani, J. I., Eckert, B., Dobereiner, J., Hartmann, A., 1997. Occurrence, physiological and molecular analysis of endophytic diazotrophic bacteria in gramineous energy plants. *Plant and Soil* 194, 45–55.

Kong, L.-Y., Min, Z.-D., Shi, J.-X., Rui, F., 1996. Chemical constituents from roots of *Jatropha curcas*. *ACTA BOTANICA SINICA-CHINESE EDITION-* 38, 161–166.

Martinez-Herrera, J., Siddhuraju, P., Francis, G., Davila-Ortiz, G., Becker, K., 2006. Chemical composition, toxic/antimetabolic constituents, and effects of different treatments on their levels, in four provenances of< i> *Jatropha curcas*</i> L. from Mexico. *Food Chemistry* 96, 80–89.

Mbagwu, J. S.C., 1992. Improving the productivity of a degraded ultisol in Nigeria using organic and inorganic amendments. Part 2: Changes in physical properties. *Bioresource Technology* 42, 167–175.

Misra, M., Misra, A. N., 2010. Jatropha: the biodiesel plant biology, tissue culture and genetic transformation–a review. *Int. J. Pure Appl. Sci. Technol* 1, 11–24.

Mohanty, S. R., Kollah, B., Banerjee, K., Dubey, G., Rao, A. S., 2012. Methane consumption and dynamics of microbial population in the rhizosphere of bioenergy crop *Jatropha curcas* in tropical Indian Soil. *Proceedings of Biofest* 12–13.

Norman, M. J. T., Pearson, C. J., Searle, P. G. E., 1995. The Ecology of Tropical Food Crops. Cambridge University Press, Cambridge.

Pasha, C., Balakrishna, K., Hanumalal, N., Srinivas, B., Chandrasekhar, B., 2013. Evaluation of toxins, antinutrients and nutrients of Indian cultivating varieties of *Jatropha curcas*. *Asian Journal of Chemistry* 25, 1638–1642.

Paulillo, L., de Sousa Lessa, G., Isaacson, J., Mo, C., Brotto, L., Abreu, E., Brotto, M., and Lima, J. G. M., 2014. Pinhão-manso (*Jatropha curcas*) demonstrates potent antibacterial properties in a rat model of third degree burns (1180.18). *The FASEB Journal* 28, 1180.18.

Pérez, Y., Schenck, N. C., 1990. A unique code for each species of VA mycorrhizal fungi. *Mycologia* 82, 256–260.

Sabandar, C. W., Ahmat, N., Jaafar, F.M., Sahidin, I., 2013. Medicinal property, phytochemistry and pharmacology of several< i> Jatropha</i> species (Euphorbiaceae): A review. *Phytochemistry* 85, 7–29.

Santos, A. F. dos, Fonseca, S. A., César, F. A., de Azevedo Albuquerque, M. C. P., Santana, J. V., Santana, A. E. G., 2014. A penta-substituted pyridine alkaloid from the rhizome of Jatropha elliptica (Pohl) Muell. Arg. is active against Schistosoma mansoni and Biomphalaria glabrata. *Parasitology Research* 113, 1077–1084.

Setha, B., Laga, A., Mahendradatta, M., 2014. Phytochemical screening and antibacterial activity of crude extracts from *Jatropha curcas*, Linn against histamine-forming bacteria.

Sharath, B. S., Mohankumar, B. V., Somashekar, D., 2014. Bio-detoxification of phorbol esters and other anti-nutrients of *Jatropha curcas* seed cake by fungal cultures using solid-state fermentation. *Applied Biochemistry and Biotechnology* 172, 2747–2757.

Singh, K., Singh, B., Tuli, R., 2013. Sodic soil reclamation potential of *Jatropha curcas*: a long-term study. *Ecological Engineering* 58, 434–440.

Singh, R. N., Vyas, D. K., Srivastava, N. S. L., Narra, M., 2008. SPRERI experience on holistic approach to utilize all parts of< i> *Jatropha curcas*</i> fruit for energy. *Renewable Energy* 33, 1868–1873.

Sinha, A., Srivastava, P. K., Singh, N., Sharma, P. N., Behl, H. M., 2011. Optimizing organic and mineral amendments to jatropha seed cake to increase its agronomic utility as organic fertilizer. *Archives of Agronomy and Soil Science* 57, 193–222.

Venkatesh, L., Naik, S. T., Suryanarayana, V., 2009. Survey for occurrence of arbuscullar mycorrhizal fungi associated wtih *Jatropha curcas* (L). and Pongamia pinnata (L), Pierre in three agroclimatic zones of Karnataka. *Karnataka Journal of Agricultural Sciences* 22, 373–376.

Wani, S. P., Chander, G., Sahrawat, K. L., Srinivasa Rao, C., Raghvendra, G., Susanna, P., Pavani, M., 2012. Carbon sequestration and land rehabilitation through< i> *Jatropha curcas*</i>(L.) plantation in degraded lands. *Agriculture, Ecosystems and Environment* 161, 112–120.

In: The Promising Future of *Jatropha curcas*
Editors: Maira R. Segura-Campos et al.

ISBN: 978-1-63484-989-0
© 2016 Nova Science Publishers, Inc.

Chapter 12

AGRIBUSINESS POTENTIAL USE OF *JATROPHA CURCAS* IN THE WORLD: A REVIEW

***Betty Jarma Arroyo*[1,*] *and Alfredo Jarma Orozco*[2,†]**
[1]Agroindustrial Engineer, Special Consultant
[2]Agricultural Engineer, University of Cordoba, Montería, Córdoba, Colombia

ABSTRACT

Jatropha curcas L. is a tropical plant presently used as a source of biodiesel; the cake, after oil extraction, is rich in protein and is a potential source for livestock feed. However, this resultant cake is compounded by a group of toxic substances, including phorbol esters (PMA), and some other antinutrients, and a good detoxification is necessary before use as food. The possibilities of potential exploitation of this plant for various applications have been explored. The *Jatropha* species has been used to remedy large areas of land that have been affected by large plantations; also it is gaining a lot of importance for the production of biodiesel around the world. These issues are dealt herewith to observe its future scope to mitigate energy crisis, environmental management and sustainable productions. The information about the detoxification methods, the cake, phytoremediation and some other topics are collected and discussed.

Keywords: renewable energy, phorbol esters, phytoremediation

INTRODUCTION

The food industry must initiate a profound change and innovation to meet the increasingly diverse and more precise demands of consumers. Food products of today for both human and animal consumption must submit a faultless sanitary guarantee, to meet the needs created by the evolution of lifestyle.

---

* Email: betyca1@hotmail.com.
† Email: ajarma@correo.unicordoba.edu.co.

Researchers consider the *J. curcas* as an alternative to fossil fuels, and to reduce emissions of greenhouse gases (Behera et al., 2010), as it is not toxic to nature when compared to petro-diesel (Pandey et al., 2012). *Jatropha curcas* L. (JCL) has been publicized as the only tropical plant with the potential to increase renewable energy sources due to its many merits; therefore, it deserves to be considered as the best candidate for the tangible and intangible benefits of ecology and the environment. The species has been recommended for extensive plantations in abandoned and degraded lands worldwide (Pandey et al., 2012).

Then, *Jatropha* could be doubly exploited in agribusiness, both in the production of Biodiesel in arid lands (Pandey et al., 2012), and obtaining a high-protein feed for livestock as an alternative food supply in times of drought, to be cultivated on a large scale (Makkar et al., 2011).

This crop also has certain limitations, especially when it comes to consumption. *J. curcas* species have several toxic principles, most of which have been reduced by heat treatment. After oil extraction, the resulting cake is rich in protein and is a potential source of livestock feed, however, it consists of a group of toxic substances, including phorbol esters (PMA) and some other anti-nutrients (Jarma et al., 2014).

The oil content in seeds is about 30 - 40% and the cake can contain up to 60% of fatty acids in similar patterns to edible oils (Pompelli et al., 2010). There are non-toxic varieties which have been found in Mexico and are consumed by the population after heat treatment (cooking). Heat treatments have proven very useful for reducing some of the toxic components of *J. curcas* L, however, there are other components that have been identified as key participants in the high toxicity of this culture; these are esters phorbol (Goel et al., 2007).

JATROPHA IN TERMS OF AGRIBUSINESS

J. curcas L., also known as the physics nut, belongs to the Euphorbiaceae family (Basili and Fontini, 2012). It is a drought resistant shrub and is well adapted to arid and semi-arid conditions (Achten et al., 2008). It has been propagated as a unique and potential tropical plant for augmenting renewable energy sources due to its several merits for which it deserves to be considered as the sole candidate in the tangible and intangible benefits of ecology and the environment. The species has been advocated for extensive plantations on degraded wastelands throughout the world (Pandey, et al., 2012).

Although this crop has a high potential for the production of biodiesel and concentrate for animals, it also has certain limitations, especially when it comes to consumption. *J. curcas* species has several toxic principles, most of which can be reduced by heat treatment. However, there are non-toxic varieties which have been found in Mexico and are consumed by the population after cooking or roasting the kernels (Makkar et al., 2011).

Heat treatment has been very useful for the reduction of some of the toxic components of JCL to aceptable levels, however, some of these species are known to contain a lethal cocktail of toxins including phorbol esters (Siam et al., 2009) that have been identified as the principal factor of the toxicity of this species (Goel, et al., 2007).

The toxicity of *Jatropha curcas* seeds can be caused by several components, including saponins, lectins (curcin), phytates, protease inhibitors, curcalonic acid, and phorbol esters.

Phorbol esters, which activate the important cellular target protein kinase C (PKC), constitute the most active components which must be removed if the oil or seed cake is to used for animal or human nutrition (Gübitz, et al., 2008; Makkar, et al., 1997).

In addition, components such as trypsin inhibitors, glucosinolates, amylase inhibitors, and cyanogenic glycosides, also can be found in *Jatropha* seeds (Rakshit, et al., 2006).

Studies in the last decade showed that *J. curcas* exhibited toxicity to a wide variety of organisms, from microorganisms to higher animals. However, the toxicity effects studied in animals are mainly by force-feeding either *J. curcas* raw or in defatted seed meals, leaves or their various organic solvent/aqueous extracts (Lin, et al., 2010).

Therefore, it is possible to claim that cultivation and detoxification of the cake can generate:

- Jobs in rural communities.
- Use of unproductive land (Achten, et al., 2008).
- Reduction of pollution of the land, from organic compounds (Hon and Joseph, 2010).
- Avoidance of using food for biofuel production.
- Participation in programs and mechanisms related to clean energy.
- Promotion of sustainability in rural areas.

DERIVATIVES OBTAINED FROM *JATROPHA*

As we have seen in this chapter, there are several products that can be obtained from *Jatropha*; some may need some complex industrial processes more than others. The understanding of the physical properties of *Jatropha* seeds (Figure 1), is necessary for harvesting, drying, cleaning, sorting, peeling and storage (Pradhan, et al., 2009;. Sirisomboon, et al., 2007).

Figure 1. *Jatropha* seed. Source: Camilo Iguarán, Universidad de Córdoba (Colombia).

The degree of difficulty of the process depends solely on the product to be obtained. For example, if the product is for livestock, then the oil must be first removed from the seeds, resulting in a defatted cake that needs to be detoxified to ensure that it is safe for consumption.

However, there are simpler processes to obtain other derivatives, it is also known that some existing *Jatropha* species are used in medicinal folklore and known as a purgative (Sabandar et al., 2013).

Bio-Fuels

On the worldwide scale, biomass is the greatest source of renewable energy (Flach et al., 2009). The amount of energy stored in the biomass is approximately 7.5-times greater than is global energy consumption. From the total technically exploitable energetic potential, the greatest share responded to biomass (Jasiulewicz and Jasiulewicz, 2006). From this issue, governments have been searching for new sources of energy that are renewable and environmentally clean. These are bio-fuels, which could be liquid, gaseous or solid, and are generated from various types of raw materials such as sugarcane, oil plants, biomass forests, and other sources of organic matter (Nass et al., 2007).

Bio-fuels are considered environmentally friendly, every time as that they release 50% less particulate matter and 98% less sulfur, as well as being biodegradable and non-toxic (Nass et al., 2007 and Demirbas, 2009). Despite the advantages of using bio-fuels, serious environmental impact can result from its production and end-usage, such as use of large amounts of water, forest destruction, reduction in food production and increased soil degradation (Escobar et al., 2009).

Therefore, energetic plants (EP) and crops are very important; the plants grown as a low cost and low maintenance harvest are used to make bio-fuels, or directly exploited for the energy content (heating or electric power production. If carbohydrate content is desired for the production of biogas, whole-crops can be made into silage and then converted into biogas (Carere et al., 2008 and Hertwich et al., 2009) (Figure 2). Energy is generated by burning plants grown for this purpose, often after the dry matter is pelletized (Masarovičová et al., 2009).

EP are typically densely planted, high yielding species cultivated for the purpose of producing (non-food) energy - burning wood or bio-fuel. There are many species used as EP, some of them are herbs (eg. *Zea mays, Brassica napus, Triticum aestivum, Helianthus annuus, Helianthus tuberosus, Sorghum bicolor, Miscanthus* spp., *Jatropha curcas*), shrubs or trees (eg *Populus* spp., *Salix* spp., *Alnus glutinosa, Ailanthus altissima, Ulmus montana*) (Johnson et al., 2007).

Bio-diesel production from JCL has become a booming business. With no competing food uses, and growth in tropical and subtropical climates across the developing world (Openshaw, 2000), the oil produced by this crop can be easily converted to liquid bio-fuel, which meets the American and European standards (Azam et al., 2005 and Tiwari et al., 2007).

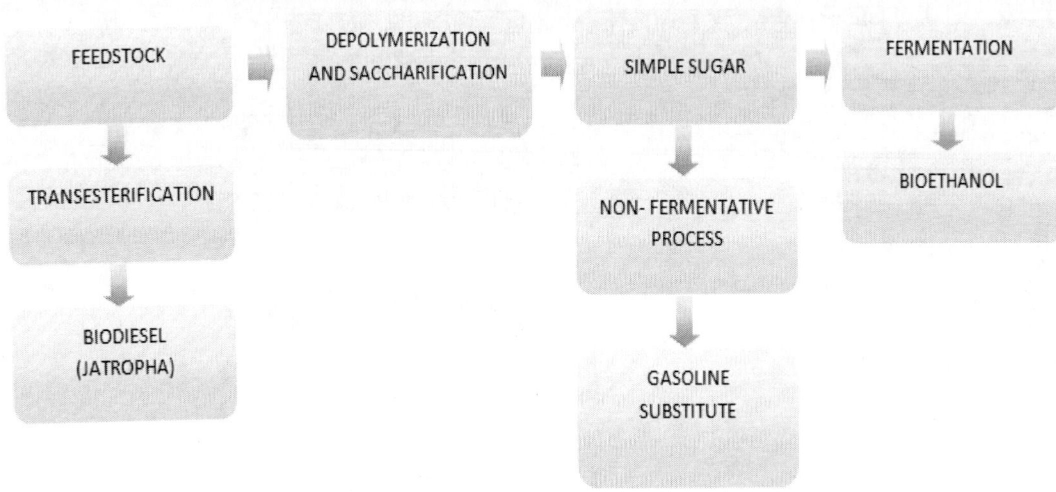

Figure 2. Pathways for bio-fuel development.

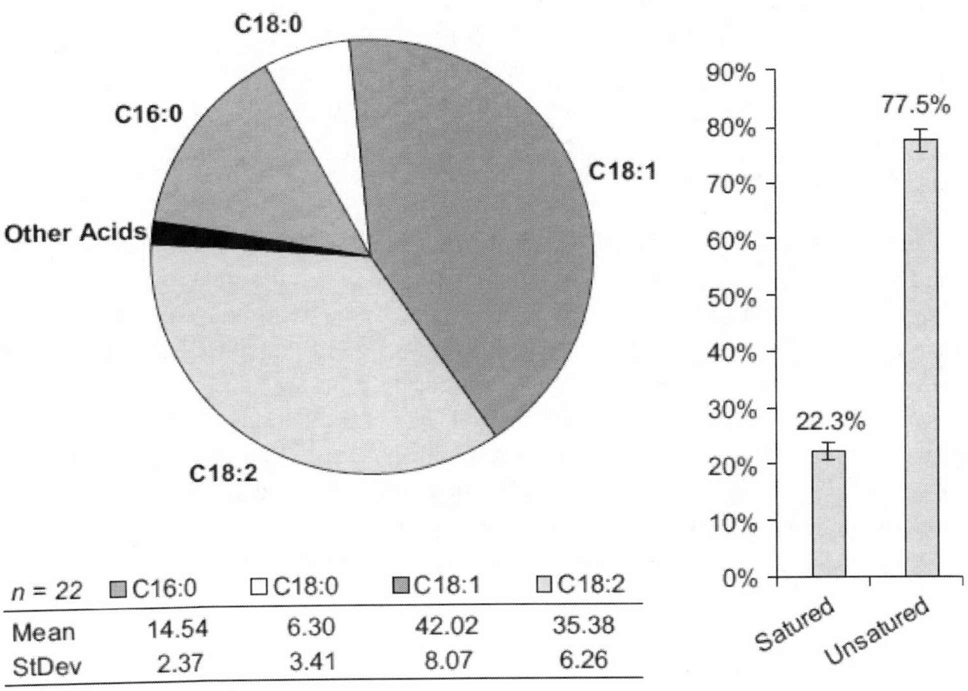

n = 22	C16:0	C18:0	C18:1	C18:2
Mean	14.54	6.30	42.02	35.38
StDev	2.37	3.41	8.07	6.26

Figure 3. Fatty acid composition (%). C16:0 ¼ palmitic acid; C18:0 ¼ strearic acid; C18:1 ¼ oleic acid; C18:2 ¼ linoleic acid. Other acids containing capric acid, myristic acid (C14:0), palmit oleic acid (C16:1), linolenic acid (C18:3), arachidic acid (C20:0), behenic acid (C22:0), cis-11-eicosenoic acid (C20:1) and cis-11,14-eicosadienoic acid (C20:2). n ¼ number of observations used. Source: Achten et al., 2008.

Jatropha proves to be very promising; the seeds contain between 25 and 32% oil (Pompelli et al., 2010), and can yield up to 1.5 t ha^{-1} oil year in plants that are at least 5 years old (Openshaw, 2000; Tiwari et al., 2007).

Table 1. JCL oil composition and characteristic (Adapted from Achten et al., 2008)

	Range	Mean	S.D	*n*
Specific gravity (g cm^{-3})	0.860-0.933	0.914	0.018	13
Calorific value (MJ kg^{-1})	37.83-42.05	39.63	1.52	9
Pour point (ºC)	-3			2
Cloud point (ºC)	-2			1
Flash point (ºC)	210-240	235	11	7
Cetane value	38.0-51.0	46.3	6.2	4
Saponification number (mg g-^{1})	102.9-209.0	182.8	34.3	8
Viscosity at 30 ºC (cSt)	37.00-54.80	46.82	7.24	7
Free fatty acids % (kg Kg^{-1}*100)	0.18-3.40	2.18	1.46	4
Unsaponifiable % (Kg Kg^{-1}*100)	0.79-3.80	2.03	1.57	5
Iodine number (mg iodine g^{-1})	92-112	101	7	8
Acid number (mg KOH g^{-1})	0.92-6.16	3.71	2.17	4
Monoglycerides % (Kg Kg $^{-1}$*100)	nd-1.7			1
Diglycerides % (Kg Kg^{-1}*100)	2.50-2.70			2
Triglycerides % (Kg Kg^{-1}*100)	88.20-97.30			2
Carbon residue % (Kg Kg^{-1}*100)	0.07-0.64	0.38	0.29	3
Sulfur content % (Kg Kg^{-1}*100)	0-0.13			2

S.D.: Standard Desviation; *n:* number of observation used; nd: no data.

The JCL oil contains more than 75% unsaturated fatty acid, which is reflected in the pour and cloud point of the oil. The fatty acid composition of JCL oil is dominated by oleic acid (C18:1) and linoleic acid (C18:2) (Figure 3). The maturity stage of the fruits at the moment of collection is reported to influence the fatty acid composition of the oil (Raina et al., 1987).

The composition and characteristics of the crude JCL oil are given in Table 1. The JCL oil meets the quality standard of rapeseed as a fuel (FACT Fundation, 2006). It is important to note that the values of the free fatty acids, unsaponifiables, acid number and carbon residue show a very wide range, although it is a small data set; this indicates that the oil quality is dependent on the interaction of environment and genetics.

Project developers and decision makers should be aware of these wide ranges, because these aforementioned characteristics are important properties for the further processing of the oil into bio-diesel (Ma, 1999 and Van Gerpen 2005).

JATROPHA CURCAS CAKE

The amino acid composition, the proportion of essential amino acids, and mineral content of the paste resulting from the extraction of oil can be compared to similar pastes used as fodder (Makkar et al., 2008).

An alternative to feedstock production of multi-nutrient and balanced blocks for cattle food consumption is *Jatropha* cake, because the seeds can contain up to 60% fatty acids in edible oils similar to patterns, and the amino acid composition essential protein, except lysine, is comparable to soybean meal (Makkar, et al., 1998; and Devappa and Bhagya, 2008).

Most clinical trials regarding the use of *Jatropha curcas* L have shown severe pathological symptoms. Several important symptoms were observed, transient loss of body mass and macroscopic/ microscopic changes from mild to severe kidney, lung, heart, liver and spleen in a dose dependent manner (Adam, 1974, Liberalino et al., 1988 and Rasshit, et al., 2008).

Toxic effects in humans are manifested in gastrointestinal disorders, liver disorders, parasitosis, and secondary skin lesions: abrasions, ulcers, etc. At high doses, the oil produces alterations in the gastrointestinal tract, manifested by nausea, vomiting and sweating; these reactions can lead to death. Alcoholic drinks are the antidote to the toxic effects (Siamazonia, 2008).

To decrease the toxicity of *J. curcas*, different detoxification methods have been created such as heat treatment, extraction with organic solvents (tetrahydrofuran, dichloromethane, methanol); extraction with ionizing radiation, among others (Rakshit, et al., 2010;. Kumar and Sharma, 2008; King, et al., 2009).

The heat treatment is not considered an effective method for decreasing the concentration of anti-nutritional factors, so a combination of temperature with strong bases has been opted for, thus obtaining a decrease of 86% of such compounds (Kumar and Sharma, 2008).

Additionally, this process improves the organoleptic properties of food, reducing the loss of body mass and delaying mortality in rats (Rakshit, et al., 2008).

Another method is extraction with ethanol, obtaining good results, removing phorbol esters and lectin, but not phytic acid, trypsin inhibitor and saponins (Saetae and Worapot, 2011). However, these detoxification processes have only been successful on a laboratory scale and not on a large scale because of the economic cost (Parawira, 2010; Kumar and Sharma, 2008).

In addition to chemical methods, biological methods have also been evaluated to reduce the toxic effect of the plant. In this regard, Joshi, et al., 2011 conducted a study to test the efficacy of *Pseudomonas aeruginosa* bacteria to degradated phorbol esters, obtaining a complete degradation during the solid state fermentation of defatted JCL seeds.

Several authors, as Aregheore et al. (2003), Martinez et al. (2006) and Rakshit et al. (2008) have used ethanol, methanol, alkali [NaOH and $Ca(OH)_2$], alkali- alcohol combinations, ethanol-NaOH-Na- $NaHCO_3$ and ClO, reporting decreases above 40%, the phorbol ester content.

Table 2. Average values (± standard deviation) of lipids, fiber, nitrogen (%), total soluble proteins and total soluble carbohydrates of *Jatropha curcas* seed cake, subjected to chemical /physical detoxification treatments (Adapted from Jarma et al., 2014)

Methodology	Lípids	Fibers	N* (%)	TSP ($g.kg^{-1}$ DM)	TSC (mmol/kg DM)
Control	15.50 ± 0.25[a]	37.36 ± 0.16[a]	44.83 ± 0.72[a]	12.08 ± 0.33[a]	383.22 ± 42.58[a]
Ethanol	6.36 ± 0.14[b]	41.30 ± 1.85[a]	36.33 ± 0.88[b]	7.82 ± 0.08[b]	163.59 ±7.00[b]
Methanol	7.06 ± 0.85[b]	18.70 ± 0.81[b]	13.33 ± 0.33[c]	4.10 ± 0.28[c]	102.23 ± 10.86b[c]
NaOH - NaClO	1.50 ± 0.05[c]	16.63 ± 1.50[b]	12.00 ± 1.52[c]	2.26 ± 0.11[d]	34.00 ± 5.42[c]

* N: Nitrogen; TSP: total soluble protein; TSC: total soluble carbohydrates; S.D: Standard Deviation; $P \leq 0.00001$ samples. Results with the same letter are statistically equal. * DM = Dry Mass.

Table 3. Phorbol esters, tannins and saponins content in cake *Jatropha curcas* seed, subjected to chemical/physical detoxification treatments (Adapted from Jarma et al., 2014)

Methodology	Phorbol esters Content (mg/Kg)	Saponins	Tannins
Control	204.60[a]	Positive	Negative
Ethanol	0.07[b]	Negative	Negative
Methanol	0.40[b]	Negative	Negative
NaOH-NaClO	0.29[b]	Positive	Negative

* Results with the same letter are statistically equal.

Table 2 shows the chemical composition of a detoxified *Jatropha* cake with three different methods, it is possible to observe that after being subjected to detoxification treatments, the cake is able to maintain acceptable values that makes its use as cattle feed possible; however, Table 3 shows that the content of phorbol esters decreases significantly with either treatment, beating the other two in significant quantities.

It is important to remember that a viable option for the use of this plant without being affected by the toxic effects is the use of non-toxic species JCL, as they are found in Mexico.

MEDICINAL PROPERTIES

The genus name *Jatropha* derives from the Greek giatros (doctor) and trophe (food) which implies medicinal uses (Pandey et al., 2012).

Jatropha species have been used as medicinal plants by native people in the tropical and subtropical countries (Openshaw, 2000). JCL contains many medicinal values for human and veterinary purposes and has a vast potential for deriving therapeutical values for usage in various segments. All parts of *Jatropha* (seeds, leaves and bark, fresh or as a decoction) have been used in traditional medicine and for veterinary purposes for a long time (Duke, 1985 and Duke, 1988). Table 4 lists the medicinal usage of *Jatropha*.

Jatropha species are famous for the purgative effect of the seed oil. This purgative effect has been directed to cure digestive system symptoms i.e., diarrhoea, dysentery, vomiting, retching and stomachache (Sabandar et al., 2013). In contrast to its purgative effect, Lioglier (1990) reported that seeds of *Jatropha* species are highly toxic and advised that they not be used in herbal medicine.

Nath and Dutta, 1998, demonstrated the wound-healing properties of curcain, a proteolytic enzyme isolated from latex. Latex has antimicrobial properties against *Staphylococcus aureus*, *Escherichia coli*, *Klebsiella pneumonia*, *Streptococcus pyogenus* and *Candida albicans* (Thomas, 1989). The latex itself has been found to be strong inhibitors to watermelon mosaic virus (Tewari and Shukla, 1982), also has been investigated as a prospective haemostatic agent (Odusote, et al., 1999).

The methanol extract of *Jatropha* roots exhibited systemic and significant anti-inflammatory activity in acute carrageenan-induced rat paw edema (Mujumda and Misar 2004, and Muanza et al., 1995), and it was found that a methanol extract of physic nut leaves afforded moderate protection for cultured human lymphoblastoid cells against the cytopathic effects of human immunodeficiency virus. Extract of the leaves showed potent cardiovascular action in guinea pigs and might be a possible source of beta-blocker agent (Fojas et al., 1986). All parts of the plant are used in traditional medicine and active components are being investigated in scientific trials. Several ingredients appear to have promising applications both in medicine as well as a plant protectant (Pandey et al., 2012).

Table 4. Various medicinal uses of *Jatropha curcas* (Adapted from: Kaushik et al., 2007; Osoniyi and Onajobi 2003; Samy et al., 1998; Mujumda and Misar 2004; Coelho-Ferreira, 2009)

S.no.	Usable plant parts	Diseases curing
1	Seeds	To treat gout, arthritis and jaundice, wound-healing, fractures, burns, purge
2	Seed oil	Eczema, skin diseases, soothe rheumatic pain, purgative action
3	Stem	Toothache, gum inflammation, gum bleeding, pyorrhoea
4	Stem bark	Infectious diseases, including sexually transmitted diseases
5	Plant sap	Dermatomucosal diseases
6	Water extract of branches	HIV, tumour
7	Plant extract	Wound healing, allergies, burns, cuts and wounds, inflammation, leprosy
8	Leaves and latex	Refractory ulcers, septic gums, styptic in cuts and bruises
9	Latex	Reduce the clotting time of human blood, sore mouth, oral thrush, fish barb wounds, snake-bites, infected sores, treating newborns' umbilical cords, coughs, mouth and throat sores
10	Root powder	In the treatment of inflammation
11	Leaf	Scabies, Eczema, Syphilis, blood cleansing, headache, flu, cough, congestion, evil eye, cleansing house during the pregnancy
12	Fruit	Stroke, toothache, numbness after bug sting, to clean mother´s and baby blood

LATEX

One of the reported traditional uses of *Jatropha curcas* leaves the latex as a haemostatic or styptic: for example, when the latex or the crushed leaf of this plant is applied directly to cuts and bleeding wounds, the bleeding soon stops (Dalziel, 1955; Watt and Breyer-Brandwijk, 1962; Neuwinger, 1996). Furthermore, Villegas et al., (1997) have demonstrated a significant wound-healing activity in *Jatropha curcas* extracts.

Osoniyi and Onajobi, (2003), demonstrate that the diluted latex can be used as an anticoagulant, the integrity of the red cell membrane of blood to which latex has been added, under the non-clotting condition, is presumably intact as, they shown how the ethanolic extracts of *Jatropha curcas* latex stabilizes the erythrocyte membrane they affirm that even the crude non-purified anticoagulant could be useful clinically, for *in vitro* assays when blood has to be obtained and kept for some time under circumstances in which chemical anticoagulants are not readily available, such as in villages and other remote places.

In nanotechnology, lead nanoparticles have received considerable attention because of their unusual properties and potential applications in diverse fields especially chemical reactions (Lu et al., 2000). The various synthetic procedures for their synthesis includes micro emulsion, (Qi et al., 1997) reverse micelles, (Lisiecki et al., 1995) reduction of aqueous lead and palladium salts, and (Ren et al., 2005) UV-light irradiation. Particularly the use of *Jatropha curcas* latex, proteins and phytochemical for synthesis of metal nanoparticles has dual advantage that, they not only act as a reducing agent but also as capping agents and deter the particle aggregation (Ravindra, 2009 and Das et al., 2010).

Joglekar et al., (2011), demonstrated that use of latex eliminates the use of synthetic reducing and capping agents previously reported and could find potential applications in lead batteries, catalysis etc.

JATROPHA AS PHYTOREMEDIATION AGENT

"The plant itself is believed to prevent and control soil erosion or can be used as a living fence or to reclaim wasteland" (Openshaw, 2000)

Marginal lands can be managed for the production of renewable energy source as their soil quality and impoverished fertility would not be able to sustain field crops and their rehabilitation, particularly in dry land degraded ecosystems could be possible through *Jatropha* plantations (Pandey et al., 2012).

An alternative to environmental pollution is phytoremediation; this technique employs photosynthetic plant systems and its microbiota as remediation agents. Phytoremediation is known as a technique that uses plant systems as agents to recover the environment (Gouveia et al., 2014).

Global agro-ecological zoning estimated some 16% of total global land area to be under risk of soil erosion. The percentage of areas under potential erosion risk varies from Europe (19%) to North Africa and the Near East (10%), so, mitigation of soil erosion with using JCL plantations is necessary for sustainability of land (Koohafkan, 2000).

JCL develops a deep taproot and initially four shallow lateral roots (Reubens et al., 2011). The taproot may stabilize the soil against landslides while the shallow roots are alleged to prevent and control soil erosion caused by wind or water, but this potential has not been investigated scientifically (Atchen et al., 2008), species are also capable to sequestrates carbon (Ogunwole et al., 2007). JCL shows high initial establishment success and survival (Zahawi et al., 2005). Living fences can be established very quickly by planting cuttings (vegetative propagation) directly in the field, while taproot is absent in propagated cuttings of JCL, which makes the plants more susceptible to uprooting by wind. Recently, Reubens et al., 2011, demonstrated that the lateral roots of *J. curcas* could decrease soil erodibility through additional soil cohesion, whereas its taproot and sinkers may facilitate exploitation of subsurface soil moisture and thus be able to increase vegetative cover, even in very dry environments.

No information is available on nutrient cycles and the impact on soil biological live. The growing concern on these issues must be checked by focused research (Atchen et al., 2008).

References

Adam, S. E. I. 1974. Toxic effects of *Jatropha curcas* in mice. *Toxicology,* 2, 1: 67–76.

Achten, WMJ., Vercho, T L., Franken, YJ., Mathijs, E., Singh, VP., Aerts, R., Muys, B. (2008). Jatropha bio-diesel production and use. *Biomass Bioenergy, 32:*1063–1084.

Aregheore, E. M., Becker, K. and Makkar, H. P. S. 2003. Detoxification of a toxic variety of *Jatropha curcas* using heat and chemical treatments and preliminary nutritional evaluation with rats. *South Pacific journal of natural science,* 21: 50-56.

Azam M. M., Waris A., Nahar N. M. 2005. Prospects and potential of fatty acid methyl esters of some non-traditional seed oils for use as biodiesel in India. *Biomass and Bioenergy,* 29:293–302.

Basili M, Fontini F. (2012) Biofuel from *Jatropha curcas*: Environmental sustainability and option value. *Ecol Econ,* 78: 1–8.

Behera, S., Srivastava, P., Tripathi, R., Singh, J. and Singh, N. (2010). Evaluation of plant performance of *Jatropha curcas* L. under different agro-practices for optimizing biomass a case study. *Biomass and Bioenergy, 34:* 34-30.

Carere C. R., Sparling R., Cicek N. and Levin D. B. 2008: Third generation biofuels via direct cellulose fermentation. *Int. J. Mol. Sci.,* 9: 1342-1360.

Coelho-Ferreira M. 2009. Medicinal knowledge and plant utilization in an Amazonian coastal community of Maruda, Para State (Brazil). *J Ethnopharmacol,* 126:159–75.

Devappa, R. K. and Bhagya, S. 2008. Biochemical and nutritional evaluation of Jatropha protein isolate prepared by steam injection heating for reduction of toxic and antinutritional factors. *J. Sci. Food Agric,* 88: 911-919.

Das RK, Borthakur BB, Bora U. 2010. *Mat Lett,* 64: 1445–7.

Dalziel, J. M., 1955. The Useful Plants of West-Tropical Africa. Crown Agents for Oversea Governments and Administration, London, p. 147.

Demirbas, A. 2009. Biofuels: green energy and technology. *Springer,* Londres.

Duke JA. 1985. Medicinal plants. *Science,* 229:1036.

Duke JA. 1988. CRC handbook of medicinal herbs. Boca Raton: CRC Press.

Escobar, J. C., E. S. Lora, O. J. Venturini, E. Yanez and E. F. Castillo. 2009. Biofuels: environment, technology and food security. Renew. *Sustain. Energ. Rev.* 13, 1275-1287.

FACT Foundation. Jatropha handbook. Eindhoven: 2006 /http://www.fact-fuels.org/images/ jatropha%20handbook%20march%202006.pdfS.

Flach B. et al: Netherlands-Germany EU-27/ EU-27 Biofuels Annual. Annual Report 2009. Gain Report Number: NL9014, Date 6/15/2009 http://gain.fas.usda.gov/Recent% 20GAIN%20Publications/General%20Report_The%20Hague_Netherlands-Germany% 20EU-27_6-15-2009.pdf.

Fojas FR, Garcia LL, Venzon EL, Sison FM, Villanueva BA, Fojas AJ, et al., 1986. Pharmacological studies on *Jatropha curcas* as a possible source of anti-arrhythmic (beta-blocker) agent. *Philippine J Sci,* 115:317–28.

Goel, G., Makkar, Hps., Francis, G. and Becker, K. (2007). Phorbol esters: Structure, biological activity, and toxicity in animals. *International Journal of Toxicology, 26:* 279-288.

Gouveia, A de F., Macruz, P. D., Araújo, J. H. B. 2015. "Fitorremediação de solos contaminados com chumbo utilizando *Jatropha curcas* L.," p. 8213-8219. In: Anais do

XX Congresso Brasileiro de Engenharia Química - COBEQ 2014 [Blucher Chemical Engineering Proceedings, v.1, n.2]. São Paulo: Blucher, ISSN 2359-1757, DOI 10.5151/chemeng-cobeq2014-1105-20957-149786.

Gübitz, G., Mittelbach, M. and Trabi, M. (2008). Biofuels and industrial products from *J. curcas* L. *Biological activities and potential application*, 4: 160-166.

Hon, L. M. and Joseph. (2010). Jatropha in South East Asia. *SEGi Review* 3, ISSN, 1985-5672.

Hertwich E. G., and Zhang X. 2009. Concentrating-solar biomass gasification process for 3rd generation biofuels. *Environ. Sci. Technol.* 43: 4207-4212.

Jarma, B., Vanegas, y., Pompelli, M., Garrido, C., Neto, E., Jarma, A. (2014). Detoxification of *Jatropha curcas* L. seed meal as possible alternative livestock feed in the colombian caribbean. *U.D.C.A Act. and Div. Cient, 17(1):* 171-178.

Jasiulewicz M. and Jasiulewicz R. 2009. Biomass from energetic plants as a renewable energy source. *Stow. Ekonom. Roln. Agrobiol,* 7: 48-51.

Joglekar S., Kisan K., Mayur D., Manish H. Novel route for rapid biosynthesis of lead nanoparticles using aqueous extract of *Jatropha curcas* L. latex. *Materials Letters,* 65: 3170-3172.

Joshi, C., Mathur, P. and Khare, S. 2011. Degradation of phorbol esters by Pseudomonas aeruginosa PseA during solid-state fermentation of deoiled *Jatropha curcas* seed cake. *Bioresour Technol, 102,* 7: 4815-4819.

Johnson J. M. F., Coleman M. D., Gesch R., Jaradat A., Mitchell R., Reicosky D. and Wilhelm W. W. 2007. Biomass- Bioenergy Crops in the United States: A Changing Paradigm. *Am. J. Plant Sci. and Biotech.,* 1: 1-28. http://www.globalsciencebooks.info/JournalsSup/images/SF/AmJPSB_1(1)1-28.pdf.

King, A., He, W., Cuevas, J., Freudenberger, M., Ramiaramanana D. and Graham, I. 2009. Potential of *Jatropha curcas* as a source of renewable oil and animal feed. *J. Experimental Botany, 60,* 10: 2897-2905.

Koohafkan AP. 2000. Land resources potential and sustainable land management: An overview. *Nat Resour Forum,* 24: 69–81.

Kaushik N, Kumar K, Kumar S, Kaushik N, Ray S. 2007. Genetic variability and divergence studies in seed traits and oil content of *Jatropha* (*Jatropha curcas* L.) accessions. *Biomass Bioenergy,* 31:497–502.

Kumar, A. and Sharma, S. 2008. An evaluation of multipurpose oil seed crop for industrial uses (*Jatropha curcas L*). *Indus Crops Prod,* 28: 1-10.

Lin, J., Zhou, X., Wang, J., Jiang, P. and Tang, K. (2010). Purification and characterization of curcin, a toxic lectin from the seed of *Jatropha curcas*. *Preparative Biochemistry Biotechnol,* 40 (2): 107-18.

Lu L., Sui M. L, Lu K. 2000. *Science,* 287:1463–6.

Lisiecki M., Biorling L., Motte B., Ninham MP. 1995. *Langmuir,* 11: 2385–92.

Liberalino, A. A. A., Bambirra, E. A., Moraes-Santos, T. and Viera, C. E. 1988. *Jatropha curcas* L. seeds. Chemical analysis and toxicity. *Braz. Arch. Biol. Technol,* 31: 539-550.

Lioglier, H. A., 1990. Plantas Medicinales de Puerto Rico y del Caribe. *Iberoamericana de Ediciones Inc.,* San Juan, PR.

Ma F., Hanna M. A. 1999. Biodiesel production: a review. *Bioresource Technology,* 70:1–15.

Masarovičová E., Kráľová K., and Peško M. 2009. Energetic plants -cost and benefit. *Ecological chemistry and engineering vol. 16,* no. 3: 263-275.

Martinez, J., Siddhuraju, P., Francis, G., Davila, G. and Becker, K. 2006. Chemical composition, toxic/antimetabolic constituents, and effects of different treatments on their levels, in four provenances of *Jatropha curcas* L. from Mexico. *Food Chemistry, 96,* 1: 80-89.

Makkar, H., Kumar, V., Oyeleye, O. O., Akinleye, A. O., Angulo, M. A. and Becker, K. (2011). Jatropha platyphylla, a new non-toxic Jatropha species: Physical properties and chemical constituents including toxic and antinutritional factors of seeds. *Food Chemistry, 125:* 63-71.

Makkar, H. P. S., and K. Becker. (1997). Studies on nutritive potential and toxic constituents of different provenances of *Jatropha curcas. J. Agric. Food Chem. 45:*3152–3157.

Makkar, H. and Becker, K. 1998. Potential of *Jatropha curcas* seed meal as a protein supplement to livestock feed, constraints to its utilisation and possible strategies to overcome constraints. En: Gübitz, G. M; Mittelbach, M and Trabi, M. Biofuel and Industrial Products from *Jatropha curcas. Dbv-Verlag Univ. Graz,* 190-205.

Makkar, H. P. S., Martinez-Herrera, J., Becker, K., 2008. Variations in seed number per fruit, seed physical parameters and contents of oil, protein and phorbol ester in toxic and non-toxic genotypes of *Jatropha curcas. J. Plant Sci. 3*, 4: 260–265.

Mujumdar AM, Misar AV. 2004. Anti-inflammatory activity of *Jatropha curcas* roots in mice and rats. *J Ethnopharmacol,* 90:11–5.

Muanza DN, Euler KL, Williams L, Newman DJ. 1995. Screening for antitumor and anti-HIV activities of nine medicinal plants from Zaire. *Int J Pharmacol,* 33:98–106.

Nass, L. L., P. A. A. Pereira and D. Ellis. 2007. Biofuels in Brazil: an overview. *Crop Sci.* 47: 2228-2237.

Nath LK, Dutta SK. 1992. Wound healing response of the proteolytic enzyme curcain. *Indian J Pharmacol,* 24:114–5.

Neuwinger, H. D., 1996. African Ethnobotany: *Poisons and Drugs.* Chap-man and Hall, New York, 500–509.

Odusote M. O., Abioye A. O., Coker H. A. B. 1999. The Latex Of Jatropha curcas Linn (Euphorblaceae): A Prospective Haemostatic Agent. Nigerian Quarterly Journal of Hospital Medicine, Vol. 9, 2: 158-166.

Openshaw, K. 2000. A review of *Jatropha curcas*: an oil plant of unfulfilled promise. *Biomass Bioenerg.* 19: 1-15.

Ogunwole J. O, Patolia J. S, Chaudhary D. R, Ghosh A, Chikara J. 2007. Improvement of the quality of a degrade dentisol with *Jatropha curcas* L. under an Indian semiarid condition. In: FACT seminar on *Jatropha curcas* L. agronomy and genetics, Wageningen, The Netherlands, March 26–28. Wageningen: FACT Foundation; article no.10.

Osoniyi O, Onajobi F. 2003. Coagulant and anticoagulant activities in *Jatropha curcas* latex. *J Ethnopharmacol,* 89:101–5.

Pandey, V. C., Singh, K., Singh, J. S., Kumard, A., Singhb, B. and Singha, R. P. (2012). *Jatropha curcas*: A potential biofuel plant for sustainable environmental development. *Renewable and Sustainable Energy Reviews, 16:* 2870-2883.

Parawira W. 2010. Biodiesel production from *Jatropha curcas*: A review. *Scientific Research Essays, 5,* 1: 1796-808.

Pompelli, M. F., Ferreira, D. T. R. G., Cavalcante, P. P. G. S., Salvador, T. L., Hsie, B.S. and Endres, L. (2010). Environmental influence on the physico-chemical and physiological properties of *Jatropha curcas* L. Seeds. *Australian J. Bot, 58:* 421-427.

Pradhan RC, Naik SN, Bhatnagar N, Vijay VK. 2009. Moisture dependent physical properties of Jatropha fruit. *Indus Crops Prod,* 29:341–347.

Qi L. M., Ma J. M., Shen J. L. 1997. *J Colloid Interface Sci*, 186: 498–500.

Ravindra P. 2009. *Mat Sci Eng B,* 163: 93–8.

Raina AK, Gaikwad B. R. 1987. Chemobotanyof Jatropha species in India and further characterisation of *curcas* oil. *Journal of Oil Technology of India,* 19: 81–5.

Rakshit KD, Madhvi D and Bhagya S. 2006. *Jatropha curcas* – a potential source of protein in food/feed. In: Symposium on Rural Food and Nutrition. Organized by Sole Company and Central Food Technological Research Institute, Mysore, India, March 22nd, 2005.

Rakshit, K., Darukeshwara, J., Rathina, K., Narasimhamurthy, K., Saibaba, P. and Bhagya, S. 2008. Toxicity studies of detoxified Jatropha meal (*Jatropha curcas*) in rats. *Food Chemical Toxicol, 46,* 12: 3621-3625.

Rakshit, K., Makkar, H. and Becker, K. 2010. Optimization of conditions for the extraction of phorbol esters from Jatropha oil. *Biomass and bioenergy,* 34: 1125 -1133.

Reubens B., Achten W. M. J., Maes W. H., Danjon F., Aerts R., Poesen J., et al., 2011. More than biofuel? *Jatropha curcas* root system symmetry and potential for soil erosion control. *J Arid Environ*, 75: 201–5.

Ren X, Chen D, Tang F. 2005. *J Phys Chem,* 109: 15803–7.

Sabandar C. W., Norizan A., Faridahanim M. J., I. Sahidin. 2013. Medicinal property, phytochemistry and pharmacology of several Jatropha species (Euphorbiaceae): A review. *Phytochemistry*, 85:7–29.

Saetae, D. and Worapot, S. 2011. oxic compound, anti-nutritional factors and functional properties of protein isolated from detoxified *Jatropha curcas* seed cake. *International. J Molecular Sciences, 12,* 1: 66-77.

Samy RP, Lgnacimuthu S, Sen A. 1998. Screening of 34 Indian medicinal plants for antibacterial properties. *J Ethnopharmacol,* 62:173–81.

Siamazonia: Sistema de información de la diversidad biológica y ambiental de la Amazonía peruana (2008). EUPHORBIACEAE *Jatropha curcas* L. http://www.siamazonia.org.pe/Archivos/Publicaciones/Amazonia/libros/28/28000008.htm.

Sirisomboon P, Kitchaiya P, Pholpho T, Mahuttanyavanitch W. 2007. Physical and mechanical properties of *Jatropha curcas* L. fruits, nuts and kernels. *Biosyst Eng*, 97: 201–207.

Siam, P., Diego, B., Sujatha, M., Makkar, H. P. S., Manish, R. L., Reddy, A. R., et al., 2009. Narrow genetic and apparent phenetic diversity in *Jatropha curcas*: Initial success with generating low phorbol ester interspecific hybrids. Nature precedings. hdl:10101/npre.2009.2782.1.

Tiwari A. K., Kumar A., and Raheman H. 2007. Biodiesel production from Jatropha (*Jatropha curcas*) with high free fatty acids: an optimized process. *Biomass and Bioenergy,* 31:569–75.

Tewari JP, Shukla IK. 1982. Inhibition of infectivity of two strains of watermelon mosaic virus by latex of some angiosperms. *Geobios,* 9:124–6.

Thomas OO. 1989. Re-examination of the antimicrobial activities of Xylopia aethiopica, Carica papaya, Ocimum gratissimum and *Jatropha curcas*. *Fitoterapia*, 60:147–55.

Van Gerpen J. 2005. Biodiesel processing and production: biodiesel processing and production. *Fuel Processing Technology,* 86:1097–107. –88.

Villegas, L. F., Fernandez, I. D., Maldonado, H., Torres, R., Zavaleta, A., Vaisberg, A. J., Hammond, G. B., 1997. Evaluation of the wound-healing activity of selected traditional medicinal plants from Peru. *Journal of Ethnopharmacology* 55, 193–200.

Watt, J. M., Breyer-Brandwijk, M. G., 1962. The Medicinal and Poisonous Plants of Southern and Eastern Africa, 2nd ed. Livingstone, Edinburgh, 420–422.

Zahawi RA. 2005. Establishment and growth of living fence species: an overlooked tool for the restoration of degraded areas in the tropics. *Restor Ecol*, 13: 92–102.

In: The Promising Future of *Jatropha curcas*
Editors: Maira R. Segura-Campos et al.
ISBN: 978-1-63484-989-0
© 2016 Nova Science Publishers, Inc.

Chapter 13

CAN *JATROPHA* HAVE A SECOND CHANCE? INSIGHTS BASED ON PREVIOUS EXPERIENCES IN MÉXICO

***Arturo Pérez-Vázquez*[1], *Jorge Martínez-Herrera*[2], *Ofelia Andrea Valdés-Rodríguez*[3,*], *Leobigildo Córdoba-Tellez*[4] *and Odilon Sánchez-Sánchez*[5]**

[1]Colegio de Postgraduados, Veracruz, Ver., México
[2]INIFAP-C. E. Huimanguillo, Tabasco, México
[3]Colegio de Veracruz. Xalapa, Veracruz, México
[4]Colegio de Postgraduados, Campus Montecillo, Texcoco, Edo. de México, México
[5]Universidad Veracruzana, CITRO. Xalapa, Veracruz, México

ABSTRACT

This chapter presents a synthesis related to the current situation of *Jatropha curcas* projects in some Mexican states where the commercial plantations for biofuel production have been established. The experiences mentioned here relates to the states of Chiapas, Michoacán, Tabasco and Veracruz. The findings indicate that *J. curcas* have failed as a cash crop due to the lack of technical support and the development of a value chain to commercialize and process the seeds and to achieve a real investment to produce biodiesel. Yet, in México there is a long tradition of using the toxic and non-toxic *J. curcas*, both of which have a high potential as food, medicine, and for ornamental purposes. These venues represent an important key to give a second chance to *J. curcas* as a commercial crop.

Keywords: biofuels, México, agroenergy, food, alternate energy

---

* Corresponding Author: valdesandrea@yahoo.com.mx.

INTRODUCTION

In early 2007 the Mexican Congress drafted the Law of Promotion and Development of Bioenergy (DOF, 2008), which began into function on February 2008. The current lines of development for biofuels in México are the use of sweet sorghum, cassava, sugar cane, *Jatropha* and *Ricinus* as well as green algae.

Jatropha curcas is a native species of México and Central America, commonly called "piñón" in México, although it has many other local names. There are 45 *Jatropha* species in the country, 39 of them are endemic. These findings suggest that México is a centre of diversity and endemism for this genus (Fresnedo-Ramirez and Orozco-Ramirez, 2013), also the non-toxic, and therefore edible genotype of *J. curcas*, has been found only in México (Makkar et al., 1998, Schmook and Sánchez-Sánchez, 2000). The main purpose of this document is to present, in a very simple way, what has happened in México with the use, growth and commercial plantation of *J. curcas*, based on previous authors´ experience and research. We present the current situation of *J. curcas* after the worldwide boom of it. Jatropha plantations have been used as a social project to intend help improving well-being of Mexican rural farmers and as a primary resource to be included in the energy official programme for biofuel production (SENER, 2012). In fact, the Inter-secretarial Commission for the Development of Bioenergy settled a programme for the sustainable production of bioenergy in 2008 as part of its national strategy for alternative energies.

The worldwide boom of *J. curcas* started with the increasing need to produce biofuel from non-edible plants, as biofuel production received a bad press due to the worldwide food insecurity, increasing worldwide food prices and land use change issues. However, none of these effects can be fully attached to *J. curcas* at least not in México, since this plant has no industrial use as a raw material for the elaboration of edible products in Mexican market. Besides these facts, due to the presence of the non-toxic *J. curcas* in México, it is widely recognized that this tropical plant has an undeveloped potential not only for the production of environmentally friendly biofuel, but also as a source of food for humans or domestic animals, or as a medicinal plant and a resource for pest control.

This essay pointed out the background research results on *J. curcas* plantations in México, which was sponsored by the government through official agricultural programmes. In order to achieve this, we created a team comprised of the leading researchers on *J. curcas* in México to produce this document. The main problems and issues that came out from the last actions in *J. curcas* agricultural projects were highlighted, and an analysis of the facts on *J. curcas* plans into policies was also presented. As a final perspective, we looked for possible solutions on the future viability of *J. curca*s as a raw material for biofuel production and also as an edible raw material for human and animal consume.

The objective of this essay is to highlight the main issues that refer to *J. curcas* in México and discuss key and research actions to promote *Jatropha* as raw material for biofuel and food proposes, as well as for other eventual additional uses.

PREVIOUS EXPERIENCES

J. curcas is a shrub that grows naturally in many states of Mexico, such as Chiapas, Tabasco, Veracruz, Oaxaca and Yucatan, among others. Farmers have traditionally used it as a natural fence on the border of their farms because it is quite easy to propagate due its rapid growth and for being unpalatable to cattle.

The commercial plantation of *J. curcas* started in México with the boom of *Jatropha* cultivation as a feedstock for biofuel production and in many cases it was used as a political dissertation on behalf of the government under the promise of being a new alternative to generate jobs and income for the rural families. Based on this outlook, most *Jatropha* projects started with lack of validated technical information, without funds and reliable technology to process the production. There was also a deficient delivery of subsidiary funds, agroecological zoning to define potential areas for this crop was not properly defined, and there was an absence of a well-integrated chain value (Solis, 2011). Therefore, the cost of production was high and demanded an elevated amount of labour due to pruning and uneven ripening fruits, particularly for harvesting. This picture is very similar in other countries, due to the lack of crop improvement, field validation, standardization of agricultural practices and the absence of a high quality certified planting material and seed. Consequently, the seed yield and oil content of the provided genetic material were not validated before field introduction; nevertheless this topic has been widely covered in the investigations on this field (Valdes et al., 2014). Usually one would expect the government or private sector to create business schemes to help people improve their income and well-being. However, these schemes were not well set up when the first Jatropha plantations started, therefore Mexican farmers could not find any market to sell their production.

There are about 2.6 million hectares (ha) with high potential for *J. curcas* cultivation in México, with an altitude from 0 up to 1000 m above sea level (asl), a temperature between 18 and 28°C and an annual rainfall between 600 and 1200 mm. The included Mexican states within the medium potential for growing *J. curcas* are Tamaulipas (442, 935 ha), Veracruz (336, 314 ha), Sonora (348, 446 ha), Guerrero (283, 191 ha) and Jalisco (261, 989 ha) according to Zamarripa and Diaz (2008). However, after the boom of Jatropha there are scarcely 8,000 ha cultivated in México. Yet, media reports –based on official information-about more than 6,000 hectares cultivated only in the state of Veracruz (Torres, 2010).

Chiapas

The cultivation of *J. curcas* in the state of Chiapas began in 2007, mostly as a feedstock for biofuel production to be used as a fuel for public transportation in Tuxtla Gutiérrez and Tapachula cities (Ando et al., 2013). Therefore, the "Instituto para la Reestructuración Productiva y Bioenergía" (Institute for the Productive Restructuration and Bioenergy) was created with the aim of establishing a plantation area of 10,000 ha, approved by a federal programme of the National Forestry Commission (CONAFOR) named ProArbol, which offered financial subsidy for its cultivation. To promote the Bioenergy Programme, the Government of Chiapas created the "Instituto de Bioenergía y Energías Alternativas" (IBEA) (Institute for Bio and Alternative Energies) in 2007, which no longer existed two years later.

Consequently all the activities were transferred to the "Instituto para la Reconversión Productiva de la Agricultura Tropical" (IRPAT) (Institute for The Productive Reconversion of the Tropical Agriculture) and civil associations called "Unión de Sociedades Bioenergéticas de Chiapas" (USB) (Union of Chiapas Bioenergetic Societies) that joined about 3,000 farmers in the state to establish the 10,000 ha of *J. curcas* committed by the government for biodiesel feedstock. Official reports indicated that the IBEA established 5000 ha of *J. curcas* in 2008 and had the goal of establishing further 5,000 ha in 2009 (Notimex, 2009). The fact is that no more than five thousand ha were established under this programme. According to information disclosed in the "1st National Integration and Planning Support Schemes for Jatropha Producers" (held on September 25, 2014 in Veracruz), the INER (Institute of Renewable Energy of Chiapas), mentioned that Jatropha plantations in Chiapas counted for less than 3,000 ha, and having some of them a density of less than 500 plants per ha.

The farmers of the Central Region in Chiapas began growing "piñon" under rainy season and irrigation systems, in good soil quality, as monoculture or intercropped with crops such as maize, watermelon, peanut and beans, and rarely with fruit trees (Solis, 2011). The genetic material used to establish the first plantation consisted of seeds imported from India, which had low germination rate. Therefore, seeds from local plants were collected to germinate and these seedlings were planted to meet the required number of plants per hectare according to the CONAFOR rules in order that farmers would be able to receive the government subsidy payment. Although nearly 92% of farmers have received explicit training by the IBEA (Solis, 2011), much of the training given to them was based on information obtained through the Internet and not based on scientific and validated research or on the technological developments conducted in México or in the state of Chiapas itself.

The biggest obstacle to keep growing *J. curcas* in Chiapas was the lack of an established market and settled low prices for its sale. According to Solis (2011) the best cost-benefit analysis was maize-piñón system, under rainy season and irrigated condition, as maize contributes to local food security and has the lowest cost-benefit in the first two years ($ 0.13 for each Mexican peso invested). It raises farmer's income from maize sales (97.84% of total revenue) during the first two years. This system could at least give an opportunity to the farmers to keep planting Jatropha until its profit can be reflected in their economy when proper market and a value chain are well developed.

Michoacán

In Michoacán, given the good agroecological conditions (high potential) for growing *J. curcas*, there are nearly 45,000 ha in marginal areas of rainy season and with assisted irrigation (Teniente et al., 2011). The suggested regions to establish this oily crop were areas with precipitation ranging from 600 to 1000 mm/year, free of frost, temperatures averaging between 20 to 30°C, low fertility soils, with low organic matter content and medium depth.

At the beginning of 2008, a business company started piñón plantations in Michoacán to achieve the establishment of 10,000 ha. For this reason, this company supported farmers who had already economic support from CONAFOR, which gave a subsidy support of 500 USD/ha. Then the company contributed with an additional $2,700 Mexican pesos (about 180 USD / ha, at currency change rate of $ 15 pesos/dollar) as a loan to start land preparation and crop establishment, until CONAFOR could give farmers the first payment. In addition, the company gave them seedlings without any cost. During the first three years, the company managed to establish about 1,000 ha, in the region known as "Tierra Caliente" that comprises 10 municipalities (Tepalcatepec, Nueva Italia, Apatzingán, Lázaro Cárdenas, Tumbiscatío, Huetamo, Buenavista, Parácuaro, Arteaga, and Turicato). The yield obtained in the first two years was less than 600 kg per ha, with a density of 1,600 plants/ha. However, some of the seed harvested was used to extend the cultivated area in Michoacán and the other part was sold to other states for propagation. One cause for the low yield was poor fertilization, as just only one dose of sulphate ammonium was applied to the plantations, adding the fact that fruiting occurs only from July up to October and sometimes in November. It is important to mention that in experimental plots a foliar fertilizer was applied and it was possible to obtain in the first year up to 1 kg of seed per plant, as it helped to tie fruits.

Unfortunately the support from CONAFOR for bioenergy crops was suspended, and this caused farmers to become demoralized and give up to continue keeping their *J. curcas* plantations, coupled with the outbreaks of violence and insecurity prevailing in the region of Tierra Caliente. Then the few companies that remained withdrew from the project and the few remaining plantations were abandoned or have been replaced by cash crops. This trend has also taken place in countries such as India, where 30% of the mono crop *J. curcas* plantations were removed and 50% completely abandoned (Ariza-Montobbio and Lele, 2010).

Tabasco

The state of Tabasco has not escaped to the boom of *J. curcas*. In 2012-2013 a group of farmers were encouraged to sow this plant in different municipalities such as Cardenas, Cunduacán, Comalcalco and Huimanguillo. Currently there are nearly 300 ha of *J. curcas* planted under rainy condition, and acceptable soils and good production level. However, until now, farmers have not had any economic support, due to the lack of market to sell the production; therefore this situation has triggered to eliminate the *J. curcas* plantations.

In spite of the market problem, farmers are still looking for the full use of the seed and the residual cake, due to the high oil contents at rates of 40-48% (Martinez-Herrera et al., 2006) and potential as food for livestock (sheep, cattle), chickens and even fish. Certainly, *J. curcas* could be an economic alternative as a protein supplement for cattle or other animals (for its protein contents above 35%), without ruling out the possibility to associate the piñón with other crops such as habanero pepper, beans, and cassava. Nevertheless, the lack of a secured market places the farmers on a difficult position. With no guaranteed profit given, there is not motivation to keep growing this plant or increasing the existing plantations.

Recently, two seed samples were analyzed from two plantations of Cunduacán, Tabasco quantifying protein content in whole seed of 18.4 and 28%, and oil content of 35.6 and 35.5%. Although, there are reports of toxic and non-toxic native plants in the state of Tabasco, phorbol esters were not detected (Martinez-Herrera, 2013). The exact origin of the seeds is unknown, however, there is some evidence that they come from the state of Morelos. The oil content in these seeds is among the ranges reported for other seeds on the states of Veracruz, Puebla, Michoacán and Morelos.

It is important to highlight the study of González-Mancillas et al. (2014) regarding the edaphoclimatic zoning of *J. curcas* in the state of Tabasco. This plant presents edaphoclimatic fitness ranges covering an area of 833,181 ha, and it is distributed on the following municipalities according to their respective suitable areas for *J. curcas* plantation: Balancán (256,201 ha), Huimanguillo (131,596 ha), Tenosique (130,708 ha), Cardenas (68,267 ha) and Teapa (38 ha). This information provides us with an approach to make the right selection for the land area for the establishment of *Jatropha* plantation and to prevent future failures from the inclusion of a non-compatible edaphoclimatic area.

There has been some interest from United States (USA) companies to settle down piñón plantations for biodiesel production and to use it directly in trailers and trucks (Nature Fuels de México and Pop Diesel, among others). Undoubtedly, it is necessary that the proposed business plan set a suitable market and value chain to provide an economic benefit for farmers and a regularization of the *J. curcas* plantation, recollection and distribution. Otherwise, even with the best of seed, technology and land it would not be possible to achieve a successful project. Therefore, the use of *Jatropha* must be under integral approach, so that the oil, the residual cake and fruit peel can respectively be used as raw material for biodiesel, balanced pellets for animal feeding or for the production of biogas and biofertilizer. On the other hand, there are many other alternative uses for *J. curcas*, for instance bushes can be planted for reforestation purposes, preventing soil erosion, or even as living fences. The oil may have other applications as a lubricant, dielectric compound, biopesticide, among many other uses that have not been yet explored (Parawira, 2010).

Veracruz

INVERBIO (the Veracruz Institute of Bioenergy Crops) reported that there are 400 hectares of *J. curcas* for biodiesel production in Veracruz, mainly in the central and northern part of the state and it is expected to reach 1,200 hectares. However, there is not any official structured programme to achieve this goal and it is used more as a political rhetoric than as a fact. Actually some plantations established since 2008-2009 are being abandoned or switched to other crops as there is not market for *Jatropha* seeds and nor any economic support.

So without any attractive economic plan it is not possible that *J. curcas* would have a second chance as a crop for biofuel production. Indeed, in most official projects there is not any formal agreement signed to buy the production and the purchase price has never been established or agreed. So, there is not any guarantee that the production will be purchased at any settle time. Just in Chiapas, in 2011 the production was bought as a political measure to show that the programme was successful. So until now, a formal purchase of the product has never happened again despite several seasons of production. Therefore most farmers do not keep any hope for *Jatropha* as a cash crop.

On the other hand, the Swiss crop science company, JATROPOWER has focused on the research and development of non-toxic *J. curcas* varieties (found naturally in some regions of México) for commercial cultivation. The company denotes that the non-toxic edible varieties of *J. curcas* are naturally available and represent a lower investment risk and relatively immediate solution for cultivating *J. curcas* for both oil and kernel meal. Non-toxic varieties produce non-toxic harvests, even when pollinated by common toxic strains. The progeny of these pollinations may however produce toxic harvests (Hardman and Co, 2015).

Currently a research center of the Mexican Department of Agriculture (SAGARPA) promotes the use of three "toxic" genetic materials, two with female flowers and one with male flowers, which planted together are expected to have a greater pollination and therefore high fructification and yield (Zamarripa, 2014). However, for the current year (2015), there is not enough seed to propagate new plantations in Mexican states which still may have some plans to establish *J. curcas* for biofuel production. This careless selection of a toxic variety reduces the utility of the plantation since it might not be used for food consumption or animal feeding. Although the proposal of detoxification considers this measure, it is unlikely to be cost effective from the economic point of view. Therefore, the utilization of the non-toxic accession must be promoted, in order to be able to use the oil and the residual cake (seed cake) directly (Martinez-Herrera et al., 2006).

Overall, the current lack of planning to set up plantations in the states of Chiapas, Michoacán, Tabasco and Veracruz, led to their establishment far from cities or industrial areas (more than two hours away of distance), raising transportation costs. Moreover, the official institutions did not have a proper plan related to the market. The production and the necessary requirements for the selected farmers (e. g. agroecological conditions, proper handling training, land access and communication, and high quality seed) were lacking. All these facts led to poor production and left the farmers to lose the opportunity cost for other cash crops that could provide them with a real profit.

Currently, some Mexican states are still promoting the establishment of *Jatropha* plantations, such as Jalisco, Morelos, Sinaloa, Puebla, Veracruz, Yucatan, Campeche and most of them with the intention to use non-toxic and edible seed. In addition, some business companies have established *J. curcas* plantations. For example LODEMO Group have planted 1500 ha and Global Energy 2200 ha. The states where *J. curcas* projects were established from 2007 to 2013 can be seen in Figure 1.

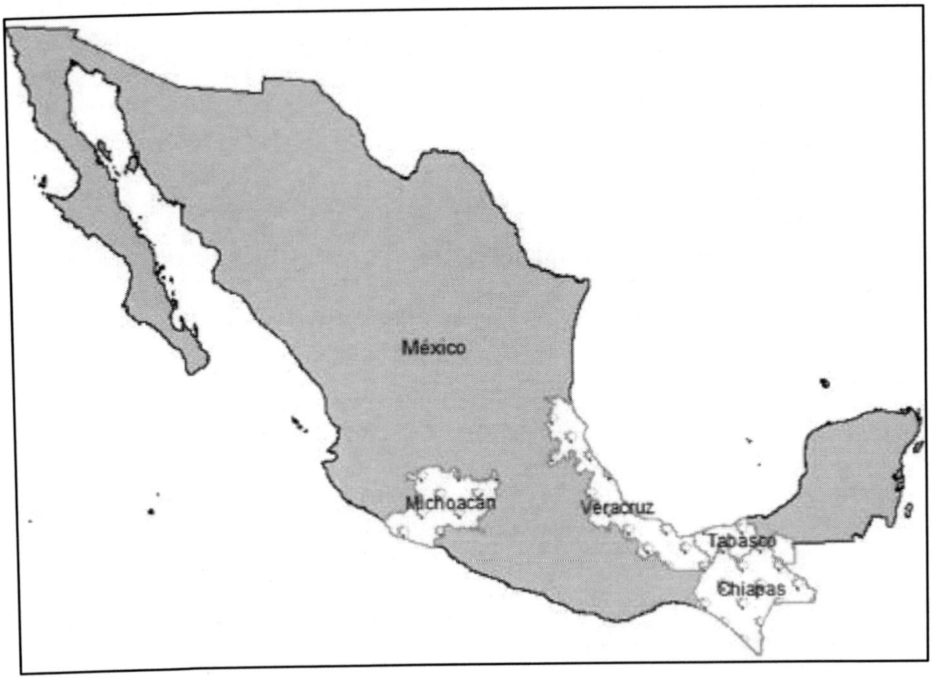

Figure 1. Some Mexican states where main projects of *Jatropha curcas* have been established during 2007-2014.

New Ways of Looking at *Jatropha curcas*

Diversity of the Genus

From the 45 species of *Jatropha* found in México, 77% of them are endemic (Martínez et al., 2002) and 79% can be found in eight regions of genetic diversity, being one of these regions the valley of Tehuacan-Cuicatlan among the states of Puebla and Oaxaca. It has been spotted in this region the existence of 11 species of *Jatropha*, being *J. neopauciflora*, *J. websteri*, *J. spathulata*, *J. vernicosa*, *J. dioica*, *J. pseudocurcas*, *J. rzedowskii*, *J. olivacea*, *J. ciliate*, *J. standleyi* and *J. curcas*. The most common are: *J. ciliate*, *J. neopauciflora*, *J. rzedowskii*, *J. oaxacana*, and *J. rufescens* (Hernández and Cordoba, 2012). The current distribution of *Jatropha* species in the Reserve Tehuacan-Cuicatlan was mainly determined by altitude, temperature and precipitation.

In the year 2009, the Colegio de Posgraduados (CP) started to collect toxic and nontoxic accessions of *J. curcas* in Veracruz, Puebla and Chiapas, as a part of the J*atropha* spp. Network run by the National System of Genetic Resource (SINAREFI) of SAGARPA. As a result, a germoplasm bank with 116 provenances has been established in the Campus Veracruz, Colegio de Postgraduados (19°11’39.32” LN, 96°20’08.59” LW at 31 m asl), being 72 non-toxic provenances from Veracruz and Puebla.

The main uses recorded in Veracruz for *J. curcas* are as an edible (46%), living fence (41%) and medicinal plant (10%) (Valdes et al., 2014b). In fact, the word of the genus *Jatropha* derived from the Greek *iatrós* (doctor) and *trophe* (food), that means that the main uses of this plant has been for medicinal and food purpose. The main use of *J. curcas* depends on the region and on the presence or absence of phorbol esters in the seed. The use of *Jatropha* as a food was first registered in the northern regions of Veracruz, where no phorbol esters were detected in seeds, while in central and southern regions the main use is as a hedge or living fence, since the concentration of phorbol esters ranged from 0.12 to 0.66 mg/g (Zavala del Angel et al., 2014).

Jatropha curcas as Food

The extended argument that *J. curcas* compete for food security due the reduction of edible plantations is not longer true when the non-toxic variety of *J. curcas* is introduced or planted; besides it can also be grown with other edible crops. Consumption of *J. curcas* seeds is possible due to non-toxic germplasm with very low or no detected content of phorbol esters. The presence and origin of this germplasm is located mainly in the region of Totonacapan in México, and it is supposed that the Totonacs domesticated this plant. In fact, one of the first studies on *J. curcas* in Mexico has relationship with food consumption (Makkar and Becker, 1997; Makkar et al., 1998; Smook and Sánchez, 2000).

References on food uses of this plant have not been found for pre-Columbian times due to lack of texts and paleobotanists reports in the area where the nontoxic ecotype is located. However, *J. curcas* has been used as a traditional food in many ethnic and rural communities in México. Mainly in the Totonaca region that includes part of the states of Veracruz, Puebla and San Luis Potosi. Therefore, it is quite possible that the Totonaca culture of the central

Gulf of México domesticated the non-toxic genotype, since oral traditions have reported that *J. curcas* seeds have been consumed during many generations by the population of this region (Valdes-Rodriguez et al., 2013).

Based on the empirical knowledge, seeds can be used to prepare different dishes with a good nutritional value. Therefore, a number of food products from seeds and flour have been elaborated with beneficial impact on society, particularly in highly marginalized communities where non-toxic *J. curcas* is grown. Apparently, there are not morphological differences among toxic and non-toxic accessions. Therefore, the introduction of toxic-genotypes in the Totonacapan region would put on risk an ancestral legacy as well as the loss of biodiversity and the possible poisoning of the residents of this region by eating toxic seeds, since traditional dishes made of *Jatropha* are consumed one or twice a week by the traditional families of the Totonacapan. During the festivity day called "*Día de los muertos*" (Day of the Dead), dishes prepared with *J. curcas* are commonly consumed on a large scale. After roasting the seeds, farmers add salt, tomato sauce, and dry chilli to prepare a rich snack. The traditional preparation of "*pulacles*" consist of grinding the *J. curcas* seeds which are then mixed with onion, zucchini, chayote, baked beans, and finally wrapped in dried corn husks and cooked. To prepare other dish called "*pipian*", *J. curcas* seeds are roasted, grounded and then combined with sesame and pumpkin seeds. To enhance flavour, chopped onions and chilli paste is added to the chicken and the preparation is ready for eating.

The high quality protein, sulphur amino-acids and easy protein digestibility, provides a wide range of uses for *J. curcas* in the food sector, which not only can generate additional jobs, but can also make a significant positive impact on the rescue process of the traditional Mexican food knowledge, which has been replaced by fast food with little nutritional value and with a consequent detriment on people´s health promoting obesity and diabetes (Martinez et al., 2013).

On the other hand, we have found that the tradition of *J. curcas* as a food has different uses, as a dish, a kind of drink called *atole*, candy, and as a snack (as toasted seed). To demonstrate this, SINAREFI-SAGARPA held a public festival on October 2013 in México City. It was set up a Food Display with diversity of food products made of *J. curcas*. This exhibit had an important role to let society know in extent that a native crop such as *J. curcas* could be used to evoke the food traditions and by some additional creativity to use this crop for food purposes into a commercial and common food products and dishes such as pizza, pastries, bread, etc. This *J. curcas* display had also the purpose to avoid the loss of the traditional use of this important native crop that still has significant importance on rural communities.

When we look at the nutritional value of *J. curcas* flour, it has significant contents of protein, oil and carbohydrates. Therefore, it has the capacity to improve the nutritious value of some common food such as "*tortillas*," an important basic food in México. It also can be used as a nut mixed with other grains and whole cereals in kind of granola (muesli).

Jatropha as Ornamental Plant

When people use the term *Jatropha*, usually they refer to *Jatropha curcas* L. which is only one of the 170 known species belonging to the largest family of Euphorbiaceae, in which, a large number of other *Jatropha* species with high potential as ornamental plants, such as *J. intergerrima*, *J. gossypifolia*, *J. neopauciflora* and *J. rzedowskii* (Figure 2), that are native from México, are included.

Figure 2. *Jatropha* species with ornamental potential.

Indeed *J. curcas* has been used as an important plant in home gardens in many rural communities in México, mainly the non-toxic one. It can be found particularly in many gardens of the poor population. It seems that for them, it is still an important plant to have in the garden as food and for medicinal use.

Medicinal Potential of *Jatropha curcas*

J. curcas was already known since pre-Columbian times, and it is possible that the Olmecs, the indigenous group who inhabited the region of Mesoamerica from 2000 up to 400 years BC, started using this plant (Dias et al., 2012). They might be using the toxic ecotype as a medicinal plant. Also short time after with the arrival of the Spaniards they were using *J. curcas* due to the antibacterial effect of the bark, leaves and roots (De la Cruz, 1964). Seeds are considered with purgative effect, but only a maximum of two seeds is recommended, after that it could became toxic (Table 1). These medicinal uses were documented on the last century in traditional markets of central México (Baytelman, 1981). The medicinal uses of *J. curcas* presented here are based on a survey conducted by the authors with the owners of plants in Veracruz state and research from Baytelman (1981). Indeed the antiviral action of the latex and its property for healing has been already mentioned by other researchers (Villegas and Fernandez, 1997; Tewari and Shukla, 1982).

Table 1. Medicinal use of *J. curcas* in México

Genotype	Part used	Use and purpose
Toxic and non-toxic	Leaf or trunk latex	Apply the latex over the skin to cure bacteria, fungus or virus diseases Healing, wounded antiseptic, anti-inflammatory, blood coagulation, anti-microbial
Toxic	Seeds	Eat two boiled seeds to purge or eliminate intestinal parasites

MAIN CONSIDERATIONS FOR INCOMING *J. CURCAS* PROJECTS

J. curcas has showed that is not just a potential crop to produce biofuel (Everson, Mengistu and Gush, 2013). Indeed, *J. curcas* has been overemphasized as a panacea plant for the global energy crisis. Even though, we have seen an extraordinary collapse of *J. curcas* official programmes that implemented it as a raw material for biofuel production. It seems that the higher cost of *J. curcas* production may have a direct negative effect by producing a very expensive biofuel. So to ensure the production of low price biomass, farmers have to make important changes in agricultural practices. It is important that the new projects related to *J. curcas* consider it not only as a biofuel, but also to produce oil and flour for human and animal feeding, among other several uses that we previously mentioned through this document.

Future *Jatropha* projects at wide scale should be set up not only by the integration of complementary process that exploits the versatility of the plant, but also taking into consideration scientific information, socio-economic analysis, a policy implementation, an evaluation on fossil and biofuel prices and the projection of the tentative markets and prices of the raw material depending on its use. Medium scale projects would be more suitable in rural communities as a business and for the rural livelihood. In other words, any project of *J. curcas* should be highly productive, competitive, profitable, sustainable and equitable. To meet those goals we must take in consideration the required research, innovation and technological development (Figure 3). So anyone must be aware that this plant is not ready for commercial cultivation, as there are not improved plantations neither certified validated high yield seeds. Also from the financial point of view it is still a not profitable crop due to the low yield obtained nowadays in México. A large quantity which has been published about Jatropha is more a speculation than real facts in terms of jobs generation, profitability and high yield under marginal land.

For a profitable market *J. curcas* projects have to consider that the yields obtained per hectare must be higher than five tonnes and the purchase prices must be at least $ 8.00 Mexican pesos / kg (40 to 50 cent of USA dollar). In addition, *J. curcas* plantations must be promoted as small or medium production units for oil extraction and using the residual cake for human and animal consumption. As many foreign "investors" are just looking to obtain financial or subsidy support from the local Governments without putting on risk their money under the promise to generate jobs, the subsidy government funding must implement measures as financial penalties for those who do not accomplish the established agreements for governmental funding. Until now, very few businessmen have invested their money to consolidate a real *Jatropha* value chain. So investors and the industry have to guarantee channels of marketing and a fair price to farmers.

Other continents such as Africa, Asia, Central and South America, are giving a second chance to *J. curcas* as a cash and biofuel crop, using elite material or highly productive seeds, compact areas to increase yields and reducing production costs. In addition, today most countries have an agronomic technology that allows more acceptable yields. Although some countries are still using toxic material, we suggest planting non-toxic materials, as after the oil extraction the residual cake can be used for animal feeding without any extra cost. We have to be aware that *J. curcas* biodiesel has a period of less than 10 years life. Even if the *J. curcas* oil has a wide application and high economic value, it also has to compete with today new alternative energies that are more efficient and profitable. Nevertheless, *Jatropha* could be an important feedstock for biofuel production, especially in countries without oil fossil reserves or keen to generate green energy.

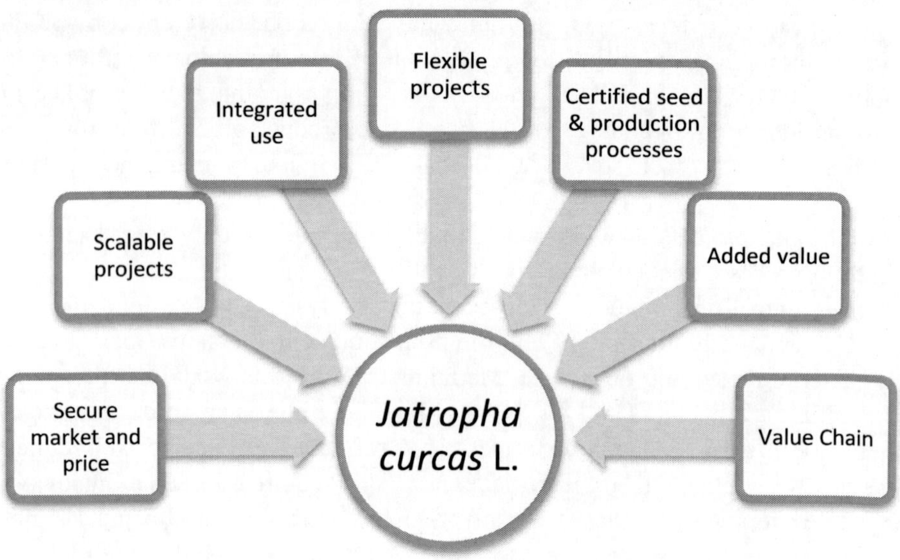

Figure 3. Main issues to include for achieving successful *Jatropha* projects.

AN APPROXIMATION TO A SUCCESSFUL *JATROPHA* PROJECT: THE SINAREFI PROGRAMME

Plant genetic resources for food and agriculture are the biological basis for food security. Thus, the Ministry of Agriculture, Livestock, Rural Development, Fishing and Food

(SAGARPA) implemented, in 2002, the programme called the National System of Plant Genetic Resources for Food and Agriculture (SINAREFI). This programme, along with four additional subsystems (Livestock, Forestry, Aquaculture and Microbial) in 2009 integrated the National System of Genetic Resources (SINARGEN).

The SINAREFI established networks (comprising researchers, farmers, farmer´s associations, etc.) to work by genus native plants or theme, using as a platform the Global Plan of Action of the Plant Genetic Resources, as well as the information from diagnostic carried out in México on Plant Genetic Resources for Food and Agriculture and published in year 2000 and a second one published in 2006. The information on both diagnostics allows the establishment of a National Plan for Action on Plant Genetic Resources, which includes activities in four strategic areas: *in situ* and *ex situ* conservation, use and empowerment and capacity building, with 20 lines of action.

The *Jatropha* spp. Network was established in 2009 under the SINAREFI programme.. Today it involves 13 researchers, from which nine belong to seven different public institutions (Colegio de Postgraduados, Campus Montecillo and Campus Veracruz; CEPROBI-IPN, Morelos; INIFAP, Experimental Research Field, Morelos; Technologic Institute Conkal, Yucatán; Universidad Tecnológica de la Huasteca, Hidalgo; UNAM; University of Veracruz; and UACh, Texcoco), as well as the association of JatroBioenergy and Oilseeds S.P.R. de R.L. This network is directed by SINAREFI representatives and a "Network Coordinator," whose activities are carried out according to a specific and strategic plan for the *Jatropha* genus and outlined under the National Plant of Action. After seven years, the Network has accomplished the following results:

1. A document called "Diagnostic and strategic plan of *Jatropha* spp." in México" which has been printed.
2. Genetic richness sites taking into account 45 species in Mexico have been found (Figure 4). Also ethno botanical and ecological information has been obtained and 15 wild species have been identified.
3. A germplasm bank has been established.
4. *J. curcas* accessions have been morphologically and chemically characterized.
5. A collecting strategy in regions with very high and excellent probability distribution of *J. curcas* has been performed (Figure 5). As a result, more than 400 accessions have been collected and kept under storage conditions in SINAREFI gene banks.
6. More than 200 accessions have been chemically assessed and more than 50% are non-toxic due to the null presence of phorbol esters.
7. Validation genotype plots have been established in Veracruz (Colegio de Postgraduados, Campus Veracruz), Espinal, Veracruz (Universidad Veracruzana), Yautepec, Morelos, Tierra Caliente, Michoacán (CEPROBI-IPN) and Sierra Norte de Puebla (Colegio de Postgraduados and UACh).
8. Non-toxic elite clones, which are under validations for registration as elite varieties, have been detected.

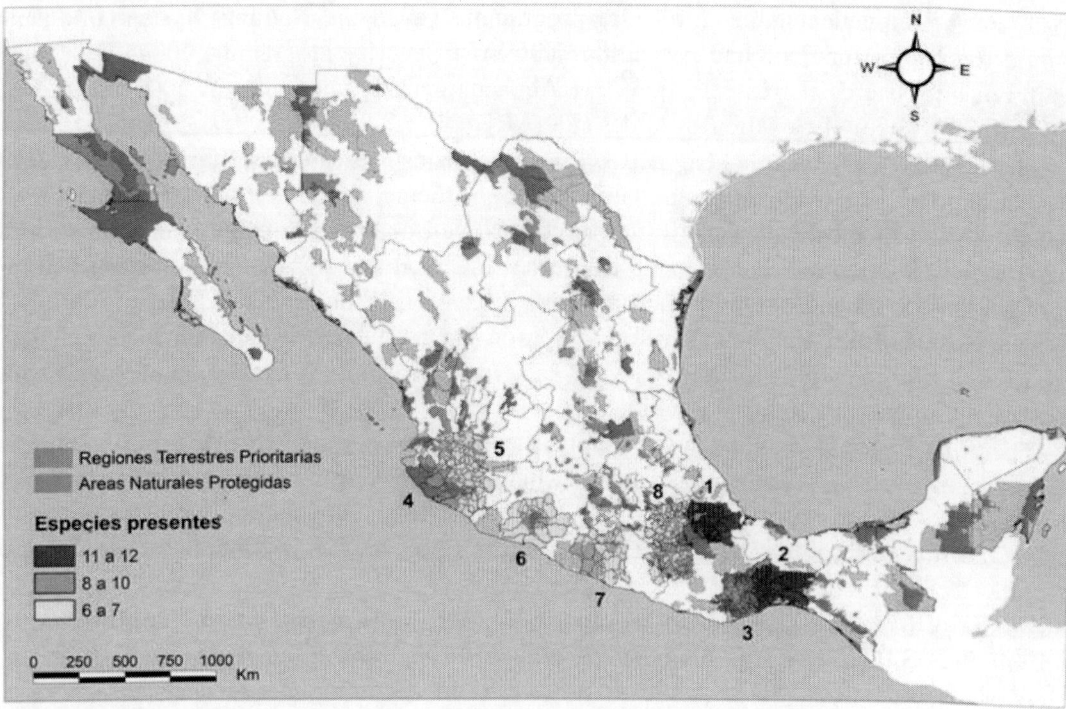

Figure 4. Genetic richness of *Jatropha* genus in México, including information from herbarium of 45 species (Cordova et al., in press).

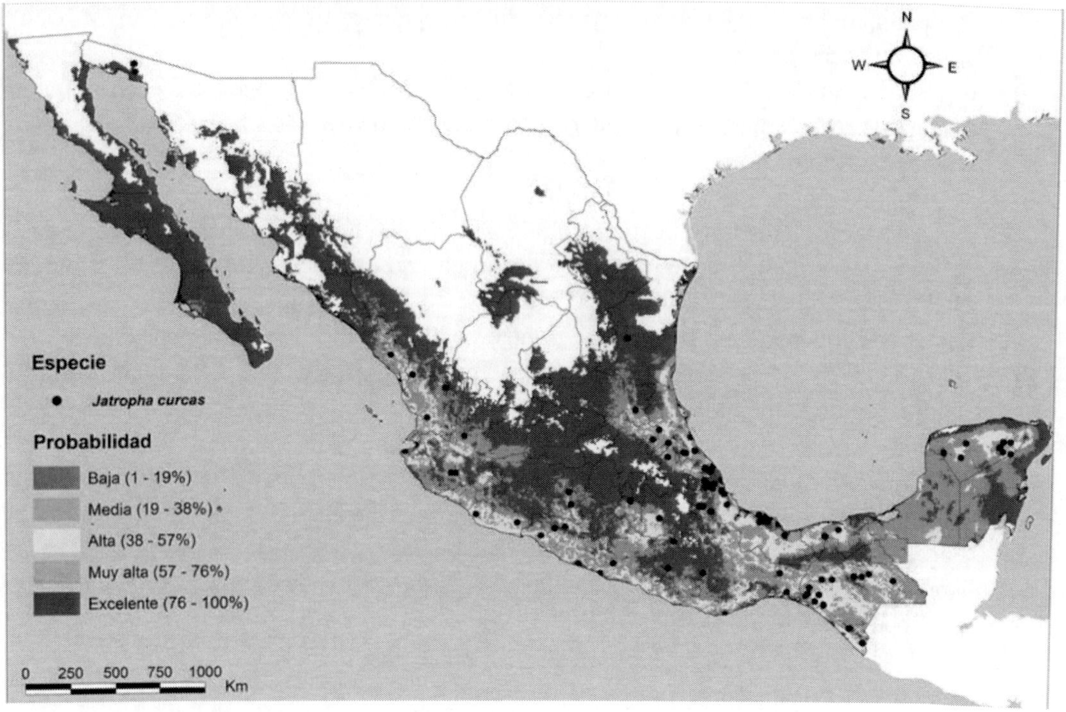

Figure 5. Potential distribution of *J. curcas* in Mexico (Cordova et al., in press).

CONCLUSION

One of the main failures of *J. curcas* in México is the way how it was introduced in the rural communities, without any previous cultivation experience, technical advice or guidelines given to the farmers, and without any formal official programme, focussed more on a political background rather than a realistic profitable business.

Since the *J. curcas* boom started, research institutions put so much emphasis and gave advice to local authorities (the Government) to not compromise farmers in a risky initiative until they were provided with the required technical information and suitable management training. However, the fever for *J. curcas* plantation as a panacea to solve some poverty problems and to produce biofuel, conducted to the use of wild *J. curcas* seeds, with negative results on the proposed projects.

As it is known, México has severe problems to achieve food security, mainly among the poor and therefore it is anticipated that using *J. curcas* as food would be more suitable than using it for biofuel production. Therefore if we want to give *J. curcas* plantations a second chance, we must look at a national and local strategy to increase its yields based on a friendly and sustainable agricultural practice, and to establish a sustainable value chain for its products and by-products.

As a result of the present investigation –which is only our approach to a very vast and difficult issue- we conclude that there is a need to develop a national platform to support and give advice towards the start of any public or private *J. curcas* initiative to achieve success.

ACKNOWLEDGMENTS

We would like to acknowledge the economic support of LPI3-Energy Alternate and Biomaterials of Colegio de Postgraduados, and to SINAREFI (SAGARPA), which is greatly appreciated.

REFERENCES

Ando, T., Tsunekawa, A., Tsubo, M. and Kobayashi, H. (2013). Identification of Factors Impeding the Spread of *Jatropha* Cultivation in the State of Chiapas, México. *Sustainable Agriculture Research,* 2 (2), 54-59. doi:10.5539/sar.v2n2p54.

Ariza-Montobbio, P. and Lele, S. (2010). *Jatropha* plantations for biodiesel in Tamil Nadu, India: Viability, livelihood trade-offs and latent conflict. *Ecological Economics,* 70(2), 189-195. doi:10.1016/j.ecolecon.2010. 05.011.

De la Cruz, M. (1964). *Libellus de Medicinalibus Indorum Herbis*. Instituto Mexicano del Seguro Social. Edición especial. México, 1964.

Dias, L. A. S., Missio, R. F. and Dias, D. C. F. S. (2012). Antiquity, botany, origin and domestication of *Jatropha curcas* (Euphorbiaceae), a plant species with potential for biodiesel production. *Genetics and Molecular Research,* 11 (3), 2719-2728.

DOF (Diario Oficial de la Federación). 2008. Ley de Promoción and Desarrollo de los Bioenergéticos. 1ro. Febrero, 2008. Secretaría de Energía. 75-82 p.

Baytelman, B. (1981). Etnobotánica en el estado de Morelos. SEP- Instituto Nacional de Antropología e Historia. 287p.

Everson, C. S., Mengistu, M. G. and Gush, M. B. (2013). A field assessment of the agronomic performance and water use of *Jatropha curcas* in South Africa. *Biomass and Bioenergy,* 59, 59-69.

Fresnedo-Ramirez, J. and Orozco-Ramirez, O. (2013). Diversity and distribution of genus Jatropha in México. *Genetic Resource Crop Evolution,* 60,1087–1104. DOI 10.1007/s10722-012-9906-7.

González-Mancillas, R., Juárez-López, J. F., Aceves-Navarro, L. A., Rivera-Hernández, B. and Guerrero-Peña, A. (2014). Zonificación edafoclimática para el cultivo de *Jatropha curcas* L., en Tabasco, México. Investigaciones Geográficas. *Boletín del Instituto de Geografía*, UNAM. doi.org/10.14350/rig.39936.

Hardman and Co. (2015). Future Harvest 21st Century Jatropha. London, UK. www.hardmandco.com.

Hernández, N. N. Y. and Cordoba, T. L. (2012). Diversidad de *Jatropha* spp. en el valle de Tehuacán-Cuicatlán. En: 1er Feria de la Agrodiversidad y Agroproductos: Resúmenes de carteles. SINARFI-SAGARPA. P. 51.

Makkar, H. and Becker, K. (1997). Potential of *Jatropha curcas* seed meal as a protein supplement to livestock feed; constraints to its utilization and possible strategies to overcome constraints. DBV Graz: 190 - 205.

Makkar, H., Becker, K. and Schmook, B. (1998). Edible provenances of *Jatropha curcas* from Quintana Roo state of México and effect of roasting on antinutrient and toxic factors in seeds. *Plant Foods for Human Nutrition,* 52, 31-36.

Martínez, G. M., Jiménez, R. J., Cruz, D. R., Juárez, A., García, A. R., Cervantes and Mejía, H. R. (2002). Los géneros de la familia Euphorbiaceae en México. Anales del Instituto de Biología, UNAM, *Serie Botánica,* 73(2), 155–281.

Martínez-Herrera, J., Siddhuraju, P., Francis, G., Dávila-Ortíz, G. and Becker, K. (2006). Chemical composition, toxic/antimetabolic constituents, and effects of different treatments on their levels, in four provenances of *Jatropha curcas* L. from México. *Food Chemistry,* 96, 80–89. doi:10.1016/j.foodchem.2005.01.059.

Martínez-Herrera, J., Jiménez-Martínez, C. and Güemes-Vera, N. (2013). Use of *Jatropha curcas* L. (non toxic variety) as traditional food and generation of new products in México. In: *Jatropha*, Challenges for a New Energy Crop. Editors N. Carels, M. Sujatha. B. Bahadur, Springer New York Heidelberg Dordrecht London, pp. 333-341.

Notimex. (2009). Bioenergéticos se han convertido en prioridad: Antolín Morales. Cuarto Poder. Sección Municipios: B24 (Col. 1). Tuxtla Gutiérrez, Chiapas. Mayo 25.

Parawira, W. (2010). Biodiesel production from *Jatropha curcas*: A review. *Scientific Research and Essays,* 5(14), 1796-1808.

Schmook, B. and Sanchez-Sanchez, O. (2000). Usos y potencial de *Jatropha curcas* L. en la península de Yucatán, México. *Revista Foresta Veracruzana,* 2, 7-11.

SENER (Secretaría de Energía). (2012). Prospectiva de Energías Renovables 2012-2026. México, D. F. Gobierno Federal. 156 p.

Solis, G. B. F. (2011). Integración de *Jatropha curcas* L. en agroecosistemas como materia prima para biodiesel en la región centro de Chiapas, México. Tesis de Doctora en Ciencias en Agroecosistemas Tropicales. Colegio de Postgraduados, Campus Veracruz. 192 p.

Teniente, O. R., Tapia, V. L. M., Zamarripa, C. A., González, Á. A., Solis, B. J. L., Martínez, V. B., and Hernández, M. (2011). Guía técnica para la producción de piñón mexicano (*Jatropha curcas* L.) en Michoacán. INIFAP. SAGARPA. 44p.

Tewari, J. and Shukla, I. (1982). Inhibition of infectivity of two strains of watermelon mosaic virus by latex of some angiosperms. *Geobios,* 9, 124-126.

Torres, C. (2010). Biodiesel a partir de *Jatropha curcas* en México. In: Biocombustible, diodiesel. Veracruzanos (Online newspaper). Available in: http://biodiesel.com.ar/2985/ biodiesel-a-partir-de-jatropha-curcas-en-México.

Villegas, L. F. and Fernández, I. D. (1997). Evaluation of the wound-healing activity of selected traditional medicinal plants from Perú. *Journal of Ethnopharmacology,* 55, 193-200.

Valdés-Rodríguez, O. A., Sanchez, S, O., Perez-Vazquez, A. and Caplan, J. (2013). The Mexican non-toxic *Jatropha curcas,* food resource or biofuel? A reflection. *Ethnobotanic, Research and Applications,* 11, 001-007.

Valdés-Rodríguez, O. A., Perez-Vazquez, A. and Muñoz, G. C. (2014a). Drivers and Consequences of the First *Jatropha curcas* Plantations in México. *Sustainability,* 6(6), 3732-3746.

Valdés-Rodríguez, O. A., Perez-Vazquez, A., Garcia, P. E. and Inurreta, A. H. (2014b). Condiciones agroecológicas de procedencias nativas de *Jatropha curcas* L. en el estado de Veracruz. In: Energía Alterna y Biocombustibles. Perez-Vazquez, A. and Garcia-Perez, E. (Eds). Mundi-Prensa. México, D. F. pp 143-152.

Zamarripa, C. A. (2014). Nuevas variedades de piñón (*Jatropha curcas* L.) para el desarrollo de los bioenergéticos en México. Conferencia en el Primer Encuentro Nacional de Integración de productores de Jatropha. Septiembre, 2014. INIFAP-SAGARPA.

Zamarripa, C. A. and Díaz, P. G. (2008). Áreas de potencial productivo de piñón *Jatropha curcas* L., como especie de interés bioenergético en México. Boletín electrónico del Comité Nacional Sistema Producto Oleaginosas 16: 4-6. www.oleaginosas.org/cargas/ mar-abr- electronico.pdf.

Zavala del Angel, I., García-Pérez, E., Pérez-Vázquez, A., González-Hernández, D., Ávila-Reséndiz, C. and Verpoorte, R. (2014). Aspectos etnobotánicos del piñón (*Jatropha curcas* L.), contenido de ésteres de forbol y su distribución geográfica en Veracruz México.

About the Editors

Maira Rubi Segura-Campos. Dietitian Biologist Chemist (2004), Faculty of Chemistry, Universidad Autonoma de Yucatan (UADY); MSc in Food Science and Technology (2007), Faculty of Chemical Engineering (UADY); Ph.D. in Agricultural Science, Campus of Biological and Agricultural Science (2010) (UADY). She is a research professor and head of the Food Science Laboratory from Faculty of Chemical Engineering (UADY) in Mexico. She is a National Researcher Level I (2012-2018) of the National System of Researchers and has collaborations with national and international research groups. She is responsible and contributor of different research projects and author of national and international articles, books, congress memories book chapters as well as advisor of undergraduate and graduate thesis.

David Betancur-Ancona is Professor and Researcher of Food Science and technology in Faculty of Chemical Engineering at the Autonomous University of Yucatan in México. He is a graduate in Biochemical Engineering, MSc in Food Science and Technology and PhD in Food Sciences. He is also a National Researcher Level III (2014-2017) of the National System of Researchers. He is responsible and contributor of research projects and author of indexed articles, books, and book chapters. Advisor of undergraduate and graduate thesis. Since 2000, he has collaborated with national and international networks to develop multidisciplinary research projects.

INDEX

A

B

C

F

G

H

J

K

L

M

P

Q

R

S

T

Z